Green Approaches to Biocomposite Materials Science and Engineering

Deepak Verma
Graphic Era Hill University, Dehradun, India

Siddharth Jain
College of Engineering Roorkee, India and University of Alberta, Canada

Xiaolei Zhang
Queens University, Belfast, UK

Prakash Chandra Gope
College of Technology, G.B.Pant University of Agriculure and Technology, Pantnagar, India

A volume in the Advances in Chemical and Materials Engineering (ACME) Book Series

ENGINEERING SCIENCE REFERENCE
An Imprint of IGI Global

Published in the United States of America by
Engineering Science Reference (an imprint of IGI Global)
701 E. Chocolate Avenue
Hershey PA 17033
Tel: 717-533-8845
Fax: 717-533-8661
E-mail: cust@igi-global.com
Web site: http://www.igi-global.com

Library of Congress Cataloging-in-Publication Data

Names: Verma, Deepak, 1983- editor. | Jain, Siddharth, 1984- editor. | Zhang, Xiaolei, 1986- editor. | Gope, Prakash Chandra, editor.
Title: Green approaches to biocomposite materials science and engineering / Deepak Verma, Siddharth Jain, Xiaolei Zhang, and Prakash Chandra Gope, editors.
Description: Hershey, PA : Engineering Science Reference, [2016] | Includes bibliographical references and index.
Identifiers: LCCN 2016012151| ISBN 9781522504245 (hardcover) | ISBN 9781522504252 (ebook)
Subjects: LCSH: Fibrous composites. | Sustainable engineering.
Classification: LCC TA418.9.C6 G74 2016 | DDC 620.1/97--dc23 LC record available at https://lccn.loc.gov/2016012151

This book is published in the IGI Global book series Advances in Chemical and Materials Engineering (ACME) (ISSN: 2327-5448; eISSN: 2327-5456)

British Cataloguing in Publication Data
A Cataloguing in Publication record for this book is available from the British Library.

Advances in Chemical and Materials Engineering (ACME) Book Series

ISSN: 2327-5448
EISSN: 2327-5456

Mission

The cross disciplinary approach of chemical and materials engineering is rapidly growing as it applies to the study of educational, scientific and industrial research activities by solving complex chemical problems using computational techniques and statistical methods.

The **Advances in Chemical and Materials Engineering (ACME) Book Series** provides research on the recent advances throughout computational and statistical methods of analysis and modeling. This series brings together collaboration between chemists, engineers, statisticians, and computer scientists and offers a wealth of knowledge and useful tools to academics, practitioners, and professionals through high quality publications.

Coverage

- Numerical Techniques
- Nanomaterials
- Computational methods
- Fracture Mechanics
- Sustainable Materials
- Industrial chemistry
- Composites
- Materials to Renewable Energies
- Thermo-Chemical Treatments
- Coatings and surface treatments

The Advances in Chemical and Materials Engineering (ACME) Book Series (ISSN 2327-5448) is published by IGI Global, 701 E. Chocolate Avenue, Hershey, PA 17033-1240, USA, www.igi-global.com. This series is composed of titles available for purchase individually; each title is edited to be contextually exclusive from any other title within the series. For pricing and ordering information please visit http://www.igi-global.com/book-series/advances-chemical-materials-engineering/73687. Postmaster: Send all address changes to above address.

Titles in this Series

For a list of additional titles in this series, please visit: www.igi-global.com

Position-Sensitive Gaseous Photomultipliers Research and Applications
Tom Francke (Myon, Sweden) and Vladimir Peskov (CERN, Switzerland)
Engineering Science Reference • copyright 2016 • 562pp • H/C (ISBN: 9781522502425) • US $240.00 (our price)

Research Perspectives on Functional Micro- and Nanoscale Coatings
Ana Zuzuarregui (CIC nanoGUNE Consolider, Spain) and Maria Carmen Morant-Miñana (CIC nanoGUNE Consolider, Spain)
Information Science Reference • copyright 2016 • 511pp • H/C (ISBN: 9781522500667) • US $215.00 (our price)

Emerging Research on Bioinspired Materials Engineering
Mohamed Bououdina (University of Bahrain, Bahrain)
Engineering Science Reference • copyright 2016 • 441pp • H/C (ISBN: 9781466698116) • US $215.00 (our price)

Petrochemical Catalyst Materials, Processes, and Emerging Technologies
Hamid Al-Megren (King Abdulaziz City for Science and Technology, Saudi Arabia) and Tiancun Xiao (Oxford University, UK)
Engineering Science Reference • copyright 2016 • 539pp • H/C (ISBN: 9781466699755) • US $225.00 (our price)

Applying Nanotechnology to the Desulfurization Process in Petroleum Engineering
Tawfik A. Saleh (Department of Chemistry, King Fahd University of Petroleum and Minerals, Saudi Arabia)
Engineering Science Reference • copyright 2016 • 555pp • H/C (ISBN: 9781466695450) • US $235.00 (our price)

Quantitative Structure-Activity Relationships in Drug Design, Predictive Toxicology, and Risk Assessment
Kunal Roy (Jadavpur University, India)
Medical Information Science Reference • copyright 2015 • 703pp • H/C (ISBN: 9781466681361) • US $295.00 (our price)

Handbook of Research on Recent Developments in Materials Science and Corrosion Engineering Education
Hwee Ling Lim (The Petroleum Institute, UAE)
Engineering Science Reference • copyright 2015 • 493pp • H/C (ISBN: 9781466681835) • US $325.00 (our price)

Table of Contents

Preface xiv

Chapter 1
Natural Fibers for the Production of Green Composites 1
Xiaolei Zhang, Queen's University Belfast, UK
Jun Li, University of Strathclyde, UK
Siddharth Jain, College of Engineering Roorkee, India
Deepak Verma, Graphic Era Hill University, Dehradun, India

Chapter 2
Processing Technologies for Green Composites Production 24
Deepak Verma, Graphic Era Hill University, India
Garvit Joshi, Graphic Era Hill University, India
Rajneesh Dabral, Graphic Era Hill University, India

Chapter 3
Concurrent Design of Green Composites 48
Muhd Ridzuan Mansor, Universiti Teknikal Malaysia Melaka, Malaysia
S. M. Sapuan, University Putra Malaysia, Malaysia
Mohd Azli Salim, Universiti Teknikal Malaysia Melaka, Malaysia
Mohd Zaid Akop, Universiti Teknikal Malaysia Melaka, Malaysia
M. T. Musthafah, Universiti Teknikal Malaysia Melaka, Malaysia
M. A. Shaharuzaman, Universiti Teknikal Malaysia Melaka, Malaysia

Chapter 4
Effect of Bamboo Hybridization and Staking Sequence on Mechanical
Behavior of Bamboo-Glass Hybrid Composite 76
Piyush P. Gohil, The M. S. University of Baroda, India
Kundan Patel, CHARUSAT, India
Vijaykumar Chaudhary, CHARUSAT, India
Ronak Ramjiyani, CHARUSAT, India

Chapter 5
Estimation of Mechanical and Tribological Properties of Epoxy-Based Green Composites 96
Supriyo Roy, Birla Institute of Technology, India
Sumit Bhowmik, National Institute of Technology, India
J. Paulo Davim, University of Aveiro, Portugal
Kaushik Kumar, Birla Institute of Technology, India

Chapter 6
Fabrication and Processing of Pineapple Leaf Fiber Reinforced Composites 125
S. H. Sheikh Md. Fadzullah, Universiti Teknikal Malaysia Melaka, Malaysia
Zaleha Mustafa, Universiti Teknikal Malaysia Melaka, Malaysia

Chapter 7
Green Composites and Their Properties: A Brief Introduction 148
Deepak Verma, Graphic Era Hill University, India
Prakash Chandra Gope, College of Technology, India
Xiaolei Zhang, Queen's University Belfast, UK
Siddharth Jain, University of Alberta, Canada
Rajneesh Dabral, Graphic Era Hill University, India

Chapter 8
Rice Husk Reinforcement in Polymer Composites 165
Sanjay Sharma, Graphic Era Hill University, India
Deepak Verma, Graphic Era Hill University, India

Chapter 9
Techno-Economic and Life Cycle Assessment for the Production of Green Composites 192
Siddharth Jain, College of Engineering Roorkee, India
Xiaolei Zhang, Queen's University Belfast, UK

Chapter 10
Banana Fiber Reinforcement and Application in Composites: A Review 201
Abhinav Shandilya, Federal Mogul Goetze India Ltd., India
Ayush Gupta, College of Engineering Roorkee, India
Deepak Verma, Graphic Era Hill University, India

Chapter 11
Bamboo Fiber-Reinforced Composites 228
Irem Sanal, Atilim University, Turkey

Chapter 12
Coir Fiber-Reinforced Composites 247
Irem Sanal, Atilim University, Turkey

Compilation of References 276

About the Contributors 312

Index 319

Detailed Table of Contents

Preface.. xiv

Chapter 1
Natural Fibers for the Production of Green Composites.. 1
Xiaolei Zhang, Queen's University Belfast, UK
Jun Li, University of Strathclyde, UK
Siddharth Jain, College of Engineering Roorkee, India
Deepak Verma, Graphic Era Hill University, Dehradun, India

Development of green composite from natural fibers has gained increasing interests due to the environmental and sustainable benefits when compared with petroleum based non-degradable materials. However, a big challenge of green composites is the diversity of fiber sources, because of the large variation in the properties and characteristics of the lignocellulosic renewable resource. The lignocellulosic fibers/natural fibers used to reinforce green composites are reviewed in this chapter. A classification of fiber types and sources, the properties of various natural fibers, including structure, composition, physical and chemical properties are focused; followed by the impacts of natural fibers on composite properties, with identification of the main pathways from the natural fibers to the green composite. Furthermore, the main challenges and future trend of natural fibers are highlighted.

Chapter 2
Processing Technologies for Green Composites Production 24
Deepak Verma, Graphic Era Hill University, India
Garvit Joshi, Graphic Era Hill University, India
Rajneesh Dabral, Graphic Era Hill University, India

Green composites became a most important and adaptable theme of research. This area/theme not only harness the agricultural wastes such as bagasse fibres, banana fibres, etc. but also provides a new material manufactured from these wastes which are reduced weight, have low cost, and have high mechanical strength. Currently, there are various methods available for the processing or fabrication of green composites.

Some of these methods are hand layup method, injection molding method, spray-up method, compression molding, Resin-Transfer Molding (RTM), etc. In this chapter, we are discussing about the fabrication method of green composite and their important parameters. Various properties and characterization of composite materials made by these methods have also been discussed and reported here.

Chapter 3
Concurrent Design of Green Composites 48
Muhd Ridzuan Mansor, Universiti Teknikal Malaysia Melaka, Malaysia
S. M. Sapuan, University Putra Malaysia, Malaysia
Mohd Azli Salim, Universiti Teknikal Malaysia Melaka, Malaysia
Mohd Zaid Akop, Universiti Teknikal Malaysia Melaka, Malaysia
M. T. Musthafah, Universiti Teknikal Malaysia Melaka, Malaysia
M. A. Shaharuzaman, Universiti Teknikal Malaysia Melaka, Malaysia

This chapter presents the overview of concurrent design process of green composite products with special focus on conceptual design stage of natural fiber composites product development. Design of green composites product especially during the early product development stage requires three main aspects in product design which are materials, design and manufacturing process to satisfy lower cost, high quality and fast development time requirements in order to ensure successful product launch into the market. In this chapter, the concurrent design process of green composite products are discussed involving several main stages in product development such as green composite materials selection for both natural fiber and matrix constituents, conceptual design development and concept design selection of green composite products, and green composites manufacturing process selection. In addition, discussion on life cycle assessment of green composites is also included in order to provide further insight of the sustainability design requirements to the overall product development process.

Chapter 4
Effect of Bamboo Hybridization and Staking Sequence on Mechanical Behavior of Bamboo-Glass Hybrid Composite 76
Piyush P. Gohil, The M. S. University of Baroda, India
Kundan Patel, CHARUSAT, India
Vijaykumar Chaudhary, CHARUSAT, India
Ronak Ramjiyani, CHARUSAT, India

The advancement of polymer composites containing natural fibers as a manageable option material for certain designing applications, especially aviation and car applications, is a well-known area of investigation. Nevertheless, the high mechanical properties connected with synthetic fibers they are awesome and lavish contrasted with natural fibers. The utilization of natural plant fibers and mixes of natural and synthetic fibers for making ease building materials has produced much interest recently. In the present work, bamboo–glass hybrid polyester composites were produced and their mechanical properties like elasticity and flexural quality were assessed for different weight fraction and distinctive stacking sequence. The outcomes observed that bamboo–glass mixture composites offered the benefits of both natural and synthetic fibers. It is also observed that hybridization started a material with general intermediate properties between pure glass and pure bamboo. However, the significance of controlling the stacking grouping to upgrade properties was evident.

Chapter 5
Estimation of Mechanical and Tribological Properties of Epoxy-Based Green Composites..96
Supriyo Roy, Birla Institute of Technology, India
Sumit Bhowmik, National Institute of Technology, India
J. Paulo Davim, University of Aveiro, Portugal
Kaushik Kumar, Birla Institute of Technology, India

Composites based on natural fibre reinforcement have generated wide research and engineering interest in the last few decades due to their small density, high specific strength, low cost, light weight, recyclability and biodegradability and has earned a special category of 'green composite'. Here, in our proposed research, wood dust reinforced epoxy composite was processed with different % filler weight primarily. For this, natural filler based epoxy composite from wood dust is developed and its mechanical behaviour, including Tensile, Flexural, Density etc., under various testing conditions and % of filler weight were studied. These samples were simultaneously tested for abrasive wear and friction coefficient measurement. Microstructure of the composites was studied to analyze the distribution of the filler in the epoxy matrix change using scanning electron microscopy.

Chapter 6
Fabrication and Processing of Pineapple Leaf Fiber Reinforced Composites..125
S. H. Sheikh Md. Fadzullah, Universiti Teknikal Malaysia Melaka, Malaysia
Zaleha Mustafa, Universiti Teknikal Malaysia Melaka, Malaysia

There is an increasing interest worldwide in the use of Pineapple Leaf Fibers (PALF) as reinforcements in polymer composites, since this type of natural fiber exhibit attractive features such as superior mechanical, physical and thermal properties, thus offer potential uses in a spectrum of applications. PALF contains high cellulose content (between 70-82%) and high crystallinity. However, being hydrophilic, it posed a compatibility issue particularly in a hydrophobic polymeric matrix system. Thus, their shortcoming need to be addressed to ensure good interfacial bonding at the fibers/matrix interphase before their full potential can be harnessed. This chapter summarized some of the important aspects relating to PALF and its reinforced composites, particularly the main characteristics of the fiber, extraction and pre-treatment process of the fibers. Following this, discussions on the available fabrication processes for both short and continuous long PALF reinforced composites are presented.

Chapter 7
Green Composites and Their Properties: A Brief Introduction 148
Deepak Verma, Graphic Era Hill University, India
Prakash Chandra Gope, College of Technology, India
Xiaolei Zhang, Queen's University Belfast, UK
Siddharth Jain, University of Alberta, Canada
Rajneesh Dabral, Graphic Era Hill University, India

Green composites are important class of biocomposites widely explored due to their enhanced properties. The biodegradable polymeric material is reinforced with natural fibers to form a composite that is eco-friendly and environment sustainable. The green composites have potential to attract the traditional petroleum-based composites which are toxic and nonbiodegradable. The green composites eliminate the traditional materials such as steel and wood with biodegradable polymer composites. The degradable and environment-friendly green composites were prepared by various fabrication techniques. The various properties of different fiber composite were studied as reinforcement for fully biodegradable and environmental-friendly green composites.

Chapter 8
Rice Husk Reinforcement in Polymer Composites 165
Sanjay Sharma, Graphic Era Hill University, India
Deepak Verma, Graphic Era Hill University, India

Increasing concern about global warming and depleting petroleum reserves and the high cost of petroleum products had made scientists to focus more on the use of natural fibres such as rice husk, baggase, coconut husk, hemp, sisal, jute, flax, banana etc. Past decade has shown many efforts to develop composites to replace the Petroleum and other non-decaying material products. Reinforcement with natural fibre in composites has recently gained attention due to low cost, easy availability, low density, acceptable, strength full, stiffness, ease of separation, enhanced energy recovery, biodegradability and recyclable in nature. Natural fibre composites are suitable as wood substitutes in the construction sector. All these have excellent physical, thermal and mechanical properties and can be utilized more effectively in the development of composite materials. In this connection, an investigation has been carried using rice husk, a natural fibre abundantly available in India.

Chapter 9
Techno-Economic and Life Cycle Assessment for the Production of Green Composites 192
Siddharth Jain, College of Engineering Roorkee, India
Xiaolei Zhang, Queen's University Belfast, UK

Botanically, green composites belong to an economically important seed plant family that includes maize, wheat, rice, and sorghum known as Saccharum offi cinarum. There are so many natural fibers available in the environment such as rice husk, hemp fibers, flax fibers, bamboo fibers, coconut fiber, coconut coir, grawia optiva and many others also. Life Cycle Assessment (LCA) is a process to estimate the environmental feature and potential impacts related to a product, by organizing a directory of pertinent inputs and outputs of a product system, assessing the potential environmental impacts related with the said inputs and outputs, explaining the results of the inventory analysis and impact evaluation phases in connection to the objectives of the study. Particularly Bagasse, an agricultural residue not only becomes a problem from the environmental point of view, but also affects the profitability of the sugarcane industries. This chapter discusses the properties, processing methods and various other aspects including economic and environmental aspects related to green composites.

Chapter 10
Banana Fiber Reinforcement and Application in Composites: A Review......... 201
Abhinav Shandilya, Federal Mogul Goetze India Ltd., India
Ayush Gupta, College of Engineering Roorkee, India
Deepak Verma, Graphic Era Hill University, India

The growing awareness about sustainable development, environmental ecology and new legislations has led researchers to focus attention on bio fibres reinforced composites. In this field research has been done on many fibres but fibres such as banana, coir, bagasse, jute have gained importance in the recent decades. The main advantage of the natural fibre based composites materials being their low cost, easy availability, low density, acceptable specific properties, ease of separation, enhanced energy recovery, C02 neutrality, biodegradability and recyclability in nature. The attention is being given to the development of natural fibre composites is to explore value-added application avenues for their use and also for a sustainable and economical use of easily available natural material in hand. Agricultural waste is a very good example of such naturally available material and it can also be used to prepare composite materials for commercial use this has a very significant advantage over other natural fibres as its abundance and because of almost no cost.

Chapter 11
Bamboo Fiber-Reinforced Composites 228
Irem Sanal, Atlim University, Turkey

Nowadays, there has been an increased interest in the applications of bio-composites based on natural fibers, with the increasing emphasis on materials and processes which are environmental friendly and sustainable. Environmental friendly, fully biodegradable reinforced polymers or 'green' composite materials will play a major role in making the products of the future to protect our environment. The use of biodegradable and environment-friendly plant-based natural fibers has been a promising choice for polymers to make them 'greener'. In addition to being obtained from renewable sources, natural fibers suitable for composites are biodegradable and have enhanced properties. Bamboo is an excellent example for the development of sustainable natural fibers, since it can grow very fast per day, and the fibers of bamboo have excellent mechanical performance. Additionally, research in the development of bamboo-reinforced composites should be increased in the future, considering their enhanced properties, economical benefits and environmental friendly nature.

Chapter 12
Coir Fiber-Reinforced Composites 247
Irem Sanal, Atilim University, Turkey

Nowadays, fiber-reinforced polymer composites have played a significant role in many different fields of applications, regarding their high specific strength and high modulus. The fiber which serves as reinforcement mechanism in polymer composites may be either synthetic or natural. Natural fibers are not only strong and lightweight but also very economical and environmental friendly. Natural fibers as reinforcement are stated to be a major step taken in promoting environmental protection and sustainability. There are many types of natural cellulose fibers but the thickest and most resistant of all commercial natural fibers, coir/coconut fiber is a coarse, short fiber extracted from the outer shell of coconuts. Coir/coconut fibers have the highest concentrations of lignin, making it most suitable for applications where slow biodegradability is required. This chapter has been written with an aim to explore the potential of the coconut/coir fiber reinforced polymer composites in terms of their performance, surface treatments/modifications and areas of application.

Compilation of References 276

About the Contributors 312

Index 319

Preface

Natural fibres can be defined as bio-based fibres or fibres from vegetable and animal origin. This definition includes all natural cellulosic fibres (cotton, jute, sisal, coir, flax, hemp, abaca, ramie, etc.) and protein based fibres such as wool and silk. Excluded here are mineral fibres such as asbestos that occur naturally but are not bio based. Asbestos containing products are not considered sustainable due to the well known health risk, that resulted in prohibition of its use in many countries. On the other hand there are manmade cellulose fibres (e.g. viscose-rayon and cellulose acetate) that are produced with chemical procedures from pulped wood or other sources (cotton, bamboo). Similarly, regenerated (soybean) protein, polymer fibre (bio-polyester, PHA, PLA) and chitosan fibre are examples of semi-synthetic products that are based on renewable resources. In this paper also the use of fibres in food industries is excluded, where in recent years these are frequently promoted as dietary fibres or as supplements for health products.

Natural fibers are abundantly available worldwide in the form of agriculture biomass or agriculture wastes. It is seen that these fibers are another or possible source of renewable materials and can be used as prospective materials for different applications. In spite of vast production of agricultural wastes, a very little amount of the agricultural biomass or wastes is used for various applications. From the industrial point of view, they are utilizing this bio-wastes in the best possible applications. Here we are only concentrated on the term Natural Fibers which is derived from the agriculture wastes and can be used for the production of the composite materials. Natural fibres generally provide outstanding properties (specifically mechanical and thermal) and can be utilized as excellent reinforcing fillers in the matrix and can be act as an another option for bio-composites and also hybrid composites. The various Natural fibre used for the development of the polymer composites are jute, Bagasse, Coconut coir, hemp, pineapple leaf etc. have a economical market cost and are utilized by the various commercial industries such as Automobile industries, Building and construction industries, ceilings and partition boards, structural applications, aerospace, sports, boats, office products, machinery etc for their different applications. It is also observed that the future research on natural fibre-based composites is not only restricted to its automotive applications but also necessitate to broaden its application in construction industry, rural housing and biomedical applications.

The utilization of the natural waste/ fiber in polymer matrix is also reduces the environment pollution. The basic components of natural fibers are cellulose, hemi-cellulose, lignin, pectin, waxes and water soluble substances. In present scenario natural fiber composites become the important materials from the environmental point of views. These materials have potential of great mechanical strength and also have low density and are light in weight. Various industries like automative and interior decorative industries are using these materials in their industries for making automative composnents, car doors, car panels, interirors of cars, furnitures etc. Each chapter aims on the awareness and insight of the interfaces demonstrated in these green composites systems and also optimizes different parameters for the novel processing. In addition, the book also encapsulate the current progress put together in the area of the various composite fabrication techniques, chemical functions and reactions of natural fibers, novel processing of biocomposites, their applications in the various industries. Some censorious issues and recommendations for future work will be discussed in the present book. The main objective of this publication is to aware the academicians and scholars about environment concern and to develop the biocomposites by using agricultural wastes or natural wastes like plants seeds etc. The research scholars will learn from the past research in the field of green composites and then can get ideas for the further development of the biocomposites by using various polymer composites like thermosetting or thermoplastic polymers. They can also learn about the green polymer matrix for the development of bidegradable composites. This book discusses the issues related to the agriculture wastes/ residue (also natural fibers) which could be effectively utilized to make the composite materials (Polymer and biopolymer based) that ultimately solves the environmental concerns. The use of natural fibers into the composites not only solves the disposal problem but also utilizes these waste natural fibers to make some useful products. It is generally termed as a renewable source because wastes from agriculture or natural fibers are grown in a mass level and can become a great source of energy by utilizing these into a material form or some other energy source like fuel etc. The proper utilization of these residues or natural fibers in polymer matrix or biopolymer matrix has resulted in the form of biodegradable material. The term green composite material means the composite which is environmental friendly and fully sustainable. Global realization related to environmental concerns, has exposed the economic and environmental friendly biocomposite materials which are free from the side effects of synthetic fiber composites. In addition, natural fibers or green materials can be used on a sustainable basis and may create unique opportunity for regional development.

It is also seen that the Natural fibers such as Flax, Hemp, Jute, Bamboo fibers generally have higher moduli as compared to the synthetic fibers (E glass fibres). The typical market specification for the automotive applications are ultimate breaking force, flexural properties, impact strength, flammability, acoustic absorption, water absorption and crash behavior.

The researchers related to the automotive industries have kept in mind specifically the durability of the automotive components and used non degradable polymeric resins such as epoxies, polyurethane by reinforcing the man-made fibers such as aramid, glass fibers etc. Many of these polymers and fibers are derived from petroleum which is also a renewable resource.

Biocomposites (natural fiber composites) from local and renewable resources offer significant sustainability; industrial ecology, eco-efficiency, and green chemistry are guiding the development of the next generation of materials, products, and processes. Considerable growth has been seen in the use of biocomposites in the domestic sector, building materials, aerospace industry, circuit boards, and automotive applications over the past decade, but application in other sectors until now has been limited. Nevertheless, with suitable development, the potential exists for biocomposites to enter new markets and thus stimulate an increase in demand. Many types of natural fibers have been investigated with polymer matrices to produce composite materials that are competitive with synthetic fiber composites which require special attention. The agricultural wastes can be used to prepare fiber-reinforced polymer composites for commercial use and have marketing appeal. The growing global environmental and social concern, high percentage of exhaustion of petroleum resources, and new environmental regulations have forced the search for new composites, compatible with the environment. Many references to the current status of research work on the applications of biocomposites are cited in this book.

In the present era of environmental consciousness, more and more material are emerging worldwide, Efficient utilization of plant species and utilizing the smaller particles and fibers obtained from various lignocellulosic materials including agro wastes to develop eco-friendly materials is thus certainly a rational and sustainable approach. Any lignocellulosic waste matter can, therefore, be turned into composite products through appropriate R& D work and development in technological aspects. These approaches offer much simpler materials for future use in comparison to metal based composites. Resilience property which makes plastics ideal for many applications like food industries, packaging, construction field and sanitation products etc. Petroleum-derived plastics can lead to waste disposal problems, as these materials are not readily biodegradable and because of their resistance to microbial degradation, they accumulate in the environment. Biodegradable plastics and polymers were first introduced in 1980s. There are many sources of biodegradable plastics, from synthetic to natural polymers. Natural polymers are available in large quantities from renewable sources, while synthetic polymers are produced from non renewable petroleum resources.

The future outlook for development in the field of biopolymers materials is promising because of its Environment friendly behaviour. Bio-composites often lead to a reduction in weight and costs and potential applications in the fields related to environmental protection and the maintenance of physical health. For these reasons the popularity of these composites is increasing in the western countries and already a significant amount of scientific knowledge is generated. Starch is a vegetable matter renewable from carbon dioxide, water and sunshine. It is biodegradable, cheap and to be physical or chemical modified easily. This means someday it is unnecessary to rely on petroleum resources. This will ensure their continued interest in agro forestry. This is necessary for keeping alive the tempo of tree growing which is necessary for the very survival of humanity.

Biodegradation takes place through the action of enzymes and/or chemical deterioration associated with living organisms. Biodegradability depends not only on the origin of the polymer but also on its chemical structure and the environmental degrading conditions. Biobased polyesters such as PLA and polyhydroxyalkanoates (PHAs) have begun to be more accepted since they are prepared from renewable feedstock rather than petroleum and are biodegradable.

The objective of this book is to utilize the natural fibers or natural resource-based material and suggests the potential replacement of synthetic fiber composite material by green composite materials. This book elaborates the information of the various available natural fibers and use of these fibers as a sustainable option to synthetic fibers and polymers by considering the present situation and growing benefit of green materials from natural resources. This book also provides awareness on recent details and progress made in the fabrication of natural polymers or renewable materials for their variety of applications. The various types of natural polymer (or biopolymer) i.e. their manufacturing and properties is also discussed in this book. Thermal analysis of green composites is also discussed in this book.

The present book concentrates on the progress of biomaterials in the field of orthopaedics, an effort to utilize the advantages offered by renewable resources for the development of biocomposite materials based on biopolymers and natural fibers. The present book focuses on the enhanced properties of natural fiber as bone implant. It is a challenge to the creation of better materials for the improvement of quality of life. The present book proposes suggestions of using natural fiber reinforced composite as a plate material, which uses pure natural fibers that are rich in medicinal properties.

Academicians, researchers focusing on bio-materials or natural fiber reinforced composite material will find this text useful in furthering their research exposure to relevant topics in green composites and assisting in furthering their own research efforts in this field. This book discusses the importance of the natural fibers and their properties. Various factors which can be considered before reinforcing in to the matrix are also discussed in the present chapter. The research scholars who are working in the same research field can take an advantage from this book by knowing about the various types of the natural fibers available and their properties.

ORGANIZATION OF THE CHAPTERS

The book is organized into twelve chapters. Brief descriptions of each of the chapters are as follows

Chapter 1 identifies the the various types of the natural fibers available and their structures. In this chapter a classification of fiber types and sources, the properties of various natural fibers, including structure, composition, physical and chemical properties are focused; followed by the impacts of natural fibers on composite properties, with identification of the main pathways from the natural fibers to the green composite. Furthermore, the main challenges and future trend of natural fibers are highlighted.

Chapter 2 establishes the information about the various types of the processes available for the manufacturing of the green composite materials. Some of these methods are hand layup method, injection molding method, spray-up method, compression molding, Resin-Transfer Molding (RTM), etc. In this chapter authors have discussed about the fabrication method of green composite and their important parameters. Various properties and characterization of composite materials made by these methods have also been discussed and reported in this chapter.

Chapter 3 discussed about the concurrent design of the composite materials with special focus on conceptual design stage of natural fiber composites product development. The main stages for the product development used by the concurrent design process are also discussed in this chapter. In addition, discussion on life cycle assessment of green composites is also included in order to provide further insight of the sustainability design requirements to the overall product development process.

Chapter 4 presents an analysis of the bamboo–glass hybrid polyester composites and their mechanical properties like elasticity and flexural quality for different weight fraction and distinctive stacking sequence. This chapter also showed that the hybridization started a material with general intermediate properties between pure glass and pure bamboo.

Chapter 5 discussed the development of natural filler based epoxy composite from wood and its mechanical behaviour, including Tensile, Flexural, Density etc., under various testing conditions and % of filler weight. These samples were simultaneously tested for abrasive wear and friction coefficient measurement. Microstructure of the composites was studied to analyze the distribution of the filler in the epoxy matrix change using scanning electron microscopy.

Chapter 6 discussed the use of Pineapple Leaf Fibers (PALF) as reinforcements in polymer composites, since this type of natural fiber exhibit attractive features such as superior mechanical, physical and thermal properties, thus offer potential uses in a spectrum of applications. This chapter summarized some of the important aspects relating to PALF and its reinforced composites, particularly the main characteristics of the fiber, extraction and pre-treatment process of the fibers. Discussions on the available fabrication processes for both short and continuous long PALF reinforced composites are also presented in this chapter.

Chapter 7 discussed a brief introduction about the green composites and their properties. The green composite have potential to attract the traditional petroleum-based composites which are toxic and non-biodegradable. The various properties of different fiber composite were discussed as reinforcement for fully biodegradable and environmental-friendly green composites.

Chapter 8 discussed the use of ricer husk as filler in the polymer matrix composites. In this chapter authors presented a review on the past research based on the rice husk reinforcement in polymer matrix and their mechanical properties. Various types of mechanical properties such as tensile strength, flexural strength, impact strength are reviewed in the present chapter. Microstructure of the rice husk reinforced composite is shown by the scanning electron microscopy.

Chapter 9 discussed the techno economic and life cycle assessment of the green composites. Life Cycle Assessment (LCA) is a process to estimate the environmental feature and potential impacts related to a product, by organizing a directory of rpertinant inputs and outputs of a product system, assessing the potential environmental impacts related with the said inputs and outputs, explaining the results of the inventory analysis and impact evaluation phases in connection to the objectives of the study.

Chapter 10 reviews the banana fiber reinforcement and application in the composites. The attention is being given to the development of natural fibre composites is to explore value-added application avenues for their use and also for a sustainable and economical use of easily available natural material in hand. In this chapter authors presented a review on the past research based on the banana fiber reinforcement in polymer matrix and their mechanical properties. Various types of mechanical properties such as tensile strength, flexural strength and impact strength are reviewed in the present chapter. Microstructurc of the banana fiber reinforced composite is shown by the scanning electron microscopy.

Chapter 11 discussed the use of bamboo fibers in polymer composites. Bamboo is an excellent example for the development of sustainable natural fibers, since it can grow very fast per day, and the fibers of bamboo have excellent mechanical performance. Additionally, research in the development of bamboo-reinforced composites should be increased in the future, considering their enhanced properties, economical benefits and environmental friendly nature. In this chapter authors presented a review on the past research based on the bamboo fiber reinforcement in polymer matrix and their mechanical properties. Various types of mechanical properties such as tensile strength, flexural strength and impact strength are reviewed in the present chapter.

Chapter 12 discussed the use of coir fibers in polymer composites. Coir/coconut fiber is a coarse, short fiber extracted from the outer shell of coconuts. Coir/coconut fibers have the highest concentrations of lignin, making it most suitable for applications where slow biodegradability is required. This chapter has been written with an aim to explore the potential of the coconut/coir fiber reinforced polymer composites in terms of their performance, surface treatments/modifications and areas of application.

We wish to express our gratitude to all the authors/contributors from all over the world for accepting our invitations for contribution and sharing their knowledge but also for commendably integrating their skills on scattered information from various fields in making the chapters and undergo editorial suggestions to finally produce this project that will hope be a success. We greatly appreciate their commitment. We thank IGI Global team for their generous cooperation at every stage of the book production.

Deepak Verma
Graphic Era Hill University, India

Siddharth Jain
University of Alberta, Canada & College of Engineering Roorkee, India

Xiaolei Zhang
Queens University Belfast, UK

Prakash Chandra Gope
GBPUAT, India

Chapter 1
Natural Fibers for the Production of Green Composites

Xiaolei Zhang
Queen's University Belfast, UK

Jun Li
University of Strathclyde, UK

Siddharth Jain
College of Engineering Roorkee, India

Deepak Verma
Graphic Era Hill University, Dehradun, India

ABSTRACT

Development of green composite from natural fibers has gained increasing interests due to the environmental and sustainable benefits when compared with petroleum based non-degradable materials. However, a big challenge of green composites is the diversity of fiber sources, because of the large variation in the properties and characteristics of the lignocellulosic renewable resource. The lignocellulosic fibers/natural fibers used to reinforce green composites are reviewed in this chapter. A classification of fiber types and sources, the properties of various natural fibers, including structure, composition, physical and chemical properties are focused; followed by the impacts of natural fibers on composite properties, with identification of the main pathways from the natural fibers to the green composite. Furthermore, the main challenges and future trend of natural fibers are highlighted.

DOI: 10.4018/978-1-5225-0424-5.ch001

INTRODUCTION

Remarkable achievements of green technology in material science have been made in past few decades, mainly by using natural resources to develop high performance engineering products. Utilization of natural fibers, or lignocellulosic fibers, extracted from biodegradable materials to reinforce composites have attracted increasing interest (Satyanarayana, Arizaga, & Wypych, 2009). The design of novel, green, value-added products focuses on sustainable ecology and "green chemistry" with the holistic approach of life cycle assessment (Faruk, Bledzki, Fink, & Sain, 2014; Satyanarayana et al., 2009). Moreover, other driving forces for increased global attention on natural fiber reinforced composites, are strong worldwide demands for:

1. Creating a resource circulating society, and
2. Addressing the environmental challenge of using petroleum-based products (Satyanarayana et al., 2009).

Lignocellulosic materials had been used earlier around a century ago to produce composites. The first polymer composite based on natural fiber was recorded as early as in 1908. This composite was produced from cotton or paper reinforced on phenol or melamine form aldehyde resins (Bledzki & Gassan, 1999). In the 1940s, the development of reinforcement using glass fibers with lower cost and superior properties took over the competition with lignocellulosic materials (Satyanarayana et al., 2009). However, utilization of glass fibers suffers the problems of high density of glass leading to high dead-weight of the material, and disposal problem at their end life.

The development of lignocellulosic fibers restarted during the period of oil crisis. Required by the laws on reduction of environmentally unfriendly materials, great efforts had been paid by the EU to manufacture eco-friendly products based on natural resources. In 1990s, cellulosic fibers were explored extensively to replace traditional glass fibers for reinforcement composites. In 21st century, polymers reinforced with natural fibers (NFCs), commonly known as bio-composites, have been developed and successfully applied in automotive and building sectors, as well as other consumer goods.

Natural fibers are playing an important role in a number of applications due to their inherent eco-friendly advantages, which are explored to substitute traditional synthetic fibers, particularly, emphasizing green reinforcement. Consequently, natural fibers reinforced composites are emerging very rapidly as the potential substitute of the traditional synthetic fibers in automotive, aerospace, marine, sporting goods and electronic industries (Vijay Kumar Thakur & Thakur, 2014) for the following major advantages:

- **Sustainability:** The renewable nature of lignocellulosic fibers lead to the sustainable benefit for their derived materials (V. K. Thakur, Singha, & Thakur, 2012a). Lignocellulosic fibers are carbon neutral materials; it is acknowledged that they do not release excess carbon dioxide into the atmosphere after being composted or combusted. The working conditions are friendly thereby reducing the chances of dermal and respiratory irritation (John & Thomas, 2008). The increasing ecological concerns help to have greater focus on plant derived fiber and crop derived plastics as materials, with attractive benefits in CO_2 sequestration, reduce dependence on petroleum products and added opportunities for agricultural industries (Satyanarayana et al., 2009).
- **Recyclability:** The thermal decomposition and biodegradable properties of natural fibers reduce the environmental impact when compared with petroleum derived materials. It was pointed out that producing petroleum from biomass needs 10^6 years, compared to 1 to10 years of time to convert the fuels or chemicals into CO_2 (Satyanarayana et al., 2009). While the degradable time for most of the natural fibers is just1 to 6 months as opposed to 13 years for painted wood, 450 years for plastic, and for glasses and tires, the degradation need an uncertain long time (Satyanarayana et al., 2009).
- **Availability:** The existence of abundant availability of lignocellulosic resources throughout the world, with some of them being abundant typically in the tropics and agricultural crops (Satyanarayana et al., 2009), is an additional factor for the interests in natural fibers, especially with the foreseeable shortage of fossil fuels (John & Thomas, 2008).
- **Weight Reduction:** Lower mass density of natural fibers than that of glass fibers brings the benefit of reduced weight and saving in energy consumption (Faruk et al., 2014; John & Thomas, 2008).
- **Cost Reduction:** The growing interest in using lignocellulosic fibers is driven by the economic benefits. Bio-fibers are nonabrasive to mixing and molding equipment, which provide longevity to the life of the processing equipment and thereby contributes to reduced processing cost to the tune of one third of glass fibers (Faruk et al., 2014).

However, the shortcomings from some properties of natural fiber might cause serious limitations and problems in shipping, storage, processing and applications of bio-composites. A major disadvantage of natural fiber reinforced composites is the non-uniform dispersion of fibers within the matrix which impairs the efficiency of the composite. This non-uniformity is caused by the difficulties jointly associated with the inherent polar and hydrophilic nature of lignocellulosic fibers and the non-polar characteristics of most thermoplastics. Another problem is that the processing of composites is restricted to 200 °C because a higher temperature results in degradation. The high moisture absorption of natural fibers is cited as another disadvantage, which causes swelling and presence of voids at the interface, resulting in poor mechanical properties and reduction of dimensional stability of composites. Lastly, low microbial resistance and susceptibility may cause rotting, which limits the exploitation of bio-fibers in durable composite applications (John & Thomas, 2008).

The obvious advantages outweigh the limitations and drive the increasing development of natural fibers for various applications as a replacement of petroleum derived synthetic materials. Based on the challenges of applying natural fibers on the variation of types, physical and chemical properties of the fiber nature, the detailed characterization of natural fibers and their impact on the composite properties will be covered in this chapter.

CHARACTERIZATION OF NATURAL FIBERS

Fiber Type

According to the sources, natural fibers can be classified into: plant fibers, animal fibers, and minerals fibers, as shown in Figure 1. The main composition for plant fibers is cellulose thus they can also be called as cellulosic fibers. Cellulosic fibers have numerous advantages over synthetic materials, such as abundant availability and environmentally friendly. Various cellulosic fibers include non-wood based or agricultural based fibers such as stem, straw, bast, leaf, seed, fruit, and grass, and wood based fibers (V. K. Thakur et al., 2012a). The main composition of animal fibers is proteins. While for mineral fibers, two typical fiber types are asbestos and metal fibers, with the main composition of hydrous magnesium silicates and metals respectively. The pure fibers need to be extracted and separated from either plant or animal raw materials, which can be further used as reinforcement (Zini & Scandola, 2011). Plant fibers can also been classified into primary and secondary fibers, where the primary fibers are produced from the plant as the main products, while the secondary fibers are produced as by-products of the plant (Faruk, Bledzki, Fink, & Sain, 2012).

Bast fibers are collected from the inner bark which surrounding the stems of dicotyledonous plants (Faruk et al., 2012). Bast fibers provide mechanical support for the stem of a plant (V. K. Thakur, Singha, & Thakur, 2012b). The cultivation conditions significantly influence the properties of a bast fiber (Bogoeva-Gaceva et al., 2007; Eichhorn et al., 2010). Most frequently used bast fibers include flax, hemp, jute, kenaf, and Ramie (Faruk et al., 2012). Flax prefers temperate climates in the world and is extensively used in the textile markets and to reinforce the green

Figure 1. Main natural fiber types
EFB: Empty Fruit Bunch (Abdul Khalil, Bhat, & Ireana Yusra, 2012; Vijay Kumar Thakur & Thakur, 2014).

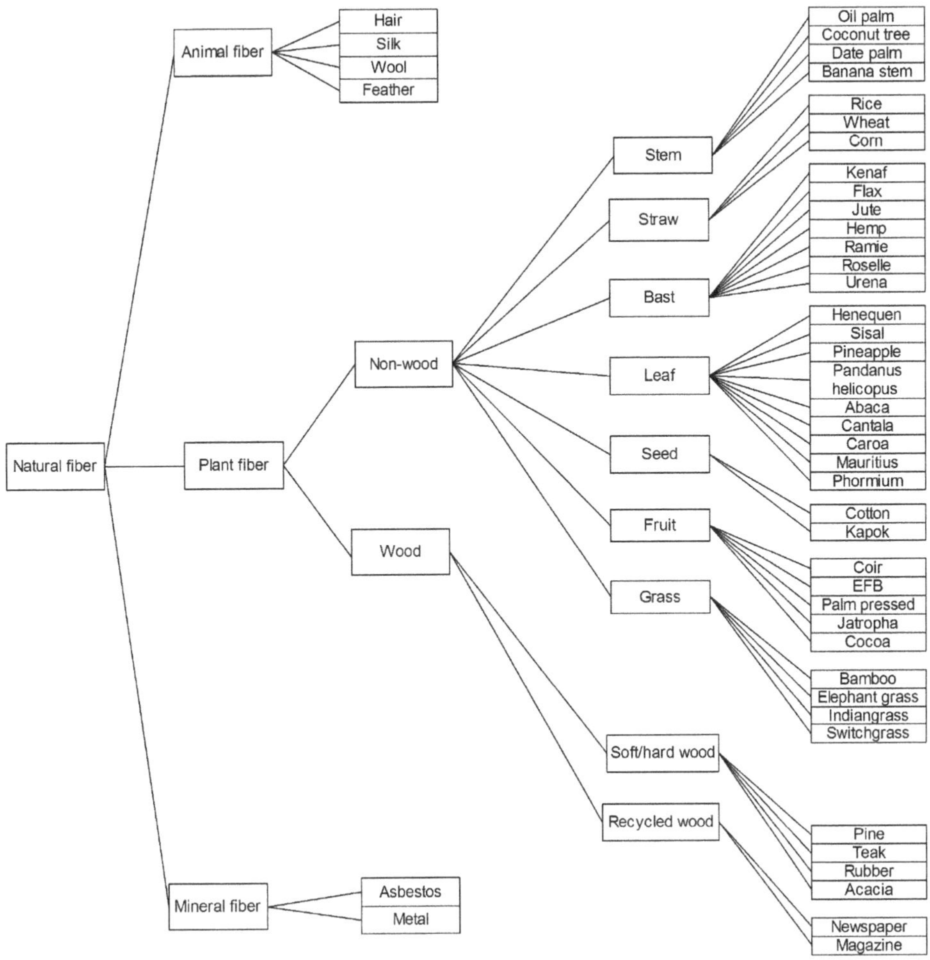

composites. Hemp is another widely used bast fiber crop that grows in temperate regions, as an important non-food agricultural fiber crop in Europe. Jute is one of the cheapest natural fibers and currently has higher production yield than other bast fiber types, with preferred growth conditions in Bangladesh, India, and China. Kenaf is a promising raw material for composite production due to the recently developed decortications equipment, which can separate the bast fiber with the core. Ramie is one of the oldest fiber crops, different with other bast fibers, a pre-treatment is needed for ramie due to its chemical composition, which limits its extensive commercialization (Faruk et al., 2012).

As opposed to bast fibers providing support to stem, leaf fibers provide the mechanical support for the leaves of a plant, mostly monocotyledonous plant (Akil et al., 2011). Leaf fibers are normally used to produce rugs, cordages, carpets, and mats, which are coarser than the products from the bast fibers (Bogoeva-Gaceva et al., 2007; Eichhorn et al., 2010). As shown in Figure 1, typical leaf fibers include henequen, sisal, abaca, and pineapple. Among which, abaca is the strongest commercial cellulosic fiber, extensively grows in Philippines, Ecuador, and Costa Rica. Sisal is a fiber plant native to southern Mexico and extensively cultivated in many other countries, with extensive application to produce rope, twine, paper, cloth, wall coverings, carpets, and dartboards. Pineapple leaf fibers have huge potential to be used as composite reinforcement due to its low-cost and abundant availability since it is a waste product from pineapple cultivation (Faruk et al., 2012).

Similar with the pineapple leaves, large amount of straw as the by-product of cereal cultivation is also considered to have huge potential to be used to reinforce composites (Li, Cai, Winandy, & Basta, 2010). Various agricultural wastes can be used to extract fibers including rice husk, wheat straw, and cornhusk (Zini & Scandola, 2011).

Seed fibers are collected from seed or seed case, which are different with bast or leaf fibers, two examples of seed fibers are cotton and kapok (Vijay Kumar Thakur & Thakur, 2014). Fruit fibers are fibers collected from fruits, with typical examples of coir. By extracting from the coconut husk, coir is collected from the material between internal shell and the outer husk. Production of coir has a significant advantage that it can be adjusted by the market demand, due to the abundant quantities of coconut husk (Faruk et al., 2012).

Different with other fiber types, grass fibers are produced from stems of monocotyledonous plants (Vijay Kumar Thakur & Thakur, 2014). Typical grass fibers include bamboo, elephant grass, indiangrass, and switchgrass. Indiangrass and switchgrass get the interests to become reinforcing agents due to the aim of land saving. Low amount of fertilizer for the growth of switchgrass is needed because of its self-seeding nature. Indiangrass is a perennial grass that can grow even with bare soil (Zini & Scandola, 2011). Bamboo is a perennial plant that grows in monsoon climates.

Animal derived protein wastes provide another source for natural fiber. Examples include hair, silk, wool, and feather. Specifically, over 4 million tons of chicken feather from poultry industry worldwide annually provide huge resource for production of natural fibers.

Figure 2. Various scale of fibers from a kenaf stem
(H P S Abdul Khalil et al., 2012).

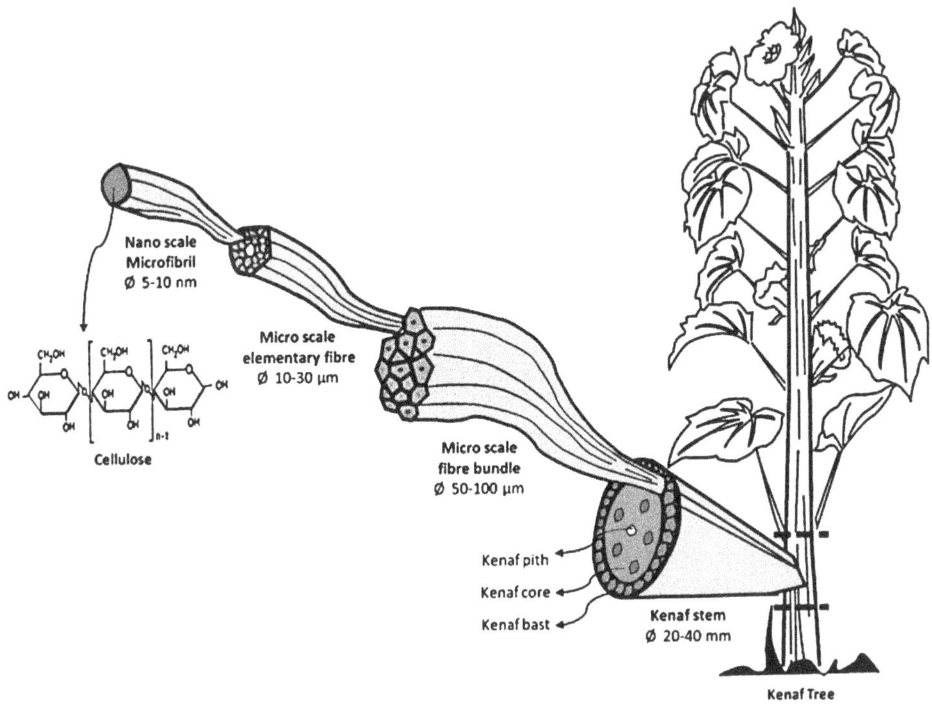

Structure and Composition

Figure 2 gives the different scales of fibers from a kenaf stem, a kenaf stem can be broke and scotched into bast fiber bundles, where one fiber bundle consist of elementary fibers. Furthermore, one elementary fiber consists of thousands of micro-fibrils. The kenaf fibers can be used to reinforce composite ranges from fiber bundles to elementary fibers, or even to micro-fibrils. The mechanical properties differ significantly with different scales of fibers. The price-performance ratio of fiber bundles is acceptable despite the fact that their lateral strength is poorer than axial strength. The tensile strength of elementary fibers is stronger than fiber bundles (H P S Abdul Khalil et al., 2012).

The structure and chemical composition of natural fibers could be influenced by climatic conditions, age and the degradation process of living plants. Water is the major chemical component of a living plant. After drying, the plant cell walls have been found to contain different amount of cellulose, hemicellulose, lignin, and also small amount of pectins and waxes (H P S Abdul Khalil et al., 2012; John & Thomas, 2008). The unidirectional cellulose microfibrils reinforce the matrix

Figure 3. The main compositions, cellulose, hemicellulose, lignin, and others, fractions of main plant fibers

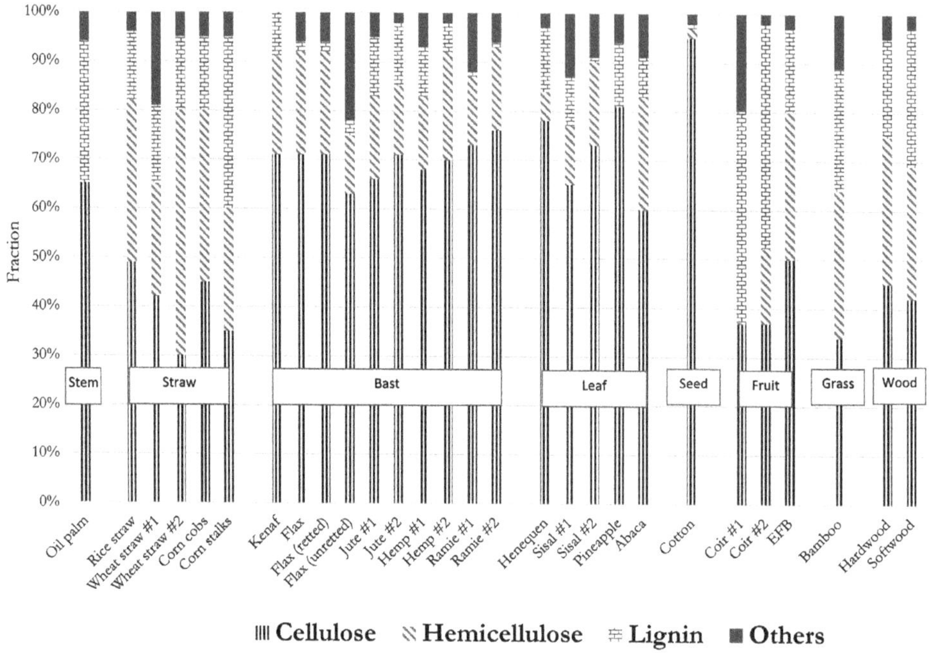

blended by hemicellulose and lignin (Gassan, Chate, & Bledzki, 2001). Pectins are a collective name for hetero-polysaccharides, serves the purpose of providing flexibility to a plant. Waxes make up the last part of fibers and they consist of different types of alcohols (John & Thomas, 2008). The chemical composition varies from plant to plant, and also fluctuates within different parts of the same plant, for example, the lignin content in the root and stalk core is higher than that in the fibers (H P S Abdul Khalil et al., 2012; Faruk et al., 2012).

The chemical composition percentage of selected natural fibers is illustrated in Figure 3. It is shown that stem, bast, leaf, and seed contain higher amount of cellulose than that in straw, fruit, grass and wood. Typically, the content of cellulose in seed is more than 95%, while the hemicellulose and lignin contents are as lower as 1 to 2%. Due to the nature of the three major components, the fibers derived from those varying sources of lignocellulosic material are quite different.

Cellulose

Generally, most of the natural cellulosic fibers contain 60 to70% cellulose, and cellulose is primarily composed of C, H, and O with general formula of $\left[C_6H_{10}O_5\right]_n$

Figure 4. Detailed chemical structure of cellulose, hemicellulose, and lignin (Zhang, Yang, & Blasiak, 2011)

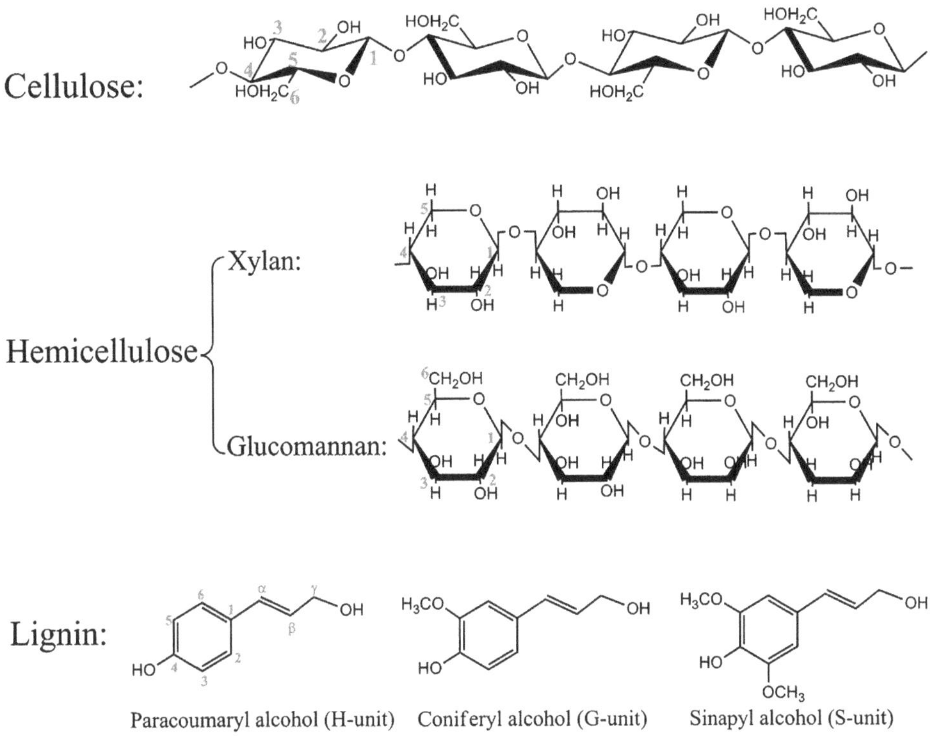

(Akil et al., 2011; Vijay Kumar Thakur & Thakur, 2014). The chemical structure of cellulose, hemicellulose, and lignin is illustrated in Figure 4.

Cellulose is considered as the main structural component that renders strength and stability to the lignocellulosic fibers. Cellulose molecules can be obtained via various stages of extraction of microscale elementary fiber bundles from various lignocellulosic fibers (H P S Abdul Khalil et al., 2012).

Cellulose is composed of polymer chains, which consists of unbranched β-(1-4)-glycosidic linked D-glucopyranose rings, while the length of the glucan chains depends on the source of cellulose. It was found that lignocellulosics has a degree of polymerization (DP) up to 10,000 (Klemm, Heublein, Fink, & Bohn, 2005; O'Sullivan, 1997). Although, the basic structure of cellulose remains the same, yet they are found to exhibit different properties in various plants, due to the differences in the degree of polymerization (Baillie, 2004).

The percentage of the cellulose in a natural fiber affects the application, properties, and cost of production (Pandey, Ahn, Lee, Mohanty, & Misra, 2010). Aside from this, many other factors can significantly affect the properties of cellulosic fibers, including the area of growth, climate, and the age of the plant, internal fiber structure, chemical composition, cell dimensions, and microfibril angle (H P S Abdul Khalil et al., 2012). The mechanical properties and the reinforcing efficiency of natural fibers are reflected by the nature of cellulose and also its crystallinity (H P S Abdul Khalil et al., 2012).

In addition, the orientation of the microfibrils determines the mechanical properties of the fibers. For example, the orientation nearly parallel to the fiber axis, gives rise to a large modulus of elasticity(Klemm et al., 2005). In consequence, cellulose has a high tensile strength due to its fibril structure and the large number of hydrogen bonds.

Hemicellulose

Hemicellulose is not a form of cellulose and the name is a misnomer. Hemicellulose comprises of a group of polysaccharides composed of a combination of 5- and 6-carbon ring sugars (Vijay Kumar Thakur & Thakur, 2014) (see Figure 4). It differs from cellulose in three major aspects:

1. Hemicellulose contains several different sugar units but cellulose contains only 1,4-linked β-D-glucopyranose units (John & Thomas, 2008);
2. Hemicellulose exhibits a considerable degree of chain branching containing pendant side groups giving rise to its non-crystalline nature, while cellulose is a linear polymer; and
3. The degree of polymerization in hemicellulose is around 50 to 300, which is 10 to 100 times lower than that of native cellulose.

In addition, hemicellulose forms the supportive matrix for cellulose microfibrils, and hemicellulose is very hydrophilic, soluble in alkali and easily hydrolyzed in acids (John & Thomas, 2008). Hemicellulose is recognized as one of the imperative constituents of natural cellulosic fibers, and hemicellulose is the second most abundant family of naturally occurring polymers (V. K. Thakur, 2013; V. K. Thakur, Singha, A. S., 2013). Within natural fibers, hemicellulose has high influence on the degradation of the fibers both from thermal and biological aspects. A higher hemicellulose content can cause a higher moisture absorption in a natural fiber (Vijay Kumar Thakur & Thakur, 2014).

Lignin

Lignin is a complex hydrocarbon polymer with both aliphatic and aromatic constituents, having very high molecular weight. Lignin is insoluble in most solvents and cannot be broken down to monomeric units. Lignin is composed of three major monolignols (see Figure 4):

1. Coniferyl alcohol (G-units),
2. Paracoumaryl alcohol (H-units), and
3. Sinapyl alcohol (S-units).

The main difficulty in lignin chemistry is that to date no method has been established to isolate lignin in its native state from the fiber (John & Thomas, 2008). Within all three main compositions of natural fibers, the lignin contribute most to the ultra-violet (UV) degradation compared to cellulose and hemicellulose (Faruk et al., 2014).

Properties of Natural Fiber

The intrinsically physical and chemical properties of natural fibers are the main driving forces for their utilization to reinforce the composite materials. These properties include relatively low cost, weightlessness, and biodegradability, with additional benefits of abundant availability (John & Thomas, 2008; Vijay Kumar Thakur & Thakur, 2014). However, except for the well accepted benefits of natural fibers mentioned above, the natural fibers are weak at the high-performance characteristics when compared with synthetic fibers (Abdul Khalil, Kumar, Asri, Nik Fuaad, & Ahmad, 2007; Khalil, Hanida, Kang, & Fuaad, 2007). Understanding the fiber characteristics is essential to extensively utilize natural fibers and improve the performance of natural fibers derived bio-composites (Baillie, 2004).

Properties of Different Fiber Types

One big difference between natural fiber and synthetic fiber is the variability (Zini & Scandola, 2011). The variation in the properties of natural fibers can be explained as an outcome of the different fiber sources, plant age, cultivation, climate conditions, and the manufacturing techniques (Baillie, 2004). For example, within different bast fibers, flax and hemp prefer temperate regions while jute and kenaf prefer tropical regions (Summerscales, Dissanayake, Virk, & Hall, 2010). As a consequence of

Table 1. Physical and chemical properties of main natural fibers

Fiber Types		Density (g/ cm^3)	Elongation (%)	Young's Modulus (GPa)	Tensile Strength (MPa)	Specific Tensile	Water Absorption (%)	Dimensions L/D (mm/ µm)	Ref
Petroleum fiber	E-glass	2.5	2.5	70	2000-3500	800-1400			(Bledzki, Reihmane, & Gassan, 1996)
	Aramide	1.4	3.3-3.7	63-67	3000-3150	2140-2250			(Bledzki et al., 1996; L. M. Lewin, 1985)
	Carbon	1.4	1.4-1.8	230-240	4000	2860			(Barkakaty, 1976; Bledzki et al., 1996)
Plant fiber	Flax	1.5	1.2-3.2	27-80	345-1500	230-1000	7	750-900/50-150	(Bledzki et al., 1996; Cheung, Ho, Lau, Cardona, & Hui, 2009)
	Jute	1.3-1.5	1.5-1.8	10--55	393-800	300-610	12	120/25-30	(Bledzki et al., 1996; Cheung et al., 2009; Mohanty, Misra, & Hinrichsen, 2000)
	Hemp	1.5	1.6	70	550-900	370-600	8		(Cheung et al., 2009)
	Kenaf		1.6	53	930				(Faruk et al., 2012)
	Ramie	1.5	1.2--3.8	24.5--128	400-938	270-620	12--17	900-1200/20-80	(Bledzki et al., 1996; Cheung et al., 2009)
	Bagasse	1.25	1.1	17--27.1	512			100-300/10-34	(Faruk et al., 2012; Satyanarayana et al., 2009)
	Cotton	1.5-1.6	3.0-10.0	5.5-12.6	287-800	190-530	8--25	35/10-45	(Bledzki et al., 1996; Chaudhary & Gohil, 2015; Cheung et al., 2009; Mohanty et al., 2000; Satyanarayana et al., 2009)
	Sisal	1.3-1.5	2.0-5.12	9.4-28	511-635	390-490	11	900/8-50	(Bledzki et al., 1996; Cheung et al., 2009; Mohanty et al., 2000; Satyanarayana et al., 2009)
	Abaca	1.5	3--10	12	400				(Faruk et al., 2012)
	Banana		2.5-3.7	27-32	700-800			300-900/12-30	(Satyanarayana et al., 2009)
	Pineapple leaf	0.8-1.6	14.5	1.44	400-627				(Faruk et al., 2012)
	Pineapple		3.2	82	180		0	900-1500	(Satyanarayana et al., 2009)
	Coir	1.2	15---51.4	04---10	131-220	110-180	10	20-150/10-50	(Faruk et al., 2012; Satyanarayana et al., 2009)
	Bamboo	0.6-1.1	0	11--17	140-230				(Faruk et al., 2012)
	Oil palm	0.7-1.55	25	3.2	248				(Faruk et al., 2012)
Wood	Softwood kraft	1.5		40	1000	670			(Bledzki et al., 1996)
Animal	Chicken feathers	0.89		42280	100-200	112-220			(Barone, Schmidt, & Liebner, 2005; Saheb & Jog, 1999)
	Silkworm silk	1.3-1.4	15	0.5			8		(Arai, Freddi, Innocenti, & Tsukada, 2003; Lee et al., 2005; Shao & Vollrath, 2002)

huge variation, the properties of natural fibers vary with different fiber types. Table 1 provides a ready glance into the various physical and chemical properties of main synthetic and natural fibers.

Currently the research on natural fibers mainly focuses on bast fibers, due to their superior properties and ease of extraction from the raw resources.

Reddy and Yang (Reddy & Yang, 2005a, 2005b) found that cornstalk fiber has lower crystallinity but similar microfibril angle like bast fibers. also compared the fibers from the leaves and stems of switchgrass with the common natural fibers extracted from cotton, linen, and kenaf and observed similar tensile properties, where the crystallinity and breaking elongation are 51%, and 2.2% respectively for switchgrass leaves, and 46%, and 6.8% respectively for switchgrass stems (Reddy & Yang, 2005a, 2005b).

The type of cellulose obtained from the fibre of water bamboo was found to be cellulose I (Shih, 2007). Recycled wood based fibers, for example, recycled newspaper, have a higher amount of reinforcement can be used based on a higher compactness and lower permeability values than that of other fibers (e.g. cellulose). It was reported that recycled paper has a permeability of 3.60×10^{-13} m^2, compared with the value of 6×10^{-10} m^2 in cellulose (Richard et al., 2002).

Increasing interests on animal fibers are due to their high performance as reinforcement for bio-composite. For example, wool keratin fibers are known to be less hydrophilic than plant fibers. The characteristic of being less hydrophilic (water friendly) can avoid the problems of water absorption and aggregation brought by the hydroxyl group. Also, the hollow structure of the feather keratin fibers results in a density lower than that of plant fibers (see Table 1). The hollow structure provides an additional benefit of low dielectric constant which promotes its application in electronic industry (Hong & Wool, 2005). As another important animal fiber, the properties of silk vary significantly with the animal species and the spinning conditions (Shao & Vollrath, 2002). Generally, silk fibers have relatively low thermal stability, with advantages of oxidation resistance, and antibacterial, and UV-resistant properties. Thus, the silk reinforced bio-composite have applications in bioengineering areas such as tissue engineering and bone fixators (Cheung et al., 2009).

Hydrophilic Nature

The hydrophilic nature is one example of the weak performance of plant fibers due to the high hydroxyl content of cellulose, which requires pre-treatment like pre-drying for the composite processing. Another issue brought from the hydroxyl content of natural fibers for the composite processing is the incompatibility between the polar

Figure 5. The equilibrium moisture content (EMC) for main plant fibers at 65% relative humidity and 21 °C
(Faruk et al., 2012).

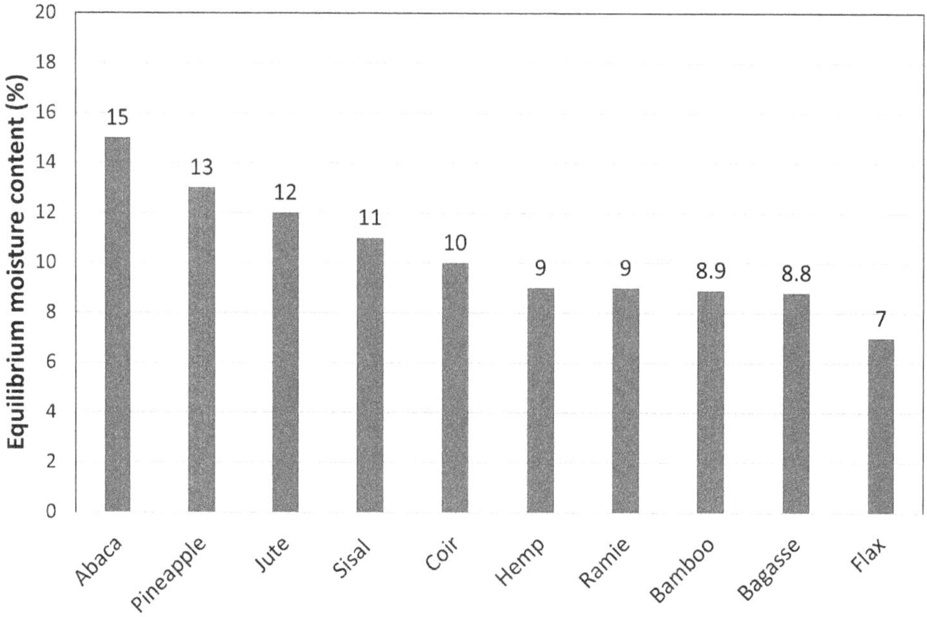

hydrophilic group from natural fibers and the non-polar hydrophobic polymer matrices. This incompatibility results in a weaker interface between the reinforcement and the matrix in a composite which can cause aggregation (H P S Abdul Khalil et al., 2012; Zini & Scandola, 2011).

The equilibrium moisture content (EMC) is the moisture content of a fiber at a dynamic equilibrium i.e. the condition when the moisture in the fiber is neither gained nor lost. EMC changes with relative humidity and temperature. EMC of certain major types of natural fibers at 65% relative humidity and 21°C are shown in Figure 5. A higher moisture content in the natural fiber leads to a high moisture content of bio-composites which is reinforced by the fiber (Faruk et al., 2012).

The moisture content of a fiber significantly influences the thermal and mechanical properties of the resulting composite. For example, for a polyvinyl chloride (PVC)/wood sawdust composite, the flexural and the tensile properties decreases and the elongation at break increases at low moisture content. Moreover, when the glass transition temperature stays steady, an increasing moisture content leads to a decreasing decomposition temperature of the composite (H P S Abdul Khalil et al., 2012).

Other Key Properties

Several physical properties of natural fiber are important for their maximal utilization. These properties include density, dimensions, defects, strength, variability, and crystallinity. From Table 1, the density of natural fibers is obviously lower than the glass fibers making them an ideal choice for various applications in the automotive industry. The length-width ratios of some main natural fibers are also shown in Table 1. For lignocellulosic fibers, the length-width ratio directly influences the strength properties of a fiber (Faruk et al., 2012). The crystallinity varies with different fiber types and also within different parts of one plant, which decreases with the increasing maturity level of a plant. Transcrystallinity on the interface in a natural fiber could possibly be induced by surface modification. The transcrystallinity can significantly strengthen the interface by increasing the interfacial shear strength in a flax fiber reinforced polypropylene (PP) composite (Faruk et al., 2012). For the specific modulus, most of the natural fibers have a higher specific modulus than glass fibers (Faruk et al., 2014). Most of the natural fibers have higher specific strength and stiffness than glass fibers (H P S Abdul Khalil et al., 2012). When compared on the same scale, the cost of most of the natural fibers was found lower than the glass fibers, even though the cost of natural fibers significantly vary at different fiber types and geographic sources (Dittenber & GangaRao, 2012).

Overall, inconsistency in the fiber qualities with various physical and chemical properties is meant to be due to the climate, growth, harvesting, and processing conditions. There exist benefits and limitations to use natural fibers as reinforcement in composites, e.g. natural fibers have advantages of low density leading to low weight; high specific strength and stiffness than glass fibers, and good thermal and acoustic insulating properties (Faruk et al., 2014)However, the utilization of natural fibers also suffers the moisture absorption which causes swelling, lower strength properties than glass fiber composite, and generates odor during the degradation process of the natural fibers.

IMPACTS OF NATURAL FIBERS ON COMPOSITE PROPERTIES

The properties of composites are influenced both by the reinforcement fibers and the polymer matrices, where the overall parameters such as shape, surface appearance, and durability are determined by the matrix, however, the material stiffness and strength are influenced significantly by the fibrous reinforcement (Faruk et al., 2012, 2014). Most common used matrices for natural fiber reinforcement include petrochemical based PP, polyethylene (PE), polystyrene (PS), and polyvinyl chloride

(PVC), polyester, epoxy resin, phenol formaldehyde, and vinyl esters; and bio-based poly (lactic acid) (PLA), polyhydroxyalkanoate (PHA), soy-based biodegradable resin, polycaprolactone (PCL), and polybutylene succinate (PBS).

Utilization of hemp fibers reinforcement on PP matrix is expected to improve the physical properties. After reinforcement, the composite exhibits higher interfacial adhesion in the PP matrix. The thermal stability and phase behavior of the composites could be significantly improved and the changes in the spherulitic morphology and crystallization behavior of PP matrix were also observed (Faruk et al., 2012). Moreover, hemp fiber reinforced PP composites exhibitgood recyclability, while remaining well preserve of the mechanical properties of hemp fiber/PP composites. Furthermore, various matrixes (e.g. epoxy resins, wheat gluten) are reinforced by hemp fiber to improve the falling weight impact properties, thermal treatment, and plasticization effect.

Faruk et al. (Faruk et al., 2012) in their review have provided a comprehensive summary on flax fiber reinforced polyester resin, jute and kenaf fiber. They have described that the Flax fiber were evaluated for their thermal degradation and fire resistance, chemical treatments on surface properties and adhesion, and the effect of chemical treatments on water absorption and mechanical properties. Similarly, jute fiber reinforced PP composites were analyzed on the aspects of matrix modification, gamma radiation, and interfacial adhesion on creep and dynamic mechanical behavior. Finally, kenaf fiber reinforcement was used to increase the strength of the PP matrix. The kenaf-maleated PP composites have a higher modulus/cost ratio and a higher specific modulus than sisal, coir, and even E-glass. Thus, kenaf-PP composites provide an option for replacing existing materials with a material having higher strength, lower cost and environmentally friendliness.

Ramie fiber reinforced PP composites exhibit to increase tensile strength and flexural strength while the impacts on strength and elongation behavior of the composites can be neglected (L. P. He, 2008).

Sisal/PP composites were investigated regarding the degradation behavior and the mechanical properties. Admicellar-treated sisal fiber reinforcement was used to improve the mechanical properties of polyester composite (Sangthong, Pongprayoon, & Yanumet, 2009). Furthermore, sisal fiber reinforced epoxy resin composites were examined regarding the influence of fiber orientation on the electrical properties. Investigations were also carried out on the sisal fiber reinforced cement composites focusing on their micro-cracking mechanisms, and the effects of accelerated carbonation on cementitious roofing (Faruk et al., 2012).

Pineapple leaf fiber is rich in cellulose, relatively cheap and abundantly available. Studies were carried out to investigate the properties of pineapple leaf fiber reinforced polycarbonate composites.It was found that the silane modified pineapple leaf fibers composite exhibits the highest tensile and impact strength, while the thermal stability decreases with increasing pineapple leaf fiber content (Threepopnatkul, 2008).

When compared to a jute-PP composite of the same density, the composite reinforced by switchgrass stems had higher modulus, flexural strength, and impact resistance (Zou, Xu, & Yang, 2010). Compared to jute and flax fiber PP composites, abaca fiber reinforced PP composites are known to have better toughness and impact properties (A. K. Bledzki, 2007).

Indiangrass fiber has been reported to improve tensile and flexural properties while having no impact on strength. In addition, the tensile, impact, and flexural properties of the soy-based resin were significantly enhanced by pretreating indiangrass fibers with an alkali solution (Liu, Mohanty, Drzal, & Misra, 2005). Bamboo fibers have been used for upgrading the ecological purpose of composites, and it was concluded that bamboo fibers (bundles) have sufficient specific strength, similar to that of conventional glass fibers (Okubo, Fujii, & Yamamoto, 2004).

CHALLENGES AND FUTURE TREND

Currently, the partial biodegradable composites utilized both in Europe and the US, are still not totally "green" i.e. even the fibers are biodegradable, the degradability of the composites is limited by the non-degradable polymeric matrices with 80% of matrix polymers coming from non-renewable resources. The latest innovation in the field of bio-composites is the substitution of oil-derived polymers with polymers from renewable resources (bio-based polymers) as the matrix component, and such bio-composites are named as 'green composites', indicating that the whole composite, both matrix and reinforcement, originates from renewable resources (Zini & Scandola, 2011).

The market economic competitive pressure for natural fibers derived composites with petroleum based composites still exists, which impedes complete replacement of petroleum based materials. One of the proposed solution could be the combination of both petroleum based and bio-based resources for material engineering (John & Anandjiwala, 2008).

Production of high quality natural fibers are limited by the inconsistency and wide variability of the fiber resources, which cause the various properties of the fiber due to the region, harvesting, and processing conditions. A multi-step manufacturing process provides a potential solution to improve the fiber quality at the expense of high economic input.

CONCLUSION

Reviewing the extant literature showed a proposed solution for the conversion of plant and animal waste to useful composite materials. This is accomplished via extraction of natural fibers from waste to reinforce bio-composites which in turn provide numerous environmental benefits including sustainability, recyclability, and additional product benefits of cost and weight reduction. This technology also gives farmers a direct benefit and profit via waste disposal.

Contrarily, the application areas are limited by several disadvantages of natural fibers such as their high moisture absorption, low strength and stiffness, and these are yet to be overcome. Moreover, despite abundant availability of natural fibers from bio-resources, changes in the growth and processing conditions result in inconsistency and wide variation in the quality of natural fibers, thereby limit their enormous applications.

The global challenge of achieving complete green composition requires handshaking of industries and academic research to direct more focused efforts to tackle the current limitations to produce biodegradable matrices.

REFERENCES

Abdul Khalil, H. P. S., Bhat, A. H., & Ireana Yusra, A. F. (2012). Green composites from sustainable cellulose nanofibrils: A review. *Carbohydrate Polymers*, *87*(2), 963–979. doi:10.1016/j.carbpol.2011.08.078

Abdul Khalil, H. P. S., Kumar, R. N., Asri, S. M., Nik Fuaad, N. A., & Ahmad, M. N. (2007). Hybrid Thermoplastic Pre-preg Oil Palm Frond Fibers (OPF) Reinforced in Polyester Composites. *Polymer-Plastics Technology and Engineering*, *46*(1), 43–50. doi:10.1080/03602550600948749

Abdul Khalil, H. P. S., Tehrani, M. A., Davoudpour, Y., Bhat, A. H., Jawaid, M., & Hassan, A. (2012). Natural fiber reinforced poly(vinyl chloride) composites: A review. *Journal of Reinforced Plastics and Composites*. doi:10.1177/0731684412458553

Akil, H. M., Omar, M. F., Mazuki, A. A. M., Safiee, S., Ishak, Z. A. M., & Abu Bakar, A. (2011). Kenaf fiber reinforced composites: A review. *Materials & Design*, *32*(8–9), 4107–4121. doi:10.1016/j.matdes.2011.04.008

Arai, T., Freddi, G., Innocenti, R., & Tsukada, M. (2003). Preparation of water-repellent silks by a reaction with octadecenylsuccinic anhydride. *Journal of Applied Polymer Science*, *89*(2), 324–332. doi:10.1002/app.12081

Baillie, C. (2004). *Green Composites: Polymer Composites and the Environment*. New York: CRC Press.

Barkakaty, B. C. (1976). Some structural aspects of sisal fibers. *Journal of Applied Polymer Science*, *20*(11), 2921–2940. doi:10.1002/app.1976.070201101

Barone, J. R., Schmidt, W. F., & Liebner, C. F. E. (2005). Compounding and molding of polyethylene composites reinforced with keratin feather fiber. *Composites Science and Technology*, *65*(3–4), 683–692. doi:10.1016/j.compscitech.2004.09.030

Bledzki, A. K. (2007). Abaca fibre reinforced PP composites and comparison with jute and flax fibre PP composites. *Express Polymer Letters*, *1*(11), 755–762. doi:10.3144/expresspolymlett.2007.104

Bledzki, A. K., & Gassan, J. (1999). Composites reinforced with cellulose based fibres. *Progress in Polymer Science*, *24*(2), 221–274. doi:10.1016/S0079-6700(98)00018-5

Bledzki, A. K., Reihmane, S., & Gassan, J. (1996). Properties and modification methods for vegetable fibers for natural fiber composites. *Journal of Applied Polymer Science*, *59*(8), 1329–1336. doi:10.1002/(SICI)1097-4628(19960222)59:8<1329::AID-APP17>3.0.CO;2-0

Bogoeva-Gaceva, G., Avella, M., Malinconico, M., Buzarovska, A., Grozdanov, A., Gentile, G., & Errico, M. E. (2007). Natural fiber eco-composites. *Polymer Composites*, *28*(1), 98–107. doi:10.1002/pc.20270

Chaudhary, V., & Gohil, P. P. (2015). Investigations on Drilling of Bidirectional Cotton Polyester Composite. *Materials and Manufacturing Processes*, 1–9. doi:10.1080/10426914.2015.1059444

Cheung, H.-y., Ho, M.-, Lau, K.-, Cardona, F., & Hui, D. (2009). Natural fibre-reinforced composites for bioengineering and environmental engineering applications. *Composites. Part B, Engineering*, *40*(7), 655–663. doi:10.1016/j.compositesb.2009.04.014

Dittenber, D. B., & GangaRao, H. V. S. (2012). Critical review of recent publications on use of natural composites in infrastructure. *Composites. Part A, Applied Science and Manufacturing*, *43*(8), 1419–1429. doi:10.1016/j.compositesa.2011.11.019

Eichhorn, S. J., Dufresne, A., Aranguren, M., Marcovich, N. E., Capadona, J. R., Rowan, S. J., & Peijs, T. et al. (2010). Review: Current international research into cellulose nanofibres and nanocomposites. *Journal of Materials Science*, *45*(1), 1–33. doi:10.1007/s10853-009-3874-0

Faruk, O., Bledzki, A. K., Fink, H.-P., & Sain, M. (2012). Biocomposites reinforced with natural fibers: 2000–2010. *Progress in Polymer Science, 37*(11), 1552–1596. doi:10.1016/j.progpolymsci.2012.04.003

Faruk, O., Bledzki, A. K., Fink, H.-P., & Sain, M. (2014). Progress Report on Natural Fiber Reinforced Composites. *Macromolecular Materials and Engineering, 299*(1), 9–26. doi:10.1002/mame.201300008

Gassan, J., Chate, A., & Bledzki, A. (2001). Calculation of elastic properties of natural fibers. *Journal of Materials Science, 36*(15), 3715–3720. doi:10.1023/A:1017969615925

He, & Wang. (2008). Study on Ramie Fiber Reinforced Polypropylene Composites (RF-PP) and its Mechanical Properties. *Advanced Materials Research, 41-42.*

Hong, C. K., & Wool, R. P. (2005). Development of a bio-based composite material from soybean oil and keratin fibers. *Journal of Applied Polymer Science, 95*(6), 1524–1538. doi:10.1002/app.21044

John, M. J., & Anandjiwala, R. D. (2008). Recent developments in chemical modification and characterization of natural fiber-reinforced composites. *Polymer Composites, 29*(2), 187–207. doi:10.1002/pc.20461

John, M. J., & Thomas, S. (2008). Biofibres and biocomposites. *Carbohydrate Polymers, 71*(3), 343–364. doi:10.1016/j.carbpol.2007.05.040

Khalil, H. P. S. A., Hanida, S., Kang, C. W., & Fuaad, N. A. N. (2007). Agro-hybrid Composite: The Effects on Mechanical and Physical Properties of Oil Palm Fiber (EFB)/Glass Hybrid Reinforced Polyester Composites. *Journal of Reinforced Plastics and Composites, 26*(2), 203–218. doi:10.1177/0731684407070027

Klemm, D., Heublein, B., Fink, H.-P., & Bohn, A. (2005). Cellulose: Fascinating Biopolymer and Sustainable Raw Material. *Angewandte Chemie International Edition, 44*(22), 3358–3393. doi:10.1002/anie.200460587 PMID:15861454

Lee, S. M., Cho, D., Park, W. H., Lee, S. G., Han, S. O., & Drzal, L. T. (2005). Novel silk/poly(butylene succinate) biocomposites: The effect of short fibre content on their mechanical and thermal properties. *Composites Science and Technology, 65*(3–4), 647–657. doi:10.1016/j.compscitech.2004.09.023

Lewin, L. M. E. M. P. (1985). Handbook of Fiber Science and Technology: Fiber Chemistry (Vol. 4). New York: CRC Press.

Li, X., Cai, Z., Winandy, J. E., & Basta, A. H. (2010). Selected properties of particleboard panels manufactured from rice straws of different geometries. *Bioresource Technology, 101*(12), 4662–4666. doi:10.1016/j.biortech.2010.01.053 PMID:20153181

Liu, W., Mohanty, A. K., Drzal, L. T., & Misra, M. (2005). Novel Biocomposites from Native Grass and Soy Based Bioplastic: Processing and Properties Evaluation. *Industrial & Engineering Chemistry Research, 44*(18), 7105–7112. doi:10.1021/ie050257b

Mohanty, A. K., Misra, M., & Hinrichsen, G. (2000). Biofibres, biodegradable polymers and biocomposites: An overview. *Macromolecular Materials and Engineering, 276-277*(1), 1–24. doi:10.1002/(SICI)1439-2054(20000301)276:1<1::AID-MAME1>3.0.CO;2-W

O'Sullivan, A. (1997). Cellulose: The structure slowly unravels. *Cellulose (London, England), 4*(3), 173–207. doi:10.1023/A:1018431705579

Okubo, K., Fujii, T., & Yamamoto, Y. (2004). Development of bamboo-based polymer composites and their mechanical properties. *Composites. Part A, Applied Science and Manufacturing, 35*(3), 377–383. doi:10.1016/j.compositesa.2003.09.017

Pandey, J. K., Ahn, S. H., Lee, C. S., Mohanty, A. K., & Misra, M. (2010). Recent Advances in the Application of Natural Fiber Based Composites. *Macromolecular Materials and Engineering, 295*(11), 975–989. doi:10.1002/mame.201000095

Reddy, N., & Yang, Y. (2005a). Properties and potential applications of natural cellulose fibers from cornhusks. *Green Chemistry, 7*(4), 190–195. doi:10.1039/b415102j

Reddy, N., & Yang, Y. (2005b). Structure and properties of high quality natural cellulose fibers from cornstalks. *Polymer, 46*(15), 5494–5500. doi:10.1016/j.polymer.2005.04.073

Richard, P. W., Shrikant, N. K., John, J. L., Shana, P. B., Jue, L., Wim, T., & George, I. W. et al. (2002). Affordable Composites and Plastics from Renewable Resources: Part II: Manufacture of Composites *Advancing Sustainability through Green Chemical Engineering (Albany, N.Y.), 823*, 205–224.

Saheb, D. N., & Jog, J. P. (1999). Natural fiber polymer composites: A review. *Advances in Polymer Technology, 18*(4), 351–363. doi:10.1002/(SICI)1098-2329(199924)18:4<351::AID-ADV6>3.0.CO;2-X

Sangthong, S., Pongprayoon, T., & Yanumet, N. (2009). Mechanical property improvement of unsaturated polyester composite reinforced with admicellar-treated sisal fibers. *Composites. Part A, Applied Science and Manufacturing*, *40*(6–7), 687–694. doi:10.1016/j.compositesa.2008.12.004

Satyanarayana, K. G., Arizaga, G. G. C., & Wypych, F. (2009). Biodegradable composites based on lignocellulosic fibers—An overview. *Progress in Polymer Science*, *34*(9), 982–1021. doi:10.1016/j.progpolymsci.2008.12.002

Shao, Z., & Vollrath, F. (2002). Materials: Surprising strength of silkworm silk. *Nature*, *418*(6899), 741–741. doi:10.1038/418741a PMID:12181556

Shih, Y.-F. (2007). A study of the fiber obtained from the water bamboo husks. *Bioresource Technology*, *98*(4), 819–828. doi:10.1016/j.biortech.2006.03.025 PMID:16759852

Summerscales, J., Dissanayake, N. P. J., Virk, A. S., & Hall, W. (2010). A review of bast fibres and their composites. Part 1 – Fibres as reinforcements. *Composites. Part A, Applied Science and Manufacturing*, *41*(10), 1329–1335. doi:10.1016/j.compositesa.2010.06.001

Thakur, V. K. (2013). *Green composites from natural resources*. CRC Press. doi:10.1201/b16076

Thakur, V. K., & Singha, A. S. (2013). *Biomass-based biocomposites*. Smithers Rapra Technology.

Thakur, V. K., Singha, A. S., & Thakur, M. K. (2012a). Green Composites from Natural Fibers: Mechanical and Chemical Aging Properties. *International Journal of Polymer Analysis and Characterization*, *17*(6), 401–407. doi:10.1080/1023666X.2012.668665

Thakur, V. K., Singha, A. S., & Thakur, M. K. (2012b). In-Air Graft Copolymerization of Ethyl Acrylate onto Natural Cellulosic Polymers. *International Journal of Polymer Analysis and Characterization*, *17*(1), 48–60. doi:10.1080/1023666X.2012.638470

Thakur, V. K., & Thakur, M. K. (2014). Processing and characterization of natural cellulose fibers/thermoset polymer composites. *Carbohydrate Polymers*, *109*, 102–117. doi:10.1016/j.carbpol.2014.03.039 PMID:24815407

Threepopnatkul, P. K., & Athipongarporn, N. (2008). Polycarbonate with Pineapple Leaf Fiber to Produce Functional Composites. *Advanced Materials Research*, *47*(50).

Zhang, X., Yang, W., & Blasiak, W. (2011). Modeling Study of Woody Biomass: Interactions of Cellulose, Hemicellulose, and Lignin. *Energy & Fuels*, *25*(10), 4786–4795. doi:10.1021/ef201097d

Zini, E., & Scandola, M. (2011). Green composites: An overview. *Polymer Composites*, *32*(12), 1905–1915. doi:10.1002/pc.21224

Zou, Y., Xu, H., & Yang, Y. (2010). Lightweight Polypropylene Composites Reinforced by Long Switchgrass Stems. *Journal of Polymers and the Environment*, *18*(4), 464–473. doi:10.1007/s10924-010-0165-4

KEY TERMS AND DEFINITIONS

Bast Fiber: Natural fibers collected from the bast surrounding the plant stem.

Equilibrium Moisture Content: The moisture content of a fiber at a dynamic equilibrium i.e. the condition when the moisture in the fiber is neither gained nor lost.

Fruit Fiber: Natural fibers collected from the fruit of the plant.

Green Composite: High quality composites that produced from natural fibers and natural resins, both of which are from the recyclable and biodegradable. Green composites are the next generation composite materials with sustainability and recyclability.

Hydrophilic: The characterization of a molecule which is attracted to water.

Leaf Fiber: Natural fibers collected from the plant leaves.

Natural Fiber: Also been called as lignocellulosic fibers, extracted from biodegradable materials to reinforce composites.

Chapter 2
Processing Technologies for Green Composites Production

Deepak Verma
Graphic Era Hill University, India

Garvit Joshi
Graphic Era Hill University, India

Rajneesh Dabral
Graphic Era Hill University, India

ABSTRACT

Green composites became a most important and adaptable theme of research. This area/theme not only harness the agricultural wastes such as bagasse fibres, banana fibres, etc. but also provides a new material manufactured from these wastes which are reduced weight, have low cost, and have high mechanical strength. Currently, there are various methods available for the processing or fabrication of green composites. Some of these methods are hand layup method, injection molding method, spray-up method, compression molding, Resin-Transfer Molding (RTM), etc. In this chapter, we are discussing about the fabrication method of green composite and their important parameters. Various properties and characterization of composite materials made by these methods have also been discussed and reported here.

DOI: 10.4018/978-1-5225-0424-5.ch002

INTRODUCTION

Green composites are a combination of bioplastics and natural fibers, which have arisen as propitious alternatives to conventional polyolefin/glass fiber composites because they offer a wide variety of advantages, such as less expensive, reduced weight, increased flexibility, renewable resource and sound insulation a certain required performance. Natural fibres are subdivided based on their origins, from plants, animals or minerals. All plant fibres are composed of cellulose while animal fibres consist of proteins. Plant fibres include bast (or stem or soft sclerenchyma) fibres found in phloem of dicotyledonous stems, leaf, seed, fruit, wood, and other grass fibres. The use of these fibers in composites has increased due to their relative cost, their ability to recycle and for the fact that they can compete well in terms of strength per weight with other material. Natural fibres can also be considered as naturally occurring composites consisting mainly of cellulose fibrils embedded in lignin matrix they are aligned along the length of the fibre, which render maximum tensile and flexural strengths, in addition to providing rigidity. The reinforcing efficiency of natural fibre is associated with the nature of cellulose and its crystallinity. The main components of natural fibres are cellulose, hemicellulose, pectins, lignin, and waxes. Currently due to increasing interest in eco-friendly, sustainability, eco-efficiency and industrial ecology materials, studies on natural fiber have been actively focused to the area of composite. In an appropriate way, it can be applied as very advantageous composite when an appropriate resin has been selected.

METHODS OF FABRICATION

There are various methods for fabricating composite components to meet specific design or manufacturing challenges. The selection of appropriate method for a particular component, therefore, will depend on the part design, materials and application. Composite fabrication molding processes is used to shape the resin and reinforcement as per design. For an overview of methods used to make mold tools.

Open Molding

Open mold processes include, hand lay-up and spray-up (chopping) (Verma, Deepak, et.al, 2015). In this process, a single-sided mold is used that acts as the form and cosmetic surface of the part. Gel coatis applied to the prepared mold surface and then reinforcements are applied either by hand and then wet-out with resin, or by the spray-up process where resin and chopped fiberglass are sprayed onto the gel coated surface. The additional laminate layer is added to build thickness and strength as

Figure 1. Hand layup fabrication method

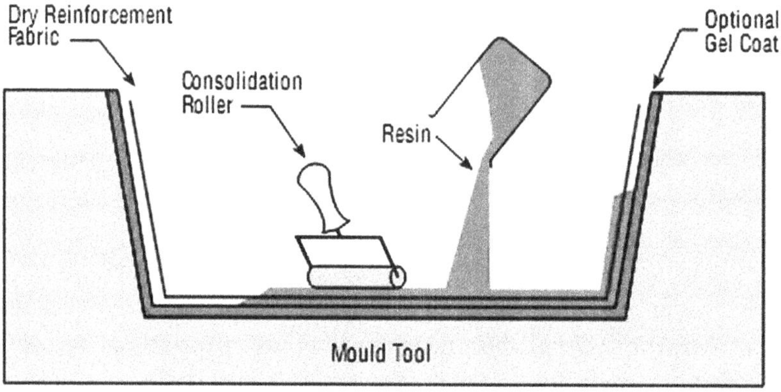

desired. Air is then rolled out of the laminate by hand and the part is left to cure. In addition to reinforcements, low density core materials such as balsa wood, foam, or honeycomb can be added to stiffen the laminate without adding significant weight.

Open molding is the most flexible of all composite fabrication processes as part size and design options are virtually limitless. Typically, the open molding process is used for a large size range of products that cannot be produced in more automated processes, or for parts that are produced in low volumes that cannot justify the higher mold costs of automated processes.

Hand Layup Process

Hand layup fabrication is very simple method and used for the thermoset composites, which typically consists of laying dry fabric layers, by hand onto a tool to form a laminate stack (Figure 1). As a part requirement for the hand lay-up process a mold must be used. The mold should be as simple as possible and can be a flat sheet or have infinite curves and edges. For few typical shapes, molds should be joined in sections/ parts so that after curing they can be taken apart for the taking away of the parts.

Resin is applied to the dry plies after layup is complete (e.g., by means of resin infusion). In a variation known as wet layup, each ply is coated with resin and "debulked" or compacted after it is placed. When heat is required for cure, the part temperature is "ramped up" in small increments, maintained at cure level for a specified period of time defined by the resin system, then "ramped down" to room temperature, to avoid part distortion or warp caused by uneven expansion and contraction. When this curing cycle is complete and after parts is demolded. A release agent can be used and apply on the surfaces of the molds so that the part will not be able stick to the mold surfaces. Some parts go through a secondary freestanding post cure, during which they are subjected for a specific period of time to a temperature higher than that of the initial cure to enhance chemical crosslink density. Electron beam curing has been explored as an efficient curing method for thin laminates.

Advantages

- Low tooling cost tooling, if cured resin is used at room-temperature.
- Widely used in many areas.
- Wide range material types are available.
- It can contain high fibre contents, and longer fibres than with spray lay-up.

Disadvantages

- Skilled person is required for Resin mixing, laminate resin contents, and laminate quality.
- Health and safety considerations are required for the proper handling of resins. The lower viscosity of the resins also leads to an increased tendency to penetrate clothing etc.
- Low resin content laminates cannot usually be achieved without the integration of large number of voids.
- Without expensive extraction systems Limiting airborne styrene concentrations to legislated levels from polyesters and vinyl esters is becoming increasingly hard.
- Low viscosity of resin generally compromises their mechanical/thermal properties due to the need for high diluent/styrene levels.

Applications

Standard wind-turbine blades, production boats, architectural mouldings.

Spray Up Molding Process

Spray-up is an open mold method that can produce complex parts more economically than hand lay-up (Figure 2). Chopped fiberglass reinforcement and catalysed resin, and in some cases, filler materials, are deposited on the mold surface from a combination chopper/spray gun. Rollers or squeegees are used to manually remove entrapped air and work the resin into the reinforcements. Woven fabric or woven

Figure 2. Spray up fabrication method

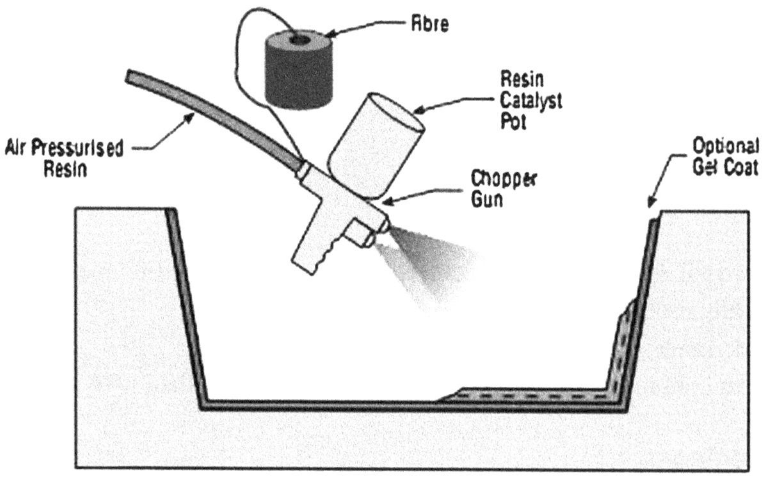

roving is often added in specific areas for greater strength. General purpose, room temperature cure polyesters are usually used to produce such parts as truck camper shells and tub & shower stalls. As in hand lay-up, gel coats are used to produce a high quality colored part surface.

In an open mold sprayup application, the mold is first treated with mold release. If a gel coat is used, it is typically sprayed into the mold after the mold release has been applied. The gel coat then is cured and the mold is ready for fabrication to begin. In the sprayup process, catalysed resin viscosity from 500 to 1,000 cps (centipoise) and glass fiber are sprayed into the mold using a chopper gun, which chops continuous fiber into short lengths, then blows the short fibers directly into the sprayed resin stream so that both materials are applied simultaneously. In the final steps of the sprayup process, workers compact the laminate by hand with rollers. Wood, foam or other core material may then be added, and a second sprayup layer imbeds the core between the laminate skins. The part is then cured, cooled and removed from the reusable mold. Balsa or foam cores may be inserted between the laminate layers in either process. Sprayup processing once a very prevalent manufacturing method has begun to fall out of favor.

Advantages

- Low cost is required for depositing fibre and resin.
- Widely used for many years.
- Low tooling cost.

Disadvantages

- Laminates tend to be very resin-rich and therefore excessively heavy.
- Only short fibres are incorporated which severely limits the mechanical properties of the laminate.
- Resins need to be low in viscosity to be sprayable. This generally compromises their mechanical/thermal properties.
- The high styrene contents of spray lay-up resins generally mean that they have the potential to be more harmful and their lower viscosity means that they have an increased tendency to penetrate clothing.
- Limiting airborne styrene concentrations to legislated levels is becoming increasingly difficult.

Applications

Simple enclosures, lightly loaded structural panels, e.g. caravan bodies, truck fairings, bathtubs, shower trays, some small dinghies.

CLOSED MOLDING

Resin Transfer Molding Process

The resin-transfer molding (RTM) is one of the most important technologies accessible for composite making. It is used for producing large complex part having great dimensional variation and good surface finish. It can fabricate 3-D complicated parts and offered fabrication of low value structural parts in moderate amounts. It is generally known as cost-effective manufacturing method for the fabrication of composites. In this method; a catalyzed thermosetting resin is injected into a confined metal mold which contains a reinforcement preform. The preform is compressed to the defined fibre volume fraction when the mating part of mold is closed. The mold is filled with resin, and fibres at that time and curing takes place inside the mold. This process is generally considered as a closed mold process which ultimately lowers the worker's subjection to pernicious volatiles, like styrene, and relates room temperature processing resins. In spite of these, RTM process is a costly process and the tool design becomes Augean when manufacturing large and complicated shaped parts. For continuous manufacturing process RTM distinctly

Figure 3. Resin transfer molding method

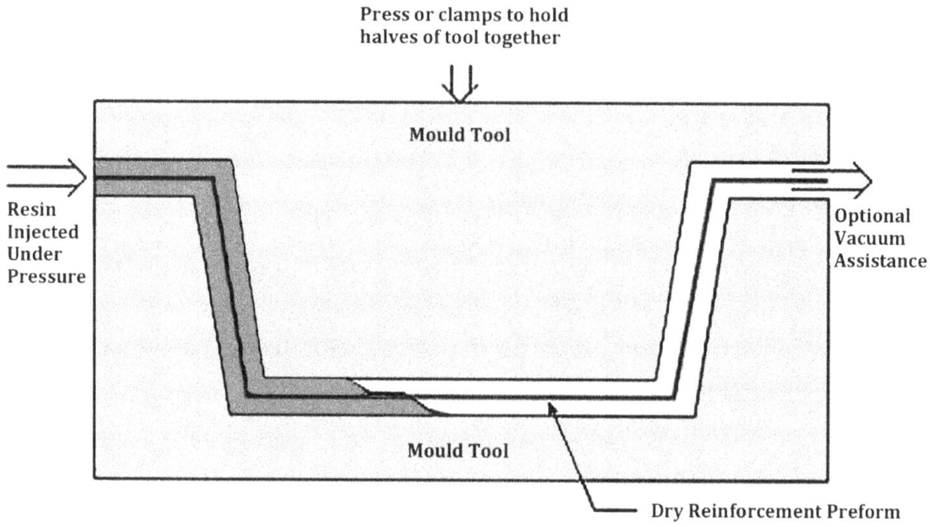

reduces the manufacturing cycle times as compared to other available composite making methods, and it is appropriate method for mass production of composites parts. The RTM process generally used a stiff closed mold. The sequence of the process is shown in Figure 3.

After this, close the mold and compress the preform and then, introduce the resin into the mold by using positive gradient pressure which replaces the air entrapped into the preform. Vacuum is also utilized at some exclusive vents and removes the entrapped air from the mold. The preform is impregnated by the resin into the mold. After that, the curing of composite is considered to start. The next and last step is to open the mold and remove the part from it. The mold closing sequence is distinguished by the compression of the reinforced fibre which permits to have the appropriate thickness. The compression changes the microstructure of the preform, developing substantial distortions and nonlinear viscoelastic effects. These effects show changes in energy of the material and generated stresses (residual or leftover) just because of the viscoelastic response of the fibres. Still, stresses released throughout the impregnation phase. The introduction of resin assured the entire impregnation of the preform, whereas in general poor impregnation of the fibres gives dry spots with disordered adherence between the layers which ultimately make the surface rough.

Advantages

- High fibre volume laminates can be obtained with very low void contents.
- Good health and safety, and environmental control due to enclosure of resin.
- Possible labour reductions.
- Both sides of the component have a moulded surface.

Disadvantages

- Matched tooling is expensive and heavy in order to withstand pressures.
- Generally limited to smaller components.
- Unimpregnated areas can occur resulting in very expensive scrap parts.

Applications

Small complex aircraft and automotive components, train seats.

Vacuum Assisted Resin Transfer Molding (VARTM)

Vacuum assisted resin transfer molding (VARTM) refers to a variety of related processes that depict the fastest growing new molding technology (Figure 4). The salient difference between VARTM processes and RTM (Resin Transfer Molding) is that in VARTM, resin is drawn into a preform through use of a vacuum only, rather than pumped in under pressure. VARTM does not require high heat or pressure. For that reason, VARTM operates with low cost tooling, making it possible to inexpensively produce large, complex parts in one shot.

Figure 4. Vacuum assisted resin transfer molding method

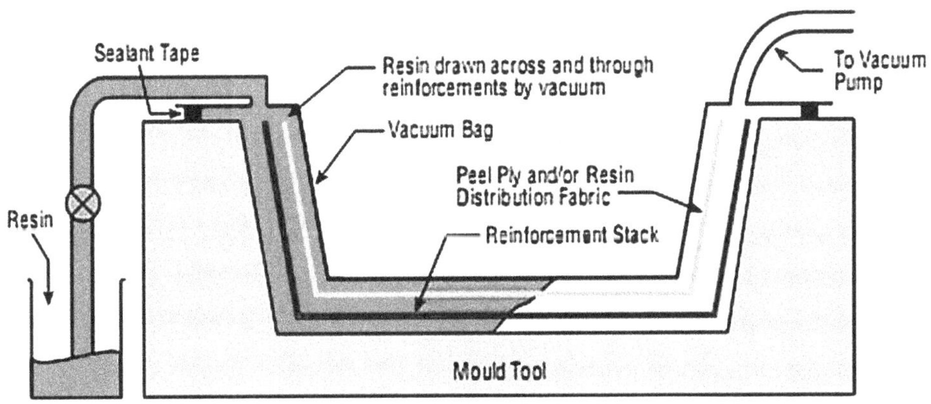

In the VARTM process, fiber reinforcements are placed in a one sided mold, and a cover (typically a plastic bagging film) is placed over the top to form a vacuum tight seal. The resin typically enters the structure through strategically placed ports and feed lines, termed as manifold. It is drawn by vacuum through the reinforcements by means of a series of designed in channels that facilitate wet out of the fibers. Fiber content in the finished part can run as high as 70 percent. A twist on the VARTM process is the use of two bags, termed double bag infusion, which uses one vacuum pump attached to the inner bag to extract volatiles and entrapped air, and a second vacuum pump on the outer bag to compact the laminate.

Advantages

- Inexpensive process.
- Can be set up without highly sophisticated equipment.
- Requires less space for set up.
- High fiber volume ratio can be achieved.

Disadvantages

- Thickness is uneven.
- Surface Quality is not good.
- Void content is high.
- No control over curing.
- Dimensional control is difficult.

Application

Parts being used in transportation, wind energy, marine, infrastructure, and aerospace applications.

HIGH VOLUME MOLDING METHODS

Compression Molding

Compression molding method for composite manufacturing needs designing for the proper routing of the excess resin. Figure 5 shows the compression molding process. It generally follows the first steps of hand lay-up. The mold for a compression molded

Figure 5. Compression molding method

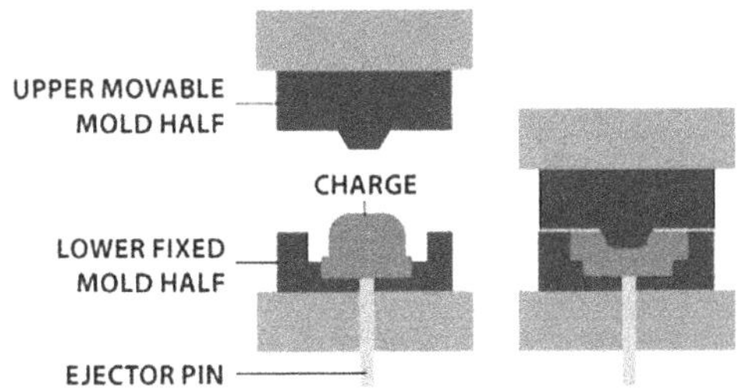

part must be designed for proper passage of excess resin. In compression molding a male and female mold will be used or a bladder mold can be used to compress into the other mold. A release agent must be applied on the mold surfaces in composite production process.

After that the fibers must be cut in accordance and placed on the surface of one of the molds. Catalyzed resin must be impregnated into the fibers with excess amount. After joining the two molds an adequate pressure will be applied either by mechanically or pneumatically. If the mold is set up in proper manner then the desired amount of resin will be throughout the fibers and out through the proper exit channels.

Advantages

- No gates, sprues or runner are required.
- Good for large parts fabrication.
- Lower cost tooling.
- Good for small productions.

Disadvantages

- Not suitable for complex moulds.
- More wastage.
- High labour cost.
- Slower process time.
- Contamination.
- Difficult to control flash.
- Moulds can be damaged.

Figure 6. Injection molding method

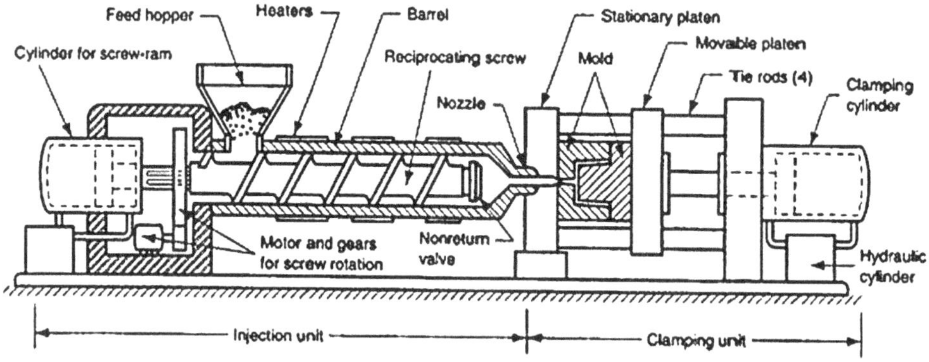

Injection Molding

Injection molding (Figure 6) is a pure plastic injection process without any fibers and has been used to make engineering thermoplastics such as polypropylene, polystyrene, and polymethylmethacrylate (PMMA) etc. Sometimes to make reinforced plastic components, short fibers such as short glass or carbon fibers can be incorporated into the thermoplastics. In this case, the fibers are mixed with the resin and injected together, rather than in the form of fiber preform.

The process cycle for injection molding is very short, and consists of the following four stages:

1. **Clamping:** The two halves of the mold must first be securely closed by the clamping unit, after that the material is injected into the mold. Each half of the mold is attached to the injection molding machine and one half is allowed to slide. These clamping units are hydraulically powered which pushes the mold halves together and exerts sufficient force to keep the mold assured closed while the material is injected. The time required to close and clamp the mold is dependent upon the machine - larger machines those with greater clamping force will require more time which is calculated from the dry cycle time of the machine.
2. **Injection:** The raw plastic material, usually in the form of pellets, is regale into the injection molding machine, and routed towards the mold by the injection unit. The material is melted by heat and pressure. The molten plastic is then injected into the mold very quickly and the build-up of pressure packs and holds the material. The shot is referred as the amount of material that is injected. The injection time is difficult to calculate accurately due to the complex and changing flow of the molten plastic into the mold but it can be estimated by the shot volume, injection pressure, and injection power.

3. **Cooling:** The molten plastic that is inside the mold begins to cool as soon as it makes contact with the interior mold surfaces and it will solidify into the desired shape of the component. However, during cooling some shrinkage of the part may occur. The packing of material in the injection stage allows additional material to flow into the mold and reduce the amount of shrinkage. It cannot be opened until the required cooling time has elapsed. The cooling time can be estimated from several thermodynamic properties of the plastic and the maximum wall thickness of the part.
4. **Ejection:** After cooling, the cooled part may be ejected from the mold by the ejection system, which is attached to the rear half of the mold. Force must be applied to eject the part because during cooling the part adheres to the mold and to facilitate this, a mold release agent can be sprayed onto the surfaces of the mold cavity before injection of the material. The time that is required to open the mold and eject the part can be estimated from the dry cycle time of the machine and should include time for the part to fall free of the mold. Once the part is ejected, the mold can be clamped shut for the next part.

After the injection molding cycle, some post processing is required. The material in the channels of the mold will solidify during the cooling process and get attached to the part. This excess material, along with any flash that has occurred, must be trimmed from the part, typically by using cutters. These scrap material can be recycled by being placed into a plastic grinder, also called regrind machines or granulators, which regrinds the scrap material into pellets and must be mixed with raw material in the proper regrind ratio to be reused in the injection molding process.

Advantages

Injection molding enables complex shapes to be manufactured, some of which might be near impossible to produce economically by any other means.

- The wide range of materials enables almost exact matching of the physical properties required from the article, and multi-layer molding enables tailoring of mechanical properties and attractive visual appearance - even in a toothbrush.
- In volume, it is a low cost process, arguably with minimal environmental impact. There is little scrap created in this process, and scrap that is produced, and be re-ground and re-used.

Disadvantages

The investment in tooling - making the mold - typically requires high volume production to recover the investment, though this does depend on the particular article.

- Producing the tooling takes development time and some parts do not readily lend themselves to a practical mold design.

Application

Wire pools, packaging, bottle caps, automotive dashboards,

Filament Winding

Filament winding is another process for the production of composite. A machine created to wind filaments in exact locations on a mandrel. Control parameters necessary to complete these tasks includes proper filament tow, longitudinal placement and mandrel winding speed. Filament winding advantages includes low material and labor costs and accuracy of product dimensions. Cost and design constraints of removable mandrels and concave surfaces are some disadvantages of filament winding. Part production by filament winding keeps increasing and only seems limited by imagination. The filament winding process is shown in Figure 7.

Figure 7. Filament winding method

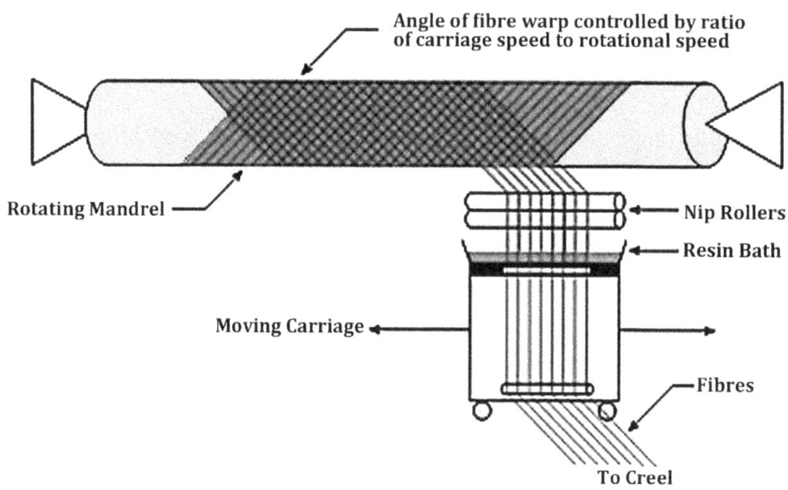

Advantages

- This can be a very fast and therefore economic method of laying material down.
- Resin content can be controlled by metering the resin onto each fibre tow through nips or dies.
- Fibre cost is minimised since there is no secondary process to convert fibre into fabric prior to use.
- Structural properties of laminates can be very good since straight fibres can be laid in a complex pattern to match the applied loads.

Disadvantages

- The process is limited to convex shaped components.
- Fibre cannot easily be laid exactly along the length of a component.
- Mandrel costs for large components can be high.
- The external surface of the component is unmoulded, and therefore cosmetically unattractive.
- Low viscosity resins usually need to be used with their attendant lower mechanical and health and safety properties.

CHEMICAL STORAGE TANKS AND PIPELINES, GAS CYLINDERS, FIRE-FIGHTERS BREATHING TANKS

Pultrusion

Pultrusion is a composite manufacturing process designed for structural shapes. Pultrusion process is only suitable and reasonable for mass production. First the fibers are drawn through a resin bath and after that through a forming block. Fast curing through steel dies is achieved by heaters and then the part is cut to proper length. The pultruded parts are stronger in the longitudinal direction because the fiber orientation in that direction leads to increase in strength. By using pultrusion process solid shapes and open sided, hollow shapes can be produced. Cores such as foam and wood can be built inside of the pultruded shapes. Due to the pressure and designs of production, protruded production can be up to 95% effective in material utilization. Figure 8 shows the pultrusion process.

Figure 8. Pultrusion

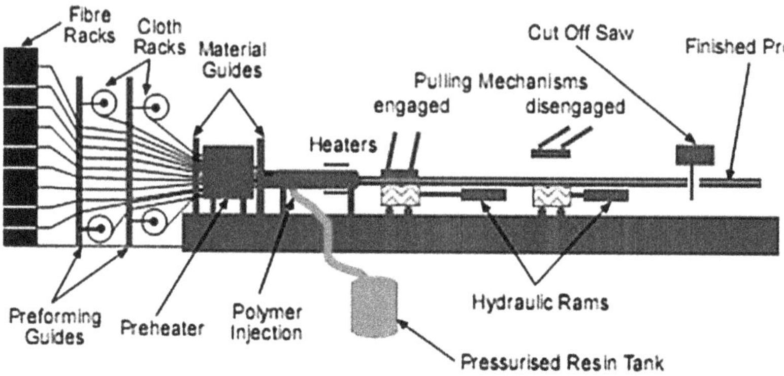

Advantages

- This can be a very fast, and therefore economic, way of impregnating and curing materials.
- Resin content can be accurately controlled.
- Fibre cost is minimised since the majority is taken from a creel.
- Structural properties of laminates can be very good since the profiles have very straight fibres and high fibre volume fractions can be obtained.
- Resin impregnation area can be enclosed thus limiting volatile emissions.

Disadvantages

- Limited to constant or near constant cross-section components.
- Heated die costs can be high.

Applications

Beams and girders used in roof structures, bridges, ladders, frameworks.

FABRICATION OF COMPOSITE: A PAST RESEARCH

Mechanical Properties

Romi Sukmawan et.al 2015 has prepared the cross-ply green composite laminates reinforced by bamboo fibers. The lignin and hemi-celluloses were removed from steam exploded bamboo fibers by treating it with alkali solution. They processed the

Figure 9. Depicts the tensile strengths of the neat PLA (poly lactic acid) and SEB/ PLA cross-ply laminate composites after hot pressing

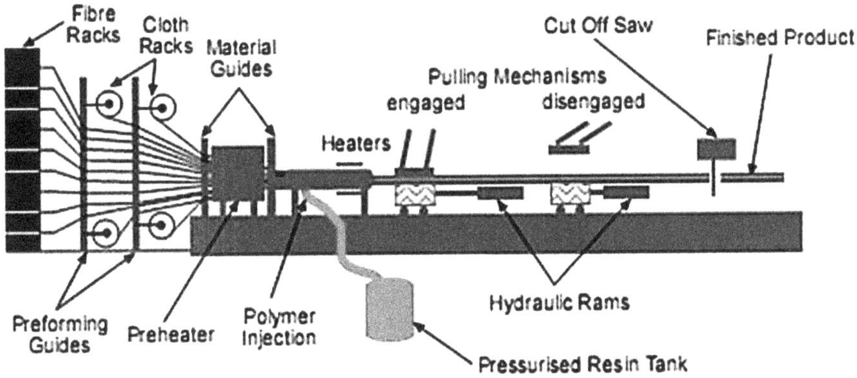

fiber by hand layup method. These bamboo fibers have average Young's modulus of 44 GPa and a width of 300 mm. The difference in strength is caused by the differences in the structure of fibers, extraction process, and also the orientation and cross section of the bamboo column.

However, it should be marked out that the bamboo fibers have a lower density of 1.15-1.3 g/cm^3, due to which their specific mechanical properties are actually comparable with those of E-glass. The bamboo fiber has almost the same average tensile strength as flax, hemp and lyocell fibers likened with other natural fibers and lyocell. The Young's modulus of bamboo fiber is slightly lower than that of flax and hemp fibers, but higher than that of lyocell fiber which means that bamboo fiber made by the steam explosion method also has great potential as reinforcing phase in composites as compared with other strong natural fibers that have been widely used in automotive applications.

The (0/90)s laminates reached its maximum tensile strength of 223 MPa and Young's modulus of 10.5 GPA, when reinforced with 68wt.% bamboo fiber. The tensile strength and Young's modulus increased linearly to approximately 7.2 and 4.9 times as higher as those of neat PLA, respectively which indicates that these laminates have the potential ability to work as reinforcement of PLA matrix. The mechanical strength and Young's modulus comparison of cross-ply (0/90)s SEB/ PLA composite with some common materials are shown in Figure 10 that are actually used in industrial applications. The tensile strength of cross-ply CFRP (carbon fiber reinforced polymer) laminate (474 MPa) and mild steel (400-500 MPa is double the tensile strength of cross-ply (0/90)s SEB/PLA composite with 68 wt.% fiber. The cross-ply (0/90)s SEB/PLA composites' specific mechanical properties was actually better than other. The Young's modulus of the composite was lower

Figure 10. Tensile strength and Young's modulus of SEB/PLA (0/90)$_s$ laminate, CFRP (0/90), mild steel Al alloy, and GFRP (0/90)
(Adapted from Romi Sukmawan et.al 2015; Nogata F, et.al 1995; Awad ZK, et.al, 2012)

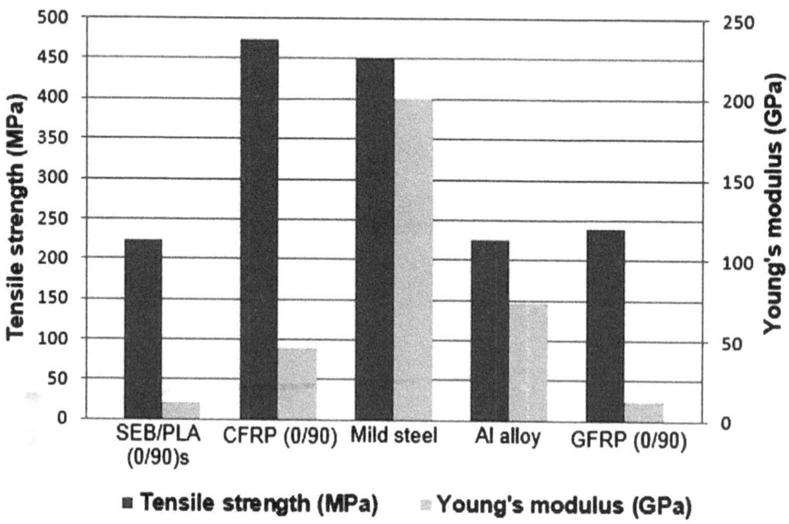

than that of carbon fiber reinforced polymer laminate (44.9 GPa), mild steel (200 GPa), and aluminum alloy (73 GPa) which show that the laminate has high strength and low rigidity. The strain at point of failure of the composites reached approximately 5.5% or 3 times of pure poly lactic acid.

Tetsuo Kikuchi et.al 2014, has used spray up fabrication method to make jute fiber composite. In this study, Jute fibers were cut into about 4 mm pieces and then sprayed with the resin onto the mold. The volume fraction of fiber of the obtained Jute-FRP (Jute-Fiber Reinforced Polymer) board was about 20%. From one Jute-FRP plate prepared using the spray up method ten specimens were extracted. The Table 1 shows the mechanical properties of Jute fiber reinforced polymer composite.

Changduk Kong et.al 2014 has prepared flax/vinyl ester natural fiber composite by vacuum assisted resin transfer molding process and examines its mechanical properties. The tests for determining mechanical properties were performed by ASTM standard. The specimen test results show the 34% fiber volume fraction of the 2D flax fabric composite and the 35% fiber volume fraction of the UD flax fiber composite. The tensile strengths of the UD flax composite is 157.5Mpa and the 2D flax fabric composite is 76.7MPa which are much higher strength than the strengths of 122.4Mpa and 62.0MPa. The mechanical properties of UD specimen and fabric specimen is shown in Table 2 and Table 3, respectively.

Table 1. Mechanical properties

Specimen No.	Tensile Strength [MPa]	Elastic Modulus [GPa]
1	17.8	-
2	18.5	4.22
3	14.4	4.07
4	16.5	4.41
5	16.2	4.01
6	16.6	4.34
7	15.7	4.47
8	18.2	4.41
9	13.2	4.22
Average	16.3	4.28
C.V	10.1%	4.4%

(Adapted from Tetsuo Kikuchi et al. 2014)

Table 2. Mechanical properties of UD flax/vinylester specimen (fiber volume fraction: 35%)

Test Institute	Jeonnam Technopark	
Test Type	**Strength (MPa)**	**Modulus (GPa)**
Tension	157.5	10.4
Compression	102.9	19.4
Flexure	188.0	9.7
In plane shear	26.24	3.8

(Adapted from Changduk Kong et.al 2014)

Table 3. Mechanical properties of 2D fabric flax/vinylester specimen (fiber volume fraction; 34%)

Test Type	Strength (MPa)	Modulus (GPa)
Tension	76.7	9.1
Compression	72.8	6.2
Flexure	108.7	6.7
In plane shear	36.1	1.6

(Adapted from Changduk Kong et.al 2014)

Rodrigo Ortega-Toro et.al 2014 has made the corn starch-glycerol (1:0.3) film by compression molding fabrication method (Rodrigo Ortega-Toro, et.al, 2014). The stress–Hencky strain curves prevail (shown in Figure 11) for the films, conditioned at 25^{0}C and 53% relative humidity for 1 week. The SG (starch-glycerol) film was the most flexible and resistant, while the addition of HPMC (Hydroxypropyl methylcellulose) brought have a small amount reduction to the film to break and to the elongation at break. This leads to a loss in the cohesion forces in the filmby a higher dispersed phase ratio and then the film resistance to break.

Saswata Sahoo et. al 2011 has made a composite from polybutylene succinate and lignin-based natural material using injection molding process (Saswata Sahoo, et.al, 2011). Table 4 shows the tensile properties and it is noticed that the tensile strength of the composites decreased firstly with 30% lignin integration and then gradually increased to 65%. Tensile strength of composites also decreased with increasing filler or agro fibre content in biodegradable polymers like PBS (Polybutylene succinate) and PLA (polylactic acid). This upshot was due to the weak interfacial bonding between the hydrophobic polymer matrix and hydrophilic filler. Nevertheless, enhancement in the tensile properties of composites was examined at higher lignin content (65%). At this, the increase in tensile strength was by a factor of 10 over the 50% lignin filled composites, was about 13% higher than the neat polymer.

Mohammad K. Hossain et.al 2011, has prepared a jute fiber reinforced biopol nanophased green composites by compression molding method. In Figure 12 and Table 5 uniaxial tensile test results of treated and untreated jute fiber bindles are

Figure 11. Typical stress–strain curves of the different films after 1 week of storage (Adapted from Rodrigo Ortega-Toro et al., 2014)

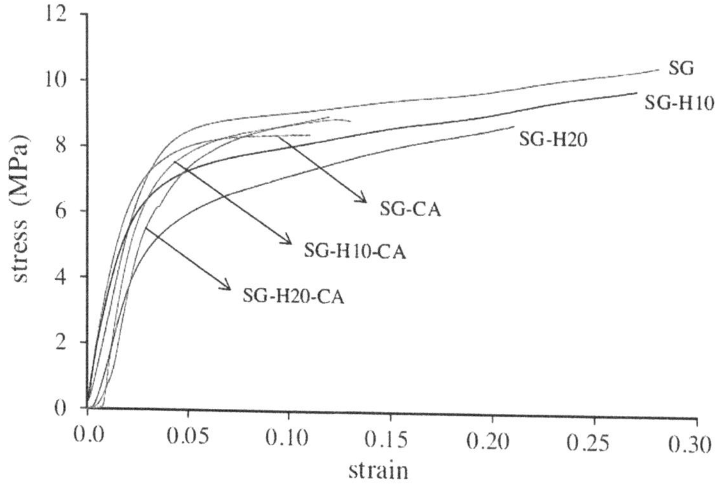

Table 4. Tensile, flexural, HDT and impact properties of composites

Specimen Label	Tensile Strength (MPa)	Tensile Modulus (GPa)	Elongation at Break (%)	Flexural Strength (MPa)	Flexural Modulus (GPa)	Impact Strength (J/M)	HDT (C)
PBS	35 ± 1.5	0.6 ± 0.01	122 ± 21	28 ± 0.4	0.6 ± 0.01	40 ± 8.4	78 ± 1.9
30% Lignin–PBS	26 ± 1.8	1.1 ± 0.03	4.6 ± 0.3	40 ± 0.5	1.1 ± 0.01	29 ± 1.0	83 ± 3.0
50% Lignin–PBS	29 ± 3.4	2.3 ± 0.35	2.0 ± 0.8	46 ± 0.3	2.2 ± 0.03	15 ± 0.9	86 ± 3.1
65% Lignin–PBS	39 ± 1.1	3.3 ± 0.04	1.5 ± 0.1	52 ± 1.1	3.8 ± 0.15	11 ± 0.9	85 ± 0.6
50% Lignin–PBS-1% PMDI	37 ± 6.1	2.0 ± 0.03	3.1 ± 1.3	68 ± 1.8	2.3 ± 0.07	29 ± 2.3	90 ± 1.9
50% Lignin–PBS-2% PMDI	42 ± 4.7	1.9 ± 0.19	4.3 ± 0.7	66 ± 0.7	2.1 ± 0.03	24 ± 3.7	94 ± 1.6

(Adapted from Saswata Sahoo, et.al, 2011)

shown. Fibers treated with NaOH for 2 h exhibited improvements in tensile strength by 13% and modulus by 17%, compared to untreated fibers. The non-cellulosic material was removed by the alkali treatment, including hemicellulose, lignin, and pectin from interfibriller regions which resulted in a higher percentage of cellulose. Higher concentrations of NaOH may weaken the fibers and make it more brittle. The optimum result is obtained by treating it for two hour with 5% NaOH.

Figure 12. Tensile stress–strain curves of treated and untreated jute fibers
(Mohammad K. Hossain et.al 2011)

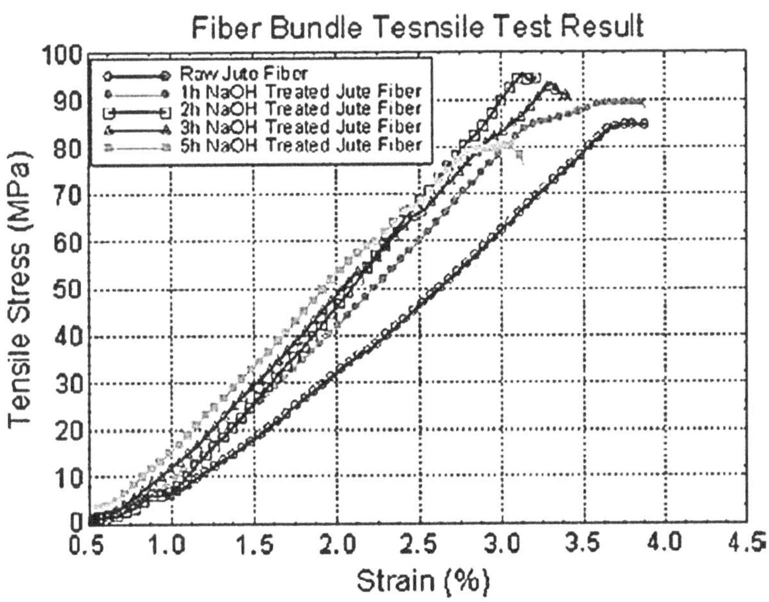

Table 5. Tensile test results of treated and untreated jute fibers

Jute Fiber Bundle	Strength (MPa)	Strength Change (%)	Strain at Failure (%)	Strain Change (%)	Modulus (GPa)	Modulus Change (%)
Untreated	81.42 ± 10	0	3.83 ± 0.62	0	1.92 ± 0.45	0
1 h NaOH treated	85.04 ± 13	4.9	3.62 ± 0.67	5	2.06 ± 0.64	7.3
2 h NaOH treated	92.54 ± 11	13	3.21 ± 0.68	16	2.25 ± 0.26	17
3 h NaOH treated	91.68 ± 10	12	3.34 ± 0.64	13	2.26 ± 0.34	17.7
5 h NaOH treated	80 ± 13	1.2	2.98 ± 0.69	22	2.28 ± 0.74	18.7

(Mohammad K. Hossain et.al 2011)

Scanning Electron Microscopy of Composities

Rodrigo Ortega-Toro et.al has made the corn starch-glycerol (1:0.3) film. Figure 1a and b shows SEM micrographs of the cross sections of films, conditioned at 25^0C and 53% RH for 1 (initial time) or 5 weeks (final time). At the initial time, the control formulations (SGi and SG-CAi) reveal a homogeneous structure with some cracks which are connected with the more brittle nature of this formulation under the analysed conditions. The addition of HPMC (Hydroxypropyl methyl-cellulose) gave rise to a two-phase structure in which this polysaccharide establishes the dispersed phase embedded in a starch continuous phase. Citric acid addition did not pitch in to avoid the formation of microcracks as can be noticed in Figure 13b. In HPMC-containing films, citric acid addition did not influence their micro-structural appearance as compared with acid-free. In spite of, a small increase in the polymer miscibility could be developed, without notably affecting the visible film microstructure.

This was especially illustrious for CA-free starch films whereas wider amorphous regions are still observable. Due to the easier water uptake from the environment and the subsequent increase in the molecular mobility the advance of the crystalline front from the film surface to the inner part of the film can be observed. The films show a more homogeneous Young's modulus (up to 3.5 GPa) at initial time, since no extreme effective modification are observed in the surface maps. Citric acid addition resulted in an increase in the surface modulus in agreement with the cross-linking effect, which is less noticeable in films containing HPMC. In samples with HPMC, more heterogeneous values of surface modulus were noticed, due to the fact that crystal growth was encumber in areas near to HPMC zones.

Figure 13. SEM micrographs of the cross sections of films without (a) and with (b) citric acid after 1 (Initial time) and 5 (Final time) weeks of storage at 53% relative humidity and 25°C
(Adapted from Toro et al.)

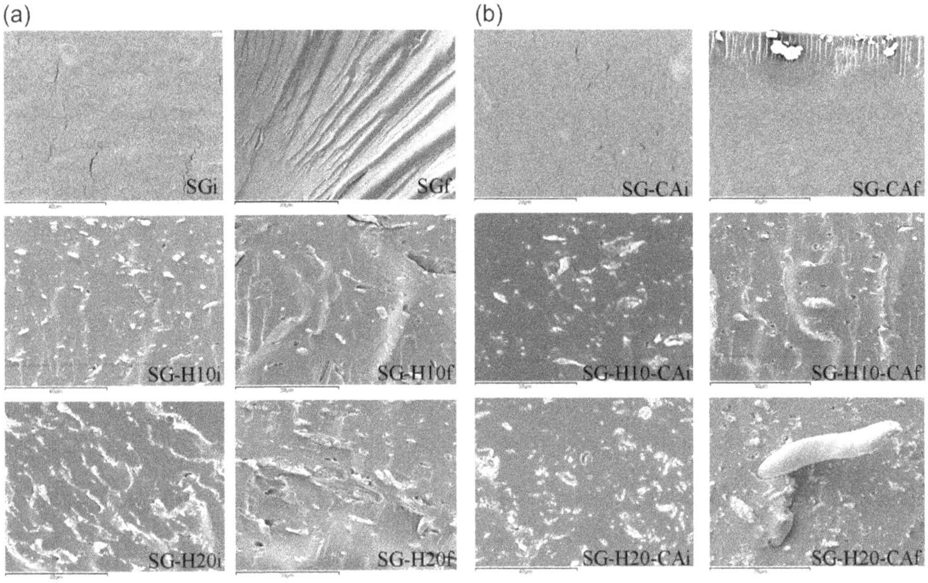

Mohammad K. Hossain et.al has prepared a jute fiber reinforced biopol nano-phased green composites by compression molding method. Surfaces of treated and untreated fibers examined divulge relatively rougher surfaces in treated fibers as similar to untreated fibers. The cleaner and rougher surfaces in finally treated fibers are a result of dispositioning of surface impurities, inorganic materials, non-cellulosic substances, and waxes. With the help of chemical treatment the mesh-like structure of fibers changes to clean and rough a single fiber which provides the higher strength to fibers.

CONCLUSION

The current chapter emphasis on the use of multifarious fabrication process for the manufacturing of natural fibre-reinforced composites. The main objective of this chapter is to examine the potential use of these processes for making polymer composites and also to study the various properties of composites. For manufacturing of composite materials both types of fibres such as chemically treated and untreated can be used by considering the various parameters of the said process. The process has good applications in the industry as well, specifically in aerospace and transportation industry for making various parts. These processes can also be found for making household equipment. The present chapter also defines the morphological study of the fractured specimen to understand the fracture mechanism of the composites.

REFERENCES

Awad, Z. K., Aravinthan, T., & Manalo, A. (2012). Geometry effect on the behavior of single and glue-laminated glass fibre reinforced polymer composite sandwich beams loaded in four-point bending. *Materials & Design*, *39*, 93–103. doi:10.1016/j.matdes.2012.02.023

Changduk, K., Park, H., & Lee, J. (2014). Study on structural design and analysis of flax natural fiber composite tank manufactured by vacuum assisted resin transfer molding. *Materials Letters*, *130*, 21–25. doi:10.1016/j.matlet.2014.05.042

Hossain, M. K., Dewan, M. W., Hosur, M., & Jeelani, S. (2011). Mechanical performances of surface modified jute fiber reinforced biopol nanophased green composites. *Composites. Part B, Engineering*, *42*(6), 1701–1707. doi:10.1016/j.compositesb.2011.03.010

Kikuchi, T., Tani, Y., Takai, Y., Goto, A., & Hamada, H. (2014). Mechanical Properties of Jute Composite by Spray up Fabrication Method. *Energy Procedia*, *56*, 289–297. doi:10.1016/j.egypro.2014.07.160

Molding, C. (n.d.). Retrieved from: http://www.coremt.com/processes/compression-molding/

Nogata, F., & Takahashi, H. (1995). Intelligent functionally graded material: Bamboo. *Compos Eng*, *5*(7), 43–51. doi:10.1016/0961-9526(95)00037-N

Rodrigo, O. T., Alberto, J., Talens, P., & Chiralt, A. (2014). Properties of starch–hydroxyl propyl methylcellulose based films obtained by compression molding. *Carbohydrate Polymers*, *109*, 155–165. doi:10.1016/j.carbpol.2014.03.059 PMID:24815412

Sahoo, S., Misra, M., & Mohanty, A. K. (2011). Enhanced properties of lignin-based biodegradable polymer composites using injection moulding process. *Composites. Part A, Applied Science and Manufacturing*, *42*(11), 1710–1718. doi:10.1016/j.compositesa.2011.07.025

Sukmawan, R., Takagi, H., & Nakagaito, A. N. (2015). Strength evaluation of cross-ply green composite laminates reinforced by bamboo fiber. *Composites. Part B, Engineering*, *84*, 9–16. doi:10.1016/j.compositesb.2015.08.072

Verma, D., Gope, P. C., Singh, I., & Jain, S. (2015). *Composites from Bagasse Fibers, Its Characterization and Applications. In Agricultural Biomass Based Potential Materials* (pp. 91–119). Springer International Publishing.

Chapter 3
Concurrent Design of Green Composites

Muhd Ridzuan Mansor
Universiti Teknikal Malaysia Melaka, Malaysia

S. M. Sapuan
University Putra Malaysia, Malaysia

Mohd Azli Salim
Universiti Teknikal Malaysia Melaka, Malaysia

Mohd Zaid Akop
Universiti Teknikal Malaysia Melaka, Malaysia

M. T. Musthafah
Universiti Teknikal Malaysia Melaka, Malaysia

M. A. Shaharuzaman
Universiti Teknikal Malaysia Melaka, Malaysia

ABSTRACT

This chapter presents the overview of concurrent design process of green composite products with special focus on conceptual design stage of natural fiber composites product development. Design of green composites product especially during the early product development stage requires three main aspects in product design which are materials, design and manufacturing process to satisfy lower cost, high quality and fast development time requirements in order to ensure successful product launch into the market. In this chapter, the concurrent design process of green composite products are discussed involving several main stages in product development such as green composite materials selection for both natural fiber and matrix constituents, conceptual design development and concept design selection of green composite products, and green composites manufacturing process selection. In addition, discussion on life cycle assessment of green composites is also included in order to provide further insight of the sustainability design requirements to the overall product development process.

DOI: 10.4018/978-1-5225-0424-5.ch003

INTRODUCTION

Concurrent design is part of the activities in the concurrent engineering (CE) approach for product development whereby it involved integration of various aspects of the product development process. Unlike traditional product development process which implemented sequential work activities, the CE approach utilized parallel process activities especially during the early design stage of the product. Also termed as Simultaneous Engineering, the CE approach removed the existing gap between design and production process in product development which has created issues regarding cost and quality due to miscommunication especially with respect to different goals and requirements between the two sequential activities (Sohlenius, 1992). Through the implementation of the CE approach, both aspects are considered in parallel as early as in the conceptual design stage which helps to produce products which able to satisfy both requirements, thus providing higher end product quality without jeopardizing the design intent, and subsequently reduced the overall product development cost due to eliminating the cost of error and repair due to the mismatch between design and production activities (Sapuan, 2005). The CE approach also helps to secure higher product development competitiveness to the market by bringing down the product development time. For green composites product developments, the CE approach is a very suitable method to be implemented considering the nature of the development process whereby green composite product designers need to consider various aspects of the product requirement such as design, materials and manufacturing process altogether during the early stage of the product development activities (Sapuan & Mansor, 2014).

Green composites products have made notable presence as substitutional materials for synthetic composites in many product applications. One of the most aggressive player using green composites is the automotive manufacturers. Driven by stringent legislations towards environmental performance for new generation of vehicles, the automotive manufacturers have applied green composites in the construction of various components such as dashboard, door trim panels, headliners and seat backs (Bismark et al., 2006). The use of green composites contributed significantly towards gaining high weight reduction, product cost and improved biodegradability of the components at the end of the life cycle (Ashori, 2008). Research in applying green composites for structural automotive components are also increasing due to similar advantages as aforementioned such as towards the development of structural frontal bonnet for buggy vehicle (Alves et al., 2010) and bumper beam (Davoodi et al., 2010).

Based on the impact and advantages of CE approach towards successful product design and development, in this chapter the concurrent design process of green composite products are discussed. Special attention is given towards the conceptual design stage of the green composites product development which involved several main activities such as green composite materials selection for both natural fiber and matrix constituents, conceptual design development and concept design selection of green composite products, and green composites manufacturing process selection. In addition, discussion on life cycle assessment of green composites products is also included in order to provide further insight of the sustainability design requirements to the overall product development process. Finally, to demonstrate the application of concurrent design for green composite products, a case study on the conceptual design development of bicycle frame using green composites is also explained in the end of this chapter.

CONCURRENT DESIGN PROCESS OF GREEN COMPOSITE PRODUCTS

Concurrent design is part of the major activities within the overall product development process. In general, systematic product development process follows Pugh product development model, termed as Total Design, which comprises of market analysis, product design specifications, conceptual design, detail design, manufacturing, and sales (Pugh, 1991). In addition, other product development models also proposed embodiment design stage to be conducted in between the conceptual design and detail design stage such as proposed by Dieter (2000) and Pahl et al. (2007). Figure 1 shows the major activities in product development process.

The product development process starts with market analysis whereby the needs of all stakeholders for the intended product are identified through several systematic methods. The stakeholders involved may comprise of end users or customers, authorities such as government, management within the company as well as internal department groups such as manufacturing, sales and procurements. Careful consideration of their needs is important towards successful development and product introduction to the market. In CE approach, several systematic market analysis

Figure 1. Major activities in product development process

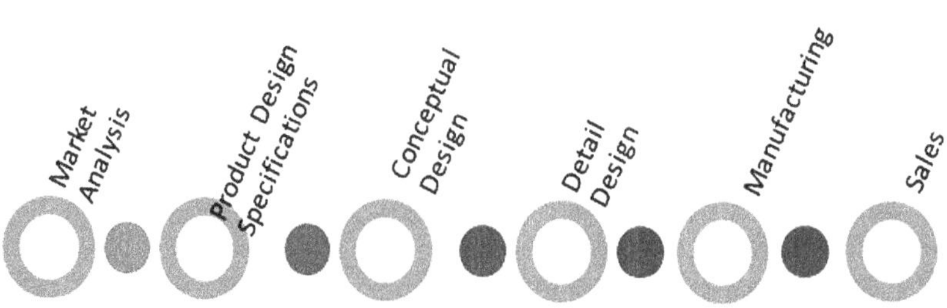

tools are available such as the Kano model and Quality Function Deployment (QFD) to determine the voice of the customer in order to tailor the desired product in fulfilling the customer preferences (Lin et al, 2011; Min et al., 2011; Yeh et al., 2011).

Based on the gathered information, the relevant attributes or factors towards successful product launch are analyze and compiled into a design guideline termed as Product Design Specifications (PDS) document. The PDS document served as the main reference which consists of all the necessary attributes that need to be considered by the new product during the whole development process. In Pugh Total Design model, the specific PDS requirements can be further group into specific elements such as cost, performance, weight and as well as standards and regulations. Similar case is applied in green composites product development whereby based from market analysis, specific PDS document are developed for the product which may includes similar elements such as cost, weight and technical performance as well as end of life requirements (recyclable and safe for disposal) and aesthetic appearance.

In the next stage of the product development process, conceptual design of the new product is developed based on the specific PDS requirements. Several tools are available within the CE approach to assist product designers in both idea brainstorming and idea generation such as the Ishikawa method or fishbone diagram method (Chandrasegaran et al., 2013), Ten Golden Rules for Ecodesign(Luttropp & Lagerstedt, 2006), Ecodesign Strategies Wheel (Brezet & Van Hemel, 1997) and Design for Assembly (DFA) (Chiu & Okudan, 2014). The design concept developed in this stage served as the founding model for the end product prior to detailed optimization process in the later stage. Information obtained at the end of the conceptual design stage are such as the overall geometry of the product, list of potential candidate materials for the embodiment of the product, the principles of operation and product working mechanism. Incorporation of decision support system are the essence of modern conceptual design activities towards getting the correct design and decision as early as possible in the overall product development process (Ghazilla et al., 2013).

In the subsequent stage termed as detail design, the rough idea and solution for the particular product is further refined towards optimizing the end product in achieving the PDS requirements more specifically such as detail product drawing with specific dimensions and tolerance, target operational performance (maximum allowable stress and deformation), safety, cost, assembly method, packaging method and end product weight. In this stage, more CE tools are applied especially with the assistance of computerized modeling and simulation applications such as Computer Aided Design (CAD), Computer Aided Analysis (CAE) and rapid prototyping method. Extensive testing and modifications of the new product are conducted to determine the prototype performance and durability in meeting the design specifications.

After the product detail design stage, the optimized product is then brought to the manufacturing stage whereby the product is produced in high volume. Similarly with the previous stage, there are many CE tools applied to ensure highest quality, lower production cost and quick time to market are achieved for the product such as through the implementation of Lean Manufacturing method, Six Sigma (Wang & Chen, 2010) as well as Failure Mode and Effect Analysis (FMEA) (Yen & Chen, 2005). Finally, the end product produced is introduced to the market which comprise of activities such as sales, marketing and after-sales. The successful launch of the product to the market is the ultimate goal for any product to achieve highest customer acceptance and subsequently contribute to higher profits and larger market share compared to other product competitors.

CONCURRENT CONCEPTUAL DESIGN OF GREEN COMPOSITE PRODUCTS

As noted in the earlier section, this chapter is focused on describing the conceptual design stage of green composite product development using the CE approach. Ulrich & Eppinger (2004) stated that, despite only 5% of total product development cost and 15% of total development time is inflicted during the conceptual design stage, the same conceptual design activities contributed to almost 70% of the total manufacturing cost of the end product. Thus, the conceptual design stage has a very significant effect to the success of the product, especially to achieve the aim of CE approach which is lower cost and faster time to market with high product quality. Decisions made during the conceptual design stage will greatly influence the subsequent product development process, and using systematic concurrent conceptual design approach, optimized and justified decisions can be made to ensure that all the PDS requirements are fulfill and the final product is able to launch successfully into the market.

Concurrent conceptual design of green composite products involved several major activities which are design concept development and selection, materials selection, and manufacturing process selection. Due to the recent environmental awareness towards product sustainability, environmental life cycle assessment is also incorporated into the conceptual design stage of the product. Figure 2 shows the major activities associated with the concurrent conceptual design of green composite products.

Under the CE approach, there are many available tools to support the implementation of all the activities in the concurrent conceptual design stage. However, for green composites product application, available literature reviews in the conceptual design stage is still limited. This is due to several factors such as the green

Figure 2. Major activities in the concurrent conceptual design of green composite products

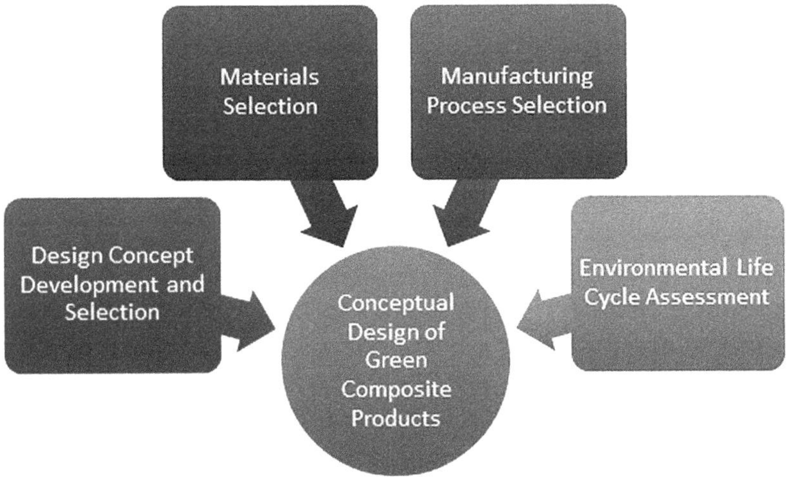

composites itself is fairly a new explored field with respect to product development compared to synthetics composites in particular and conventional engineering materials (metals, ceramics, plastics) in general. Apart from that, due to the inherent limitation of green composites properties especially structural properties (such as strength, elastic modulus, dimensional stability and thermal resistance) compared to synthetic composites, current product application of green composites are limited to non-structural products with low load bearing capability. However, by using systematic concurrent conceptual design process, it is highly possible to break the design barrier and apply green composites to products with higher load bearing requirements. In return, more environmentally sustainable product can be developed for new markets such as automotive, aerospace, building and construction using green composites to reduce the current consumption of less environmentally favorable synthetic composites. The following sections will discuss the activities related to concurrent conceptual design process of green composite products.

DESIGN CONCEPT DEVELOPMENT AND SELECTION OF GREEN COMPOSITE PRODUCTS

Design concept development and selection is conducted to obtain potential solutions in term of geometry for the green composite products. As we are aware of, green composites is inherently weak in term of structural strength and stiffness compared to synthetic composites which is made from synthetic fibers such as glass, carbon and aramid. Therefore, in order to directly substitute the use of synthetic composites with green composites, the original product geometry needs to be altered to cater for the materials property limitation. Among the possible solution is by enlarging the initial product geometry such as thickness and surface area, adding stiffeners such as ribs and wedges to the main structure, or reducing the stress concentration areas by using larger value of fillet radius and chamfer size to product sharp corners. However, these alternatives to enhancing the original green composites product structural performance is applicable at the detail designs stage of the product, while the original issue still lies in the conceptual design stage as to create the initial form of the product. In addition, the geometrical design of green composite products are more challenging due to their unidirectional force tolerance of the fibers as compared to homogenous materials such as metals which has almost equal load absorption capacity at all force direction.

In general, there are two main activities under the concurrent conceptual design approach which relates to the design concept development and selection. First is the design concept development whereby initial form of the green composites product is produced and then followed by selection of the possible design concept developed to obtain the best of final design concept for the green composites product. The following sections described both major activities related to the design concept development and selection stage.

Design Concept Development of Green Composite Products

In the design concept development process, possible solution in term of geometry for the intended product is developed. The development process normally covers idea generation and idea development. In the idea generation, based on problem statement or design requirements, possible solutions is generated which is able to address all the design intents. There are many methods available for design concept development for green composites products such as Theory of Inventive Problem Solving (TRIZ), Morphology Chart and biomimetic. One of the most important advantages of the aforementioned methods is that they provide a systematic idea generation process for users in generating ideas for the intended green composites products, which is a valuable alternative to the inherent creative thinking and experiences of the designer themselves. All the methods served as a guideline for users in producing solutions for the green composites products.

One of the available design concept development method under the concurrent design approach is the Theory of Inventive Problem Solving method (or originally term as Teoriya Resheniya Izobretatelskikh Zadach in Russian language) which is a systematic problem solving method based on study of patents. Developed in 1940s by Russian patent officer name GenrichAltshuller, the TRIZ method proposed several algorithms to identify the problem and create the final solutions based on the available solution model. The TRIZ method is reported to excel in solving problem with contradictions, which is defined as the problem which has both improving and worsening characteristics for the selected solution. The TRIZ method is able to address the worsening effect while still able to keep the improving effect both simultaneously by solving the contradiction. Among the solution model in TRIZ is the trimming tool, 40 Inventive Principles, and 76 Standard Inventive Solutions (San et al., 2009).

Among the initiative in applying TRIZ for green composites product design was the integration of TRIZ-Morphological Chart-AHP method for the development of hybrid natural fiber composites automotive parking brake lever design (Mansor et al., 2014b). The new integrated method has provided an all-round tool for composite product designers to execute many aspects of product design such performing problem identification, idea generation, concept design development and concept design selection using a single integrated and systematic approach. Apart from that, conceptual design of automotive rear spoiler component was also developed using TRIZ method which encompassed the use of kenaf fiber composites as the construction material (Mansor et al., 2015). For other product developments areas, the TRIZ method was also reported to be applied for conceptual design of notebook casing product (Yeh et al., 2011), eco-friendly bottle casing (Chen et al., 2012), vacuum cleaner (Yen & Chen, 2005) and washing machine (Yamashina et al., 2002) by integrating the TRIZ method with other CE tools such as QFD, FMEA, Eco-design and axiomatic design.

Another design concept development method under the concurrent design approach is the Morphology Chart which is a systematic idea generation method based on the function of the product. Products are in general comprised of multiple components whereby all the components are design to perform different functions. By analyzing the product functions, alternatives method to serve the functions are brainstorm and arranged into a simple matrix. Finally, the potential solutions or the potential design concepts for the product are developed by combining the individual function alternatives listed in the matrix. The advantage of this method is that it provided visual solution to aid users in developing the potential ideas for their products as well as expanding the richness of potential solutions created (Silvester et al., 2013). Despite the usefulness of the method, very limited report on the application of the morphology chart for composite product development purpose was available. One notable application was for the development of polymer composites automotive pedals (Sapuan, 2006). The morphological chart was employed to generate ideas based on the sub-functions of the component. Among the sub-functions included are pedal movement control method, cylinder or cable actuation method, geometry of the pedal pad, pedal profile, rib reinforcement pattern and location of the pivoting shaft of the component. Apart from that, the Morphological Chart was also employed by integrating it with the TRIZ method for hybrid natural fiber composites automotive parking brake lever design (Mansor et al., 2014b).The combination of TRIZ and Morphological Chart method utilized in the report was also found able to effectively produce many innovative conceptual designs using green composites as the construction material to develop the product, which can be further applied in other green composites related product for various applications

Alternatively, the biomimetic or biomimicry concept is also part of the available concept development method in concurrent design. The biomimetic concept is an idea generation model which is inspired by nature. The term was introduced by Otto Schmitt in the 1950s. Despite the recent introduction of the method, the informal application of the model for product design purpose was documented since long ago such as by Leonardo da Vinci in his idea of a mechanical flying contraption which is based on bat physical characteristics (Yurtkuran et al., 2013). The biomimetic model was revitalized by Benyus (1997) through her publication on the concept of harnessing natures inherent patterns and strategies into solution for modern products.

In general, there are two approaches in implementing biomimetic for idea generation and development in product design. First approach is termed as problem-based approach or top-down approach whereas the second approach is termed as solution-based approach or bottom-top approach (Versos & Coelho, 2011). In the problem based approach, designers seek solution for the nature characteristics, and apply relevant traits from the nature to achieve the intended product goal and design requirements. There are six (6) recommended steps associated with the implementation of the biomimetic problem-based approach which are:

1. Problem definition,
2. Reframing the problem,
3. Searching for biological solution,
4. Defining the biological solution obtained,
5. Extracting the principles or method from the biological solution and finally
6. Applying the principles to the intended design.

On the other hand, application of the biomimetic method using the solution-based approach is started the other way around whereby the designers first identify the principles of solution from nature and later apply the principles to address the design problem. The second approach is solution driven and offered the advantages of obtaining new possible applications for the nature's principle identified aside from the intended initial design requirement. In the solution-based approach, there are seven (7) implementation steps involved which started from:

1. Identification of the biological solution,
2. Defining the biological solution,
3. Extracting the principles from the biological solution,
4. Reframing the solution to explore the possible application for the biological solution principles,
5. Searching for suitable design problems relevant to the biological solution principles,
6. Defining the design problem and finally
7. Applying the biological solution principles to solve the design problem (El-Zeiny, 2012).

There are several reported case studies on the implementation of the biomimetic model for idea generation of composite products. By mimicking the cross-section of a lily plant leaf using scanning electron microscopy, designers obtained design ideas for composite sandwich panels for improved structural performance (Easterling, 1990). Furthermore, the initial compressive strength for composite laminates was also reported able to be improved by adopting nature's bamboo bast fiber structure which incorporate double helical configuration to the new composite laminates design. Similar report also found that the interlaminar shear strength of composite laminates may also be enhanced by mimicking the bamboo's inherent interlaminar transition zone model to the new composite laminates design (Li et al., 1995). In addition, another report also identified that by using biomimetic concept, the composite joining strength was able to be increased by implementing similar T-joint configuration principles occurred during growth of tree brunch (Burns et al., 2012b). The bio-inspired T-joint principle was also reported able to improve the ductility and damage tolerance of structural joint performance for aerospace composites (Burns et al., 2012a).

Design Concept Selection of Green Composite Products

After the design concept development process, normally more than one (1) possible design solutions are able to be created for the intended product. The next stage involved is to determine which solution that is most suitable to carry out the product function and selected as the final design concept. The final design concept will be the reference for the final product when it is launch into the market after enduring the optimization process later on during the detail design stage. The final design concept evaluation is conducted based on the previously determined PDS. There are many concurrent design tools for design concept selection applicable for this purpose. Among them are the Pugh concept selection method (Pugh, 1991) and the weighted objective method (Cross, 1994). For design concept selection process which involved more advanced decision making capability, the use of multi criteria decision making (MCDM) methods such as Analytic Hierarchy Process (AHP) and Technique for Order Preference by Similarity to Ideal Solution (TOPSIS) may also be employed to obtain the final decision. Mansor et al. (2014b) reported the application of the AHP method for design concept selection process of automotive parking brake lever made from hybrid green composites. Furthermore, Davoodi et al. (2011) also reported on the application of the TOPSIS method for the design concept selection process of automotive bumper beam component using hybrid green composites. In addition, the design concept selection of similar automotive bumper beam component using synthetic composites was also carried out using the AHP method as reported by Hambali et al. (2009a).

MATERIALS SELECTION OF GREEN COMPOSITE PRODUCTS

Materials selection is another important activity in the concurrent design of green composite products. Green composite materials are comprised of two main constituents, which are the reinforcing natural fibers (such as kenaf, oil palm empty fruit bunch, jute, sisal, hemp and pineapple leaf fibers) and the matrix or resin which acts as to hold all the fibers together giving the final shape to the composite materials as well as distributing the load evenly to the whole composite structure. Thus, major concern within the materials selection activity of green composite product development is to decide on the best type of natural fibers and type of matrix material to be selected to produce the green composite material (Shah, 2013). Materials selection for green composites are often challenging considering the vast amount of option available for both fiber material and matrix material, and also by the introduction of newer class of fiber and matrix materials into the market due the recent advancement in research on green composites worldwide. For instance, the green composites matrix can be choosen from two categories of polymers, which are thermosetting matrix and thermoplastic matrix. Both categories can be further diversified into matrix derived from petroleum resource and matrix material derived from natural resources such as thermoplastic starch (Hamma et al., 2014) and poly-lactic acid (PLA) (Fiore et al., 2014).

The complexity of deciding on the best fiber and matrix material are also inflicted due to the multiple stakeholder requirements that need to be satisfied simultaneously during the product development process. Among the stakeholders which have direct requirements for decision makers to fulfill in the materials selection process are the customer, product designer and product manufacturing engineer, and in most cases, their preference for the materials to be selected are conflicting between each other (Sapuan, 2010). Thus, to address this complexity, the multi criteria decision making (MCDM) method is employed during the concurrent design process to aid decision makers in getting the optimized decision for materials selection. The use of MCDM method enabled all the selection criteria from various stakeholders and pool of potential materials with varying attributes between them to be analyzed simultaneously in a systematic and quantitative ways (Behzadian et al., 2012).

On the other hand, aspects that are considered in materials selection for green composite product can be group into three (3) main categories namely functional, sensorial and intangible properties. The functional aspect is related to the functionality and technical performance of the material to suit the intended application such as mechanical properties, thermal properties and acoustic properties. Secondly is the sensorial aspect of the material which involved consideration such as visual (i.e. colour intensity and colorfulness), tactual (i.e softness-hardness), weight (heavy–light), olfactory (i.e. odorless–fragrant) and auditory (i.e. loudness-quietness) (Baxter et al., 2015). The third aspect involved is the intangible properties such as perceived values and cultural meanings, emotions, trends of the material (Karana et al., 2008). The consideration of all the mentioned aspects is crucial towards successful green composite product development whereby the material selected will also provide meaningful experience of the end product itself to the customers and acts as a linked between the user with the end product as highlighted in the Meaning of Material (MoM) model for materials selection process proposed by Karana et al. (2010).

In general, the materials selection process for green composites product can be divided into two (2) main stages, which are the screening and ranking stage (Jahan et al., 2010). The screening stage is performed to narrow down possible alternative materials to a manageable quantity based on the product design general requirements. Among the screening methods applied in materials selection are the Ashby chart (Shah, 2014), ternary diagram using three bi-dimensional factors (specific strength, specific stiffness and cost per weight) (Koronis et al., 2013) and knowledge-based system (using IF-THEN rule based technique) (Sapuan & Abdalla, 1998).

The ranking process is later performed to analyze in detail the list of potential candidate materials and select the best alternative material which is able to successfully meet all the selection criteria. In green composites materials selection, the selection criteria is extracted from the PDS developed for the intended product. In most cases, the ranking process by the majority of the available MCDM ranking methods is started by determining the weight of the selection criteria or the magnitude of preference between the selection criteria involved using method such as Digital Logic (DL) method and Analytic Hierarchy Process (AHP). However, Attri & Grover (2013) reported that there are MCDM methods which do not need the information of the weight for the selection criteria such as Preference selection index (PSI) method (Maniya & Bhatt, 2010), VlsekriterijumskaOptimizacija I KOmpromisnoResenje(VIKOR) method (Girubha & Vinodh, 2012) and Grey

Relational Analysis (GRA) method (Chan & Tong, 2007). The weight ratings are quantified based on individual preference or from group preference involving the product stakeholders. For group decision making, the average value for their preferences is calculated using either arithmetic mean or geometric mean method and applied for the respective selection criteria (Wu et al., 1998).

The next process involved during the ranking stage is analyzing the score of every candidate materials based on the respective attributes with respect to the selection criteria and weight of the criteria as defined in the previous process. Finally, the rank for every candidate materials are determined based on the individual total score value obtained whereby highest score value represents the best candidate material to fulfill the intended design requirement. In overall, selection of the MCDM method to be applied also depends on the user's preference with consideration of computational time, amount of mathematical calculation required and level of complexity of the algorithm involved (Attri & Grover, 2013). Among the ranking methods which are available for materials selection process including for green composites are Analytic Hierarchy Process (AHP) method (Akadiri et al., 2013; Mansor et al., 2013), Preferences by Similarity to the Ideal Solution (TOPSIS) method (Anupam et al., 2014; Mansor et al., 2014a), Preference Ranking Organization Method for Enrichment Evaluation (PROMETHEE) method (Çalişkan et al., 2013; Peng & Xiao, 2013), ELimination Et Choix Traduisant la REalité (ELECTRE) method (Shanian & Savadogo, 2006) and weighted property index method (Findik & Turan, 2012).

MANUFACTURING PROCESS SELECTION OF GREEN COMPOSITE PRODUCTS

Another important aspect involved in concurrent conceptual design stage is the green composite product manufacturing process selection. Similarly with the previous section, MCDM methods are normally employed to perform the selection process due to the presence of multiple selection criteria and multiple manufacturing process alternatives which needs to be analyzed concurrently. In general, the manufacturing process selection criteria for green composites product is determined based on the final composites product architecture which are the volumetric composition of the green composites in term of intended fiber volume fraction and porosity), reinforcement geometry in term of fiber length and fiber orientation, and the type of matrix chosen, whether it is thermoset or thermoplastic matrix (Shah, 2013). In addition, the criteria for the manufacturing process selection also includes end product quality and low labor cost (Singh et al., 2011). Furthermore, the manufacturing process flexibility, good precision to reach the required dimensional tolerance, low cost, low scrap and rapid production capability are also desirable traits for selection of green composites manufacturing process (Jahazi & Hossein-Nejad, 2004). Hambali et al. (2009b) demonstrated the application of MCDM method for automotive bumper composites manufacturing process selection using AHP method. In their report, among the selection criteria applied were cost (tooling, labor and equipment), production characteristics (production quantity, rate of production and processing time), geometry of the final composite product design (such as shape, wall thickness and weight), type of materials used, ease of maintenance and availability of the equipment and labor.

LIFE CYCLE ASSESSMENT OF GREEN COMPOSITE PRODUCTS

Due to the increase of awareness from consumers on the product they consumed especially on its effect to the environment, product designers are now incorporating design for sustainability tools for new product being developed. The growing trend isalso now applied for products which used green composites as the raw material. In the concurrent conceptual design stage, one of the tool applied is the life cycle assessment (LCA) method which served as to determine the potential environmental impact of the product throughout its entire life cycle (Finkbeiner & Hoffmann, 2006). The implementation of the LCA method in the conceptual design stage helps product designers especially the one dealing with green composites to quantify the resulting potential environmental burden throughout the whole product life cycle starting from raw materials extraction, production and up to the its end of life of stage. Apart from that, the LCA method will also yield quantitative measures to assess "how green" is the green composites product itself as targeted during the initial product development process. Furthermore, the information obtained through LCA method also help designers to identify key product life cycle stages which contributes to the highest environmental burden so that correction action can be undertaken as early as possible such as during the conceptual design stage of the product (Joshi et al., 2004).

There are many reported studies on the implementation of LCA method for green composites product development. Wötzel et al. (1999) applied LCA method using Eco-indicator 99 impact assessment method to assess the environmental impact of automotive side panel component developed using hemp/epoxy green composites. Schmehl et al. (2008) also reported the application of LCA method to measure the potential environmental impact of hemp/PTP thermoset green composites for passenger bus body component casing product design. Moreover, Luz et al. (2010) utilized the LCA method to assess the potential environmental burden of sugarcane bagasse/polypropylene green composites for interior component of passenger vehicle. They also analyzed the environmental performance of the similar component constructed using talc/polypropylene composites and later compared both materials in term of the environmental performance throughout the component life cycles.

CONCEPTUAL DESIGN OF GREEN COMPOSITES PRODUCT: CASE STUDY ON GREEN COMPOSITES BICYCLE FRAME DESIGN

Bicycle is among the earliest form of road transportation created by men. Among the main component of a bicycle is the frame which acts as the main load bearing structure of the vehicle which holds all the other components together to make it functional. Currently, steel is the most used material to construct the frame component due to its high strength and stiffness properties required for safe and reliable operation to the user. Synthetic composites were also used to construct the frame component especially carbon fiber composites to gain better lightweight properties compared to steel for higher cycling performance while maintaining the structural requirements for safe and functional operation. However, despite the lightweight gain by using carbon fiber composites, trade-offs were made in term of higher material cost, which limits the carbon fiber composites bicycle frame to high-end users such as competitive racing cyclists, maintaining the domination of steel for bicycle frame design. On the other hand, green composites, made from natural fiber such as kenaf, sugar palm and pineapple leaf reinforced in either thermoplastic or thermoset matrices is gaining wider attentions nowadays especially as the most potential candidate materials to replace the use of synthetic composites such as carbon fiber composites and steel for various structural applications. Green composites offer the cost and lightweight solution to synthetic composites and steel, respectively. Furthermore, green composites which used fibers made from natural resources also provide better sustainability performance such as renewable and recyclable compared to synthetic fiber made from petroleum resource. However, green composites still lag behind synthetic composites and steel especially in term of structural performance, henceforth seeing the application of green composites material to be concentrated to non-structural products. In order to gain the potential of green composites as substitution material to construct the bicycle frame, other solution to address its inherent low structural performance must be worked out, and among the solution is through applying CE approach of the product during its development process. In this case study, the integrated TRIZ-Morphological Chart method is selected to assist in developing innovative design solutions during the concurrent conceptual design stage for the new bicycle frame using green composites material. The development process of the new bicycle frame conceptual design using green composites material is divided into three stages and the explanation are described as below.

Stage 1: Problem Identification

In the first conceptual design stage, problem identification of the project was first determined using the TRIZ method. The aim of the case study was to utilize green composites for bicycle frame construction to improve the product cost, sustainability and lightweight performance compared to the current frame material, which is steel. However, applying green composites also reduced the structural performance, which is crucial to guarantee functional and safe operation of the product. Thus, a contradiction occurred in which by improving one parameter of product, another parameter worsened in the same time. According to the TRIZ method, the contradiction occurred is classified as engineering contradiction due to the type of the parameter involved. Next, the specific problem statement is matched with TRIZ 39 engineering parameters to convert it to general problem statement. TRIZ engineering parameter number 2 (weight of stationary object) was selected to represent weight of the green composites bicycle frame as the improving parameter in the design while engineering parameter number 11 (stress) was selected to represent the reduce of structural performance due to lower stress that can be absorbed by the green composites bicycle frame.

Stage 2: Solution Generation

In the second stage of the conceptual design process, solutions to solve the known contradiction were generated. Again by applying TRIZ method, general solutions were obtained using the TRIZ contradiction matrix based on the identified improving and worsening parameters as shown in Table 1. In this particular case, four potential inventive solution principles were suggested based on the contradiction matrix which are number 10 (preliminary action or prior action – "do it in advance"), number 13 (the other way around), number 18 (mechanical vibration/oscillation) and number 29 (pneumatics and hydraulics). The proposed solution principles served as guidelines in brainstorming the potential idea to solve the problem.

Table 1. Contradiction matrix for the green composites bicycle frame based on the TRIZ 39 Engineering Parameters

Improving Parameter	Worsening Parameter	TRIZ 40 Inventive Solution Principles
#2. Weight of stationary object	#11. Stress	#10. Preliminary action (prior action – "do it in advance") #13. The other way around #18. Mechanical vibration/oscillation #29. Pneumatics and hydraulics

(San et al., 2009)

Table 2. Specific solution strategy for the green composites bicycle frame based on the TRIZ solution principles

TRIZ General Solution Principles	Solution Descriptions	Specific Solution Strategy
#10. Preliminary action (prior action – "Do it in advanced")	1. Perform the required change of an object (either fully or partially) before it is needed 2. Pre-arrange objects such as that they can come into action from the most convenient place without losing time for their delivery	• Add ribbing pattern as reinforcements on the bicycle frame to increase structural strength and stiffness (Imihezri et al., 2005) • Change from multiple section frame to single piece frame to increase strength and stiffness by eliminating concentrated stress using joints as well as ease of frame manufacture using green composites material)
#13. The other way around	1. Invert the action used to solve the problem 2. Make movable parts fixed, and fixed parts movable 3. Turn the object 'upside down'	• Change from straight shape frame to arch shape frame to increase static vertical load bearing capability • Change from tubular-shape frame cross-section to beam-shape frame cross-section such as I-beam shape design to increase stiffness as well as ease of frame manufacture using green composites material (Imihezri et al., 2006)

The next step taken was to analyze each of the general solution principles obtained from the contradiction matrix to determine relevant general solution associated with the product. Based on the conducted analysis, the principle number 10 and number 13 were found to provide the most suitable solutions to solve the problem faced in the design. Thus, specific solution strategies relevant to the green composites bicycle frame were later devised based on the general solution selected. Table 2 summarizes the overall specific solution strategies based on the TRIZ general solution principles identified to be applied to the new green composites bicycle frame.

Stage 3: Conceptual Design Development

In the final stage, the classical Morphological Chart method was applied to refine the previously identified TRIZ specific solutions strategies into appropriate alternative system elements. The integration of Morphological Chart with TRIZ method enabled further identification of possible specific design features related to the recommended TRIZ principles, which in return also enhanced the effectiveness of the TRIZ method. Furthermore, the integrated TRIZ-Morphological Chart method also provided a quick way for designers to systematically translate the general TRIZ solutions into specific design idea and generate more solution ideas for the desired conceptual design. In addition, the designer may also visualize the solution idea

Figure 3. Morphological chart for green composites bicycle frame design

TRIZ solution principles and specific solution strategy	Design feature	Conceptual Design Solution			
		1	2	3	4
#10. Preliminary action (prior action – "Do it in advanced") i) Add ribbing pattern as reinforcements on the bicycle frame ii) Change from multiple section frame to single piece frame	Rib pattern	I	V	X	
	Single frame	Yes	No		
#13. The other way around i) Change from straight shape frame to arch shape frame ii) Change from tubular-shape frame cross-section to beam-shape frame cross-section	Frame shape	Flat	Arch	Mixed	
	Frame cross section	I-shape	T-shape	Rectangle	

more clearly prior to combining them together to generate new conceptual design of the product. Figure 3 shows the developed Morphological Chart for the green composites bicycle frame design.

Note: Example for development of concept design 1= Rib pattern (X) + single frame + frame shape (flat) + frame cross-section (I-shape).

Using the Morphological Chart developed in Figure 3 as guideline, four (4) new conceptual designs of the green composites bicycle frame were developed as shown in Figure 4. Each of the new conceptual designs combined individual elements related to the design feature listed in the chart. For example, concept design 1 was developed using the combination of X-type ribbing pattern as the frame reinforcements and was based according to single frame construction as opposed to the multiple frame for the current steel bicycle frame. The single frame construction design was permitted due to the capability of producing single piece component using polymer composites manufacturing technology such as injection molding and resin transfer molding process. Moreover, concept design 1 also retained the straight shape frame design for design simplification and an I-shape frame cross-section to replace the tubular steel frame cross-section applied in current bicycle frame design. The I-shape frame cross-section also helped to enhance the structural rigidity of the green composites structure while in the same time reduced the volume of material

Figure 4. New conceptual designs of green composites bicycle frame developed using integrated TRIZ-morphological chart method

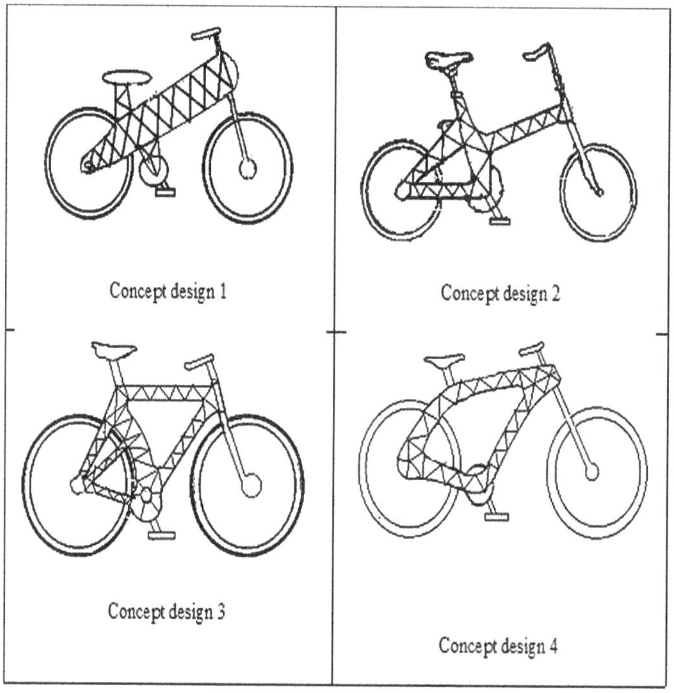

used for lightweight purpose. Utilization of the solution strategies identified using the TRIZ method for the development of new green composites bicycle frame conceptual designs have allowed innovative and practical design solutions for structural purpose to be achieved despite the initial limitation on mechanical properties of the green composite material.

CONCLUSION

The know-how of materializing green composites into workable and acceptable products both for non-structural and structural applications as substitution to conventional engineering materials and synthetic composites is an important feature required to product designers. Despite the challenging task faced due to the limitation on the structural properties of green composites, the inherent inadequacy can be addressed successfully through implementation of systematic CE approach especially during concurrent conceptual design process. Without compromising the functional requirements such as structural strength and safety, innovative solutions and optimized decision can be developed for the green composite products which also able to provide better sustainable advantages to consumers throughout the product entire life cycle especially when compared to synthetic composite products. Many CE tools are applicable to handle the product development process involving idea generation and development as well as MCDM for materials, process and design selection for green composites. The potential of success was made available for reference to designers and practitioners in the green composites field through the case study shown in the previous section using systematic CE approach during the conceptual design stage of the product.

ACKNOWLEDGMENT

The authors wish to express sincere gratitude to Universiti Teknikal Malaysia Melaka for the warm support provided throughout the preparation of this publication.

REFERENCES

Akadiri, P. O., Olomolaiye, P. O., & Chinyio, E. (2013). Multi-criteria evaluation model for the selection of sustainable materials for building projects. *Automation in Construction*, *30*(0), 113–125. doi:10.1016/j.autcon.2012.10.004

Alves, C., Ferrão, P. M. C., Silva, A. J., Reis, L. G., Freitas, M., Rodrigues, L. B., & Alves, D. E. (2010). Ecodesign of automotive components making use of natural jute fiber composites. *Journal of Cleaner Production*, *18*(4), 313–327. doi:10.1016/j.jclepro.2009.10.022

Anupam, K., Lal, P. S., Bist, V., Sharma, A. K., & Swaroop, V. (2014). Raw material selection for pulping and papermaking using TOPSIS multiple criteria decision making design. *Environmental Progress and Sustainable Energy*, *33*(3), 1034–1041. doi:10.1002/ep.11851

Ashori, A. (2008). Wood–plastic composites as promising green-composites for automotive industries. *Bioresource Technology*, *99*(11), 4661–4667. doi:10.1016/j.biortech.2007.09.043 PMID:18068352

Attri, R., & Grover, S. (2013). Application of preference selection index method for decision making over the design stage of production system life cycle. *Journal of King Saud University -. Engineering and Science*, *27*(0), 207–216.

Baxter, W. L., Aurisicchio, M., & Childs, P. R. N. (2015). (in press). Materials, use and contaminated interaction. *Materials & Design*, *0*(0). doi:10.1016/j.matdes.2015.04.019

Behzadian, M., Otaghsara, S. K., Yazdani, M., & Ignatius, J. (2012). A state-of the-art survey of TOPSIS applications. *Expert Systems with Applications*, *39*(17), 13051–13069. doi:10.1016/j.eswa.2012.05.056

Benyus, J. M. (1997). *Biomimicry: Innovation Inspired by Nature*. New York: HarperCollins Publisher Inc.

Bismarck, A., Baltazar-Y-Jimenez, A., & Sarlkakis, K. (2006). Green composites as Panacea? Socio-economic aspects of green materials. *Environment, Development and Sustainability*, *8*(3), 445–463. doi:10.1007/s10668-005-8506-5

Brezet, H., & Van Hemel, C., & the United Nations Environment Programme. Industry and Environment (Paris), Rathenau Institute (TheHague), Delft University of Technology (Delft). (1997). Ecodesign: A promising approach to sustainable production and consumption. UNEP.

Burns, L. A., Mouritz, A. P., Pook, D., & Feih, S. (2012a). Bio-inspired design of aerospace composite joints for improved damage tolerance. *Composite Structures*, *94*(3), 995–1004. doi:10.1016/j.compstruct.2011.11.005

Burns, L. A., Mouritz, A. P., Pook, D., & Feih, S. (2012b). Strength improvement to composite T-joints under bending through bio-inspired design. *Composites. Part A, Applied Science and Manufacturing*, *43*(0), 1971–1980. doi:10.1016/j.compositesa.2012.06.017

Çalişkan, H., Kurşuncu, B., Kurbanoĝlu, C., & Güven, Ş. Y. (2013). Material selection for the tool holder working under hard milling conditions using different multi criteria decision making methods. *Materials & Design*, *45*(0), 473–479. doi:10.1016/j.matdes.2012.09.042

Chan, J. W. K., & Tong, T. K. L. (2007). Multi-criteria material selections and end-of-life product strategy: Grey relational analysis approach. *Materials & Design*, *28*(5), 1539–1546. doi:10.1016/j.matdes.2006.02.016

Chandrasegaran, S. K., Ramani, K., Sriram, R. D., Horváth, I., Bernard, A., Harik, R. F., & Gao, W. (2013). The evolution, challenges, and future of knowledge representation in product design systems. *Computer Aided Design*, *45*(2), 204–228. doi:10.1016/j.cad.2012.08.006

Chen, H., Tu, J., & Guan, S. (2012). Applying the Theory of Problem-Solving and AHP to develop eco-innovative design. In M. Matsumoto et al. (Eds.), Design for Innovative Value Towards a Sustainable Society (pp. 489–494). Dordrecht: Springer Science+Business Media. doi:doi:10.1007/978-94-007-3010-6_93doi:10.1007/978-94-007-3010-6_93

Chiu, M. C., & Okudan, G. (2014). An investigation on the impact of product modularity level on supply chain performance metrics: An industrial case study. *Journal of Intelligent Manufacturing*, *25*(1), 129–145. doi:10.1007/s10845-012-0680-3

Cross, N. (1994). *Engineering Design Methods: Strategies for Product Design.* Chichester, UK: John Wiley & Sons.

Davoodi, M. M., Sapuan, S. M., Ahmad, D., Aidy, A., Khalina, A., & Jonoobi, M. (2011). Concept selection of car bumper beam with developed hybrid bio-composite material. *Materials & Design*, *32*(10), 4857–4865. doi:10.1016/j.matdes.2011.06.011

Davoodi, M. M., Sapuan, S. M., Ahmad, D., Ali, A., Khalina, A., & Jonoobi, M. (2010). Mechanical properties of hybrid kenaf/glass reinforced epoxy composite for passenger car bumper beam. *Materials & Design*, *31*(10), 4927–4932. doi:10.1016/j.matdes.2010.05.021

Dieter, G. E. (2000). Engineering Design: A Materials and Processing Approach (3rd ed.). Boston: McGraw-Hill.

Easterling, K. E. (1990). *Tomorrow's Materials.* London: The Institute of Materials.

El-Zeiny, R. M. A. (2012). Biomimicry as a problem solving methodology in interior architecture. *Procedia: Social and Behavioral Sciences*, *50*(0), 502–512. doi:10.1016/j.sbspro.2012.08.054

Findik, F., & Turan, K. (2012). Materials selection for lighter wagon design with a weighted property index method. *Materials & Design*, *37*, 470–477. doi:10.1016/j.matdes.2012.01.016

Finkbeiner, M., & Hoffmann, R. (2006). Application of life cycle assessment for the environmental certificate of the Mercedes-Benz S-Class (7 pp). *The International Journal of Life Cycle Assessment*, *11*(4), 240–246. doi:10.1065/lca2006.05.248

Fiore, V., Botta, L., Scaffaro, R., Valenza, A., & Pirrotta, A. (2014). PLA based biocomposites reinforced with Arundo donax fillers. *Composites Science and Technology*, *105*(0), 110–117. doi:10.1016/j.compscitech.2014.10.005

Ghazilla, R. A. R., Taha, Z., Yusoff, S., Rashid, S. H. A., & Sakundarini, N. (2013). Development of decision support system for fastener selection in product recovery oriented design. *International Journal of Advanced Manufacturing Technology, 70*(5-8), 1403–1413. doi:10.1007/s00170-013-5373-3

Girubha, R. J., & Vinodh, S. (2012). Application of fuzzy VIKOR and environmental impact analysis for material selection of an automotive component. *Materials & Design, 37*(0), 478–486. doi:10.1016/j.matdes.2012.01.022

Hambali, A., Sapuan, S. M., Ismail, N., & Nukman, Y. (2009a). Application of Analytical Hierarchy Process in the design concept selection of automotive composite bumper beam during the conceptual design stage. *Scientific Research and Essays, 4*(4), 198–211.

Hambali, A., Sapuan, S. M., Ismail, N., & Nukman, Y. (2009b). Composite manufacturing process selection using Analytical Hierarchy Process. *International Journal of Mechanical and Materials Engineering, 4*(1), 49–61.

Hamma, A., Kaci, M., Ishak, Z. A. M., & Pegoretti, A. (2014). Starch-grafted-polypropylene/kenaf fibres composites. Part 1: Mechanical performances and viscoelastic behaviour. *Composites. Part A, Applied Science and Manufacturing, 56*(0), 328–335. doi:10.1016/j.compositesa.2012.11.010

Imihezri, S. S. S., Sapuan, S. M., Ahmad, M. M. H. M., & Sulaiman, S. (2005). A study of the comparison of `V' and `X' ribbing in a composite pedal using mold flow analysis software. *Materials & Design, 26*(2), 157–166. doi:10.1016/j.matdes.2004.05.011

Imihezri, S. S. S., Sapuan, S. M., Sulaiman, S., Hamdan, M. M., Zainuddin, E. S., Osman, M. R., & Rahman, M. Z. A. (2006). Mould flow and component design analysis of polymeric based composite automotive clutch pedals. *Journal of Materials Processing Technology, 171*(3), 358–365. doi:10.1016/j.jmatprotec.2005.06.077

Jahan, A., Ismail, M. Y., Sapuan, S. M., & Mustapha, F. (2010). Material screening and choosing methods – A review. *Materials & Design, 31*(2), 696–705. doi:10.1016/j.matdes.2009.08.013

Jahazi, M., & Hossein-Nejad, S. (2004). The development of an optimum manufacturing and material selection process for the fabrication of labyrinth seal strips. *Journal of Materials Processing Technology, 152*(3), 272–275. doi:10.1016/j.jmatprotec.2003.11.035

Joshi, S., Drzal, L., Mohanty, A., & Arora, S. (2004). Are natural fiber composites environmentally superior to glass fiber reinforced composites? *Composites. Part A, Applied Science and Manufacturing*, *35*(3), 371–376. doi:10.1016/j.compositesa.2003.09.016

Karana, E., Hekkert, P., & Kandachar, P. (2008). Material considerations in product design: A survey on crucial material aspects used by product designers. *Materials & Design*, *29*(6), 1081–1089. doi:10.1016/j.matdes.2007.06.002

Karana, E., Hekkert, P., & Kandachar, P. (2010). A tool for meaning driven materials selection. *Materials & Design*, *31*(6), 2932–2941. doi:10.1016/j.matdes.2009.12.021

Koronis, G., Silva, A., & Fontul, M. (2013). Green composites: A review of adequate materials for automotive applications. *Composites. Part B, Engineering*, *44*(1), 120–127. doi:10.1016/j.compositesb.2012.07.004

Li, S. H., Zeng, Q. Y., Xiao, Y. L., Fu, S. Y., & Zhou, B. L. (1995). Biomimicry of bamboo bast fiber with engineering composite materials. *Materials Science and Engineering C*, *3*(0), 125–130. doi:10.1016/0928-4931(95)00115-8

Lin, C.-S., Chen, L.-S., & Hsu, C.-C. (2011). An innovative approach for RFID product functions development. *Expert Systems with Applications*, *38*(12), 15523–15533. doi:10.1016/j.eswa.2011.06.007

Luttropp, C., & Lagerstedt, J. (2006). EcoDesign and The Ten Golden Rules: Generic advice for merging environmental aspects into product development. *Journal of Cleaner Production*, *14*(15-16), 1396–1408. doi:10.1016/j.jclepro.2005.11.022

Luz, S. M., Caldeira-pires, A., & Ferrão, P. M. C. (2010). Environmental benefits of substituting talc by sugarcane bagasse fibers as reinforcement in polypropylene composites: Ecodesign and LCA as strategy for automotive components. *Resources, Conservation and Recycling*, *54*(12), 1135–1144. doi:10.1016/j.resconrec.2010.03.009

Maniya, K., & Bhatt, M. G. (2010). A selection of material using a novel type decision-making method: Preference selection index method. *Materials & Design*, *31*(4), 1785–1789. doi:10.1016/j.matdes.2009.11.020

Mansor, M. R., Sapuan, S. M., Hambali, A., Zainudin, E. S., & Nuraini, A. A. (2015). Conceptual design of kenaf polymer composites automotive spoiler using TRIZ and Morphology Chart methods. *Applied Mechanics and Materials*, *761*(0), 63–67. doi:10.4028/www.scientific.net/AMM.761.63

Mansor, M. R., Sapuan, S. M., Zainudin, E. S., Nuraini, A. A., & Hambali, A. (2013). Hybrid natural and glass fibers reinforced polymer composites material selection using Analytical Hierarchy Process for automotive brake lever design. *Materials & Design*, *51*(0), 484–492. doi:10.1016/j.matdes.2013.04.072

Mansor, M. R., Sapuan, S. M., Zainudin, E. S., Nuraini, A. A., & Hambali, A. (2014a). Application of integrated AHP-TOPSIS Method in hybrid natural fiber composites materials selection for automotive parking brake lever component. *Australian Journal of Basic and Applied Sciences*, *8*(5), 431–439.

Mansor, M. R., Sapuan, S. M., Zainudin, E. S., Nuraini, A. A., & Hambali, A. (2014b). Conceptual design of kenaf fiber polymer composite automotive parking brake lever using integrated TRIZ – Morphological Chart – Analytic Hierarchy Process method. *Materials & Design*, *54*(0), 473–482. doi:10.1016/j.matdes.2013.08.064

Min, S. H., Kim, H. Y., Kwon, Y. J., & Sohn, S. Y. (2011). Conjoint analysis for improving the e-book reader in the Korean market. *Expert Systems with Applications*, *38*(10), 12923–12929. doi:10.1016/j.eswa.2011.04.087

Pahl, G., Beitz, W., Feldhusen, J., & Grote, K. H. (2007). Engineering Design: A Systematic Approach (3rd ed.). London: Springer-Verlag London Limited.

Peng, A. H., & Xiao, X. M. (2013). Material selection using PROMETHEE combined with analytic network process under hybrid environment. *Materials & Design*, *47*(0), 643–652. doi:10.1016/j.matdes.2012.12.058

Pugh, S. (1991). *Total Design: Integrated Methods for Successful Product Engineering*. Wokingham, UK: Addison-Wesley Publishing.

San, Y. T., Jin, Y. T., & Li, S. C. (2009). *TRIZ: Systematic Innovation in Manufacturing*. Selangor: Firstfruits Sdn Bhd.

Sapuan, S. M. (2005). Concurrent design and manufacturing process of automotive composite components. *Assembly Automation*, *25*(2), 146–152. doi:10.1108/01445150510590514

Sapuan, S. M. (2006). Using the morphological chart technique for the design of polymeric-based composite automotive pedals. *Discovery and Innovation*, *18*, 311–317.

Sapuan, S. M. (2010). *Concurrent Engineering for Composites*. Serdang: UPM Press.

Sapuan, S. M., & Abdalla, H. S. (1998). A prototype knowledge-based system for the material selection of polymeric-based composites for automotive components. *Composites. Part A, Applied Science and Manufacturing*, *29*(7), 731–742. doi:10.1016/S1359-835X(98)00049-9

Sapuan, S. M., & Mansor, M. R. (2014). Concurrent engineering approach in the development of composite products: A review. *Materials & Design*, *58*, 161–167. doi:10.1016/j.matdes.2014.01.059

Schmehl, M., Mussig, J., Schonfeld, U., & Von Buttlar, H.-B. (2008). Life cycle assessment on a bus body component based on hemp and PTP. *Journal of Polymers and the Environment*, *16*(1), 51–60. doi:10.1007/s10924-008-0103-x

Shah, D. U. (2013). Developing plant fibre composites for structural applications by optimising composite parameters: A critical review. *Journal of Materials Science*, *48*(18), 6083–6107. doi:10.1007/s10853-013-7458-7

Shah, D. U. (2014). Natural fibre composites: Comprehensive Ashby-type materials selection charts. *Materials & Design*, *62*, 21–31. doi:10.1016/j.matdes.2014.05.002

Shanian, A., & Savadogo, O. (2006). A material selection model based on the concept of multiple attribute decision making. *Materials & Design*, *27*(4), 329–337. doi:10.1016/j.matdes.2004.10.027

Silvester, S., Beella, S. K., Van Timmeren, A., Bauer, P., Quist, J., & Van Dijk, S. (2013). Exploring design scenarios for large-scale implementation of electric vehicles: The Amsterdam Airport Schiphol case. *Journal of Cleaner Production*, *48*(0), 211–219. doi:10.1016/j.jclepro.2012.07.053

Singh, M., Khan, I. A., & Grover, S. (2011). Selection of manufacturing process using graph theoretic approach. *International Journal of System Assurance Engineering and Management*, *2*(4), 301–311. doi:10.1007/s13198-012-0083-z

Sohlenius, G. (1992). Concurrent engineering. *CIRP Annals-Manufacturing Technology*, *41*(2), 645–655. doi:10.1016/S0007-8506(07)63251-X

Ulrich, K. T., & Eppinger, S. D. (2004). Product Design and Development (3rd ed.). Boston: McGraw-Hill Irwin.

Versos, C. A. M., & Coelho, D. A. (2011). Biologically inspired design: Methods and validation. In D. A. Coelho (Ed.), *Industrial Design – New Frontiers* (pp. 101–120). Rijeka, Croatia: InTech. doi:10.5772/20326

Wang, F.-K., & Chen, K.-S. (2010). Applying Lean Six Sigma and TRIZ methodology in banking services. *Total Quality Management & Business Excellence*, *21*(3), 301–315. doi:10.1080/14783360903553248

Wötzel, K., Wirth, R., & Flake, M. (1999). Life cycle studies on hemp fibre reinforced components and ABS for automotive parts. *Die Angewandte Makromolekulare Chemie*, *272*(1), 121–127. doi:10.1002/(SICI)1522-9505(19991201)272:1<121::AID-APMC121>3.0.CO;2-T

Wu, W., Chiang, C., & Lin, C. (1998). Comparing the aggregation methods in the analytic hierarchy process when uniform distribution. *WSEAS Transactions on Business and Economics*, *5*(3), 82–87.

Yamashina, H., Ito, T., & Kawada, H. (2002). Innovative product development process by integrating QFD and TRIZ. *International Journal of Production Research*, *40*(5), 1031–1050. doi:10.1080/00207540110098490

Yeh, C. H., Huang, J. C. Y., & Yu, C. K. (2011). Integration of four-phase QFD and TRIZ in product R&D: A notebook case study. *Research in Engineering Design*, *22*(3), 125–141. doi:10.1007/s00163-010-0099-9

Yen, S. B., & Chen, J. L. (2005). An eco-innovative tool by integrating FMEA and TRIZ methods. In *Proceeding of the 4th International Symposium on Environmentally Conscious Design and Inverse Manufacturing* (pp. 678–683). Tokyo, Japan: IEEE.

Yurtkuran, S., Kırlı, G., & Taneli, Y. (2013). Learning from nature: Biomimetic design in architectural education. *Procedia: Social and Behavioral Sciences*, *89*(0), 633–639. doi:10.1016/j.sbspro.2013.08.907

KEY TERMS AND DEFINITIONS

Composites: Made of different parts or materials.

Conceptual Design: Outline of solutions to a design problem.

Concurrent Design: Design activities which exist, happen, or done at the same time.

Life Cycle Assessment: A tool for the systematic evaluation of the environmental aspects of a product or service system through all stages of its life cycle.

Morphological Chart: A graphical representation based on the function analysis for idea generation purpose.

Natural Fiber: Textile fibre made from natural resources as opposed to synthetic materials.

TRIZ: A structured problem solving method. It is a Russion acronym for "Teoriya Resheniya Izobretatelskikh Zadatch" which means 'Theory of Inventive Problem Solving'.

Chapter 4
Effect of Bamboo Hybridization and Staking Sequence on Mechanical Behavior of Bamboo–Glass Hybrid Composite

Piyush P. Gohil
The M. S. University of Baroda, India

Vijaykumar Chaudhary
CHARUSAT, India

Kundan Patel
CHARUSAT, India

Ronak Ramjiyani
CHARUSAT, India

ABSTRACT

The advancement of polymer composites containing natural fibers as a manageable option material for certain designing applications, especially aviation and car applications, is a well-known area of investigation. Nevertheless, the high mechanical properties connected with synthetic fibers they are awesome and lavish contrasted with natural fibers. The utilization of natural plant fibers and mixes of natural and synthetic fibers for making ease building materials has produced much interest recently. In the present work, bamboo–glass hybrid polyester composites were produced and their mechanical properties like elasticity and flexural quality were assessed for different weight fraction and distinctive stacking sequence. The outcomes observed that bamboo–glass mixture composites offered the benefits of both natural and synthetic fibers. It is also observed that hybridization started a material with general intermediate properties between pure glass and pure bamboo. However, the significance of controlling the stacking grouping to upgrade properties was evident.

DOI: 10.4018/978-1-5225-0424-5.ch004

INTRODUCTION

A composite material is made by joining two or more materials to give a special blend of properties, one of which is comprised of solid, long fibers and the other, a binder or "matrix" which holds the fibers set up. Composites help us to accomplish the coveted properties by consolidating distinctive materials in a reasonable manner. When all is said in done they have high specific strength and high specific modulus which makes them valuable in different modern applications including such attributes (Callister Jr, W.D. 2006).

Composites are extensively named Synthetic fiber reinforced composites and natural fiber reinforced composites. Synthetic fiber reinforced composites have applications in different zones because of their positive properties when contrasted and the ordinary materials; also they can be uniquely designed according to the necessities of the buyer (Nunna, S. et al. 2012). It was observed that materials made of renewable assets like natural fibers inserted in a polymer network called biopolymers give us great different options for synthetic fiber composites.

In late decades, the utilization of natural fibers as an option reinforcement material in polymer composites has pulled in the consideration of numerous analysts and researchers because of the unmistakable points of interest of these materials. Natural fibers which have favorable circumstances of being prudent to fabricate, eco-accommodating, innocuous to wellbeing, lightweight, high firmness and particular quality give a conceivable distinct option for the synthetic fibers (Athijayamani, Aet al. 2010; Ku, Het al. 2011; Malkapuram, Ret al. 2008). The sorts of natural fibers that have been concentrated on incorporate flax, hemp, jute, sisal, kenaf, coir, kapok and henequen. Natural fibers offer different favorable circumstances over man-made glass and carbon fibers, including ease and low thickness; moreover, natural fibers display particular ductile properties tantamount to those of glass fibers, are non-abrasive toward gear, are non-aggravating to the skin, show low vitality utilization, represent a low wellbeing hazard, and are renewable, recyclable and biodegradable. These composite materials are suitable for applications in the aviation, relaxation, development, games, bundling and car businesses, especially the last. The utilization of natural fiber composites has as of late experienced huge development in the car business because of the benefits of renewability, diminished outflow of toxins and enhanced fuel proficiency as a result of the light weight of the segments. In spite of their good properties, natural fibers have disservices, for example, an absence of warm solidness, quality corruption, water ingestion and poor effect properties. Attributable to the unfriendly impacts of composite materials on the earth, their high cost and other unfavorable properties, specialists have started to investigate natural fiber-based hybrid composites. Hybrid composites, which are gotten by joining synthetic and natural fibers, have been created to defeat the previously stated deficiencies. Natural fibers themselves can be dealt with as composites, which makes them harder than synthetic fibers; moreover, suitably designed superb natural fiber-reinforced hybrid composites have great quality and solidness values that inexact those of glass fiber-reinforced composites. It has been watched that the incomplete supplanting of manufactured fibers with natural fibers takes into account the development of artificial–natural mixture composites, which indicate attributes middle of the road between those of simply natural and absolutely synthetic fiber-based composites. Without a doubt, specialists have shown that enhanced properties can be accomplished by hybridizing natural fiber-based composites with glass fibers.

Scientists substantiated that improved properties can be accomplished by hybridizing natural fiber based composites with glass fibers (Venkateshwaran, N. et al. 2012; Sreekala, M. S. et al. 2002; Velmurugan, R., and Manikandan, V. 2007; Jarukumjorn K and Supakarn N. 2009). There has been an ecological mindfulness centered the regard for utilization of natural fibers as reinforcements for the polymeric network because of the natural points of interest. Biological concern and a dangerous atmospheric devation have started an impressive enthusiasm for utilizing natural materials to create green items and lessen anthropogenic carbon dioxide discharges by every conceivable mean and as written in "Kyoto Protocol", overall development is swung to CO_2 diminishment (Pervaiz, M., and Sain, M. M. 2003). In addition, the strict natural regulations have confronted the composite business to discover then again eco-accommodating reinforcements and gum frameworks to deliver ecologically amicable composite materials. The utilization of natural composites is being focused in different fields because of the ecological and temperate advantages which could be utilized as a part of car industry as inside parts and in developments area, for example, dividers and rooftops. Plant fibers are exceptionally appealing for composite materials because of the extensive qualities of these fibers, for example, biodegradability, wealth, renewability, ease, low specific gravity, high specific quality, and so forward. However certain disadvantages, for example, inconsistency with the hydrophobic polymer grids, the inclination to frame totals amid preparing, and poor imperviousness to dampness ingestion, diminish altogether the mechanical properties of the natural fibers reinforced composite materials (Mohanty, A. K. et al. 2005; Pickering, K. 2008; Saheb, D. N. and Jog, J. P. 1999).

It is clear that, in spite of the huge points of interest of natural fibers, as a result of their constrained mechanical properties, they are not supported for utilization alone in composite materials to get satisfactory reinforcement in specific applications obliging high mechanical execution. For this situation, hybridizing with synthetic fibers may prompt better results. Dissimilar to those of synthetic fibers, the mechanical properties of natural fibers demonstrate an extensive variety of variety; in this way, it is especially discriminating to focus the properties of these materials in fabric and composite structures in light of the fact that the properties of composite plates can likewise fluctuate with the properties of yarn and fabric. Therefore, the point of this study was to examine the mechanical properties of bamboo, glass and bamboo/glass composites with varied fiber weight fraction and layering arrangement.

LITERATURE SURVEY

Mechanical properties of natural fiber composites are much lower than those of synthetic fiber composites. Another burden of natural fiber composites which makes them less alluring is the poor resistance to moisture absorption. With an end goal to build up a predominant, yet prudent composite, a natural fiber can be combined with a synthetic fiber in the same matrix material in order to exploit the properties of both the fibers. This section reviews thorough literature review on effect of weight fraction / volume fraction and stacking sequence on mechanical properties of hybrid polymer matrix composite materials.

Effect of Weight Fraction/Volume Fraction

Numerous scientists have considered the behavior of the hybrid composites by differing the volume/weight fraction of the fibers and observed positive and negative impacts on mechanical properties because of that variation.

Mishra et al. (2003) prepared glass-pineapple leaf-sisal fiber hybrid composites with polyester resin. They studied mechanical properties of hybrid composites by changing treatment of fibers. They found that alkali treatment gives optimum impact and tensile strength while cyanoethylation gives maximum increment in flexural strength of hybrid composites.

John and Venkata Naidu (2004) prepared hybrid composites using sisal and glass fibers with various fiber weight fractions. They studied impact and compressive properties of hybrid composites and observed that there was insignificant change in impact strength while the compressive strength was increased slightly with increase in glass fiber weight fraction.

John and Venkata Naidu (2004) prepared hybrid composites using sisal and glass fibers with various fiber weight fractions. They studied flexural properties of hybrid composites with various treatments of fibers and observed that there was insignificant change in flexural property by saline treatment while alkali treatment slightly increased flexural property. They also observed that flexural property increases with increase in glass fiber weight fraction.

John and Venkata Naidu (2004) prepared hybrid composites using sisal and glass fibers with various fiber weight fractions. They studied tensile properties of hybrid composites and observed that there was significant change in tensile strength and modulus of elasticity.

Idicula et al. (2005) prepared hybrid composites using short banana and sisal fibers. They varied fiber length, layering arrangement and volume fraction to study flexural, impact and tensile strength. They found significant change in flexural, impact and tensile properties of hybrid composites with variation in fiber volume fraction. They found optimum tensile and flexural properties at 0.40 $V_{f.}$ They also found that composites with banana as skin material and sisal at center gives maximum elastic properties.

Dabade et al. (2006) prepared hybrid composites using palmyra and short sun hemp fibers. They studied elasticity of hybrid composites by varying fiber length and fiber weight fraction and observed the ideal fiber length and ideal fiber weight fraction.

Sabeel Ahmed et al. (2006) prepared hybrid composites using glass and woven jute fibers with isothalic polyester resin by varying fiber weight portion. They studied mechanical properties and observed that there was significant change in mechanical properties with increase in glass fiber weight fraction. The composite's behavior is investigated with the assistance of execution curves.

Abdul Khalil et al. (2007) prepared hybrid composites using oil palm and glass fibers with polyester resin by varying fiber weight fraction. They observed that the hybrid composites give good properties than the oil palm empty fruit bunch.

Laly et al. (2012) prepared hybrid composites using banana and glass fibers with varying fiber volume fraction. They observed that by adding volume fraction 0.11 of glass increases elasticity by 54.5% and impact strength by 196%. They also observed that volume fraction 0.17 of glass gives maximum tensile properties.

Girisha et al. (2012) prepared hybrid composites using areca nut husk and tamarind fibers with epoxy resin using hand lay-up method. They varied fiber weight fraction from 10% to 50%. They observed that the elasticity increases with increase in fiber weight fraction.

Madhukiran et al. (2013) manufactured and tested Banana/Pineapple hybrid composites. They presumed that the banana/pineapple hybrid composite with weight portion of 25/15 demonstrates greatest flexural strength, most extreme flexural modulus, greatest inter laminar shear quality and most extreme break load. The hybridization of these natural fibers has given significant change in flexural strength when contrasted with individual reinforcement. The work additionally exhibits the hybrid natural fiber composite materials for utilization in various consumable merchandises. Because of the low thickness of proposed natural fibers contrasted with the synthetic fibers (Glass fibers, carbon fibers, etc...), the composites can be viewed as valuable materials in light weight applications.

Bhargavi Devi et al. (2014) investigated the mechanical properties of bamboo/glass fiber reinforced polymer composites. They were supplanting routine metal compounds inferable from their eco-accommodating nature. They added bamboo fiber with glass fiber in extent of 1:1. They varied fiber length of 5 mm, 10 mm, 15 mm and weight portion of 20%, 30%, and 40%. The control parameters considered were fiber length (Fl) and fiber weight portion (W_f). In their work an endeavor has been made to show the mechanical properties through response surface methodology (RSM). Analysis of variance (ANOVA) is utilized to check the model's legitimacy.

Thomas et al. (2009) prepared hybrid composites using banana and sisal fibers by varying fiber volume fraction. The admissible fiber volume portion was specified as 40% past that the correct stress transfer from fiber to resin was not happened because of fiber agglomeration. As banana fibers have higher strength compared to sisal fibers, the flexural and tensile properties have been enhanced which demonstrated a positive hybrid impact. The impact quality of the hybrid composites was not exactly the sisal fiber polyester composites, because of permeable nature and the high winding edge of the sisal fibers.

Vara Prasad (2011) prepared hybrid composites using glass and jute fibers with polyester resin by varying fiber volume fraction. They observed that with increase in glass fiber volume fraction increases tensile properties of hybrid composites. They also observed that the Young's modulus was more for the hybrid composite with more prominent jute fiber volume portion than the glass fiber volume portion.

Yahaya et al. (2014) studied the effect of weight fraction on mechanical strength of woven kenaf kevlar hybrid composites. They prepared hybrid composites using hand layup technique with epoxy resin. They also prepared kenaf-epoxy and Kevlar-epoxy specimens to compare with hybrid composites. They observed that the mechanical strength of hybrid composite is function of fiber weight fraction.

Effect of Stacking Sequence

Stacking succession represents to the example of plan of the fiber layers in the hybrid composites. The impact of variety of the fiber layer's position on the mechanical behavior of the hybrid composites have been studied by the scientists.

Amico et al. (2010) prepared hybrid composites with sisal and glass fibers using compression molding method. They varied layering arrangement of fibers. They observed that hybrid composites give mechanical properties between pure sisal and pure glass fiber composites. They also observed that by controlling layering arrangement properties can be upgraded. They concluded that glass fibers at the top and base surface give higher flexural strength.

Gupta (2009) prepared hybrid composites using flax and glass fibers by varying layering arrangement. They studied mechanical properties of hybrid composites and observed that hybrid composites having flax fiber at center give better mechanical properties than glass fibers at center.

Jayabal et al. (2011) prepared hybrid composites using glass and woven coir fibers with polyester resin. They varied layering arrangement to study the mechanical properties. They observed that composites having glass fibers at center gives higher breaking resistance as coir fibers have less strength than glass fibers.

Jawaid et al. (2011) prepared hybrid composites using jute and oil palm fibers with epoxy resin by varying the layering arrangement of fibers. They observed that hybrid composites give mechanical properties between pure jute and pure oil palm fiber epoxy composites. They also observed that high strength fibers influence flexural and tensile properties in hybrid composites. They concluded that hybrid composites having jute fibers on the compelling parts have more noteworthy quality.

Khalil et al. (2009) prepared hybrid composites using glass and oil palm empty fruit bunch with vinylester resin. They varied layering arrangement and studied mechanical properties. They observed that glass at compelling parts created higher qualities than whatever other layered example. They also observed that glass fibers at center in hybrid composites give greater tensile properties. They concluded that natural fiber at furthest points in hybrid composites demonstrates greater impact quality.

Randjbaran et al (2014) contemplated the failure mechanism and compressive strength for hybrid composites. Static uniaxial compressive tests were performed on notched specimens prepared using two layers of carbon, glass, and Kevlar fibers and epoxy matrix combined to give six distinctive stacking hybrid composites. Stacking arrangement and orientation of them are as per the following; H1-[0K/0C/0G/0K/0G/0C]S, H2-[0G/0C/0K/0C/0K/0G]S, H3-[0K/0G/0C/0G/0C/0K]S, H4[0G/0K/0C/0C/0G/0K]S, and H5-[0K/0C/0G/0G/0C/0K]S. Strong zone model is connected to assess the compression quality. In addition stacking the first layer with glass fiber is superior to utilize the Kevlar fiber and utilizing the mix of carbon and glass is more productive than utilizing as a part of the middle layers, in addition, utilizing the carbon fiber is not prescribed at the last layer.

Mohanta and Acharya (2015) prepared hybrid composites using luffa cylindrical and glass fibers with epoxy resin. They used hand lay-up method for preparing composite specimens. They varied the layering arrangement and observed that the ideal flexural and impact properties accomplished by setting luffa cylindrical fibers at center upheld by glass fibers on each side.

Yahaya et al. (2015) studied the effect of fiber orientation on mechanical properties of aramid kenaf hybrid composites. They prepared hybrid composites using hand layup technique. They observed that woven kenaf hybrid composites have tensile strength of around 20.78% greater than unidirectional composites and 43.55% more than mat samples. They also observed that the impact strength of woven kenaf hybrid composites gives higher strength than unidirectional and mat samples.

MATERIALS AND METHODS

Specimen Preparation

To prepare the composite specimen for testing purpose, a mould of 13"x13" was prepared. Composites are made from glass fiber and bamboo fiber. Figure 1 shows bamboo yarn. Figure 2 shows glass fiber used for the reinforcement and the matrix was polyester resin. The specimen plates are cut into same pieces of dimension 12"x12".

Composite specimens were prepared by varying weight fraction which is shown in Table 1. Figure 3 shows specimens prepared with different weight fraction in unidirection. Figure 4 shows specimens prepared with different weight fraction in bidirection.

Figure 5 shows the prepared composite specimens for checking the effect of stacking sequence. The fiber weight fraction was kept same for all specimen and the stacking sequence as well as the fiber weight fraction detail is shown in Table 2.

MECHANICAL TESTING

Mechanical testing was carried out using Universal Testing machine (Tinius Olsen/L-Series H50KL).

Figure 1. Bamboo yarn

Figure 2. Glass fiber

Figure 3. Specimens prepared with different weight fraction in unidirection

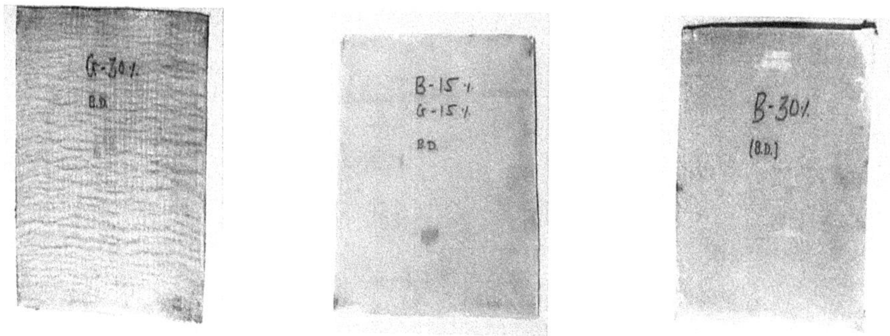

(a) (b) (c)

Table 1. Details of fiber weight fraction (Unidirectional and bidirectional)

Sr. No.	Weight Fraction	Direction of Fiber	Weight of Bamboo (gm and Layer)	Weight of Glass (gm and Layer)	Weight of Polyester (gm)	Total Weight (gm)
1	B 30%	UD	120gm, 4 L	-	280	400
2	B 20%, G 10%	UD	80gm, 5 L	40gm, 1 L	280	400
3	B 15%, G 15%	UD	60gm, 4 L	60gm, 2 L	280	400
4	B 10%, G 20%	UD	40gm, 3 L	80gm, 2 L	280	400
5	G 30%	UD	-	120gm, 3 L	280	400
6	B 15%, G 15%	BD	60gm, 4 L	60gm, 2 L	280	400
7	B 30%	BD	120gm, 4 L	-	280	400
8	G 30%	BD	-	120gm, 3 L	280	400

Table 2. Details of fiber weight (gm) and stacking sequence

Sr. No	Composite Plate No.	Stacking Sequence	Weight of Bamboo (gm, Layer)	Weight of Glass (gm, Layer)	Weight of Polyester (gm)	Total Weight (gm)
1	H 1	GGGGBBBB	60gm & 4 L	220gm, 4 L	280	400
2	H 2	GBGBGBGB	60gm & 4 L	220gm, 4 L	280	400
3	H 3	GGBBBBGG	60gm & 4 L	220gm, 4 L	280	400
4	H 4	BBGGGGBB	60gm & 4 L	220gm, 4 L	280	400
5	H 5	BGBGGBGB	60gm & 4 L	220gm, 4L	280	400
6	H 6	GBGBBGBG	60gm & 4 L	220gm, 4 L	280	400

Figure 4. Specimens prepared with different weight fraction in bidirection

(a) (b) (c)

Figure 5. The prepared composite specimens for checking the effect of stacking sequence

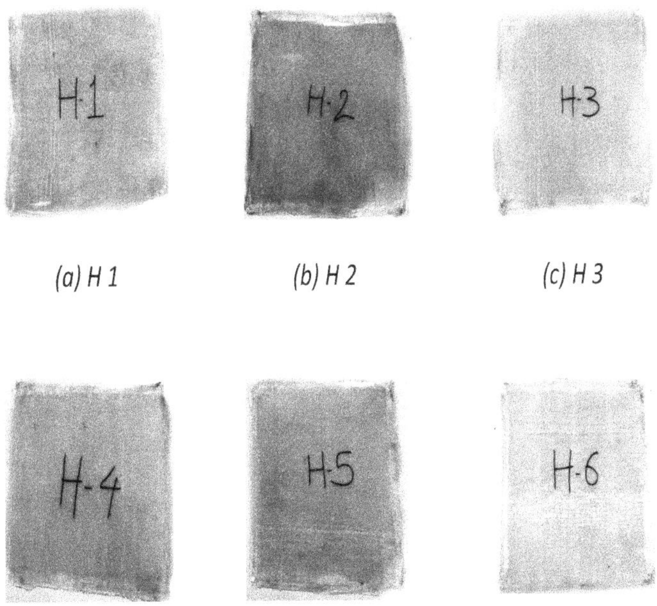

(a) H 1 (b) H 2 (c) H 3

(d) H 4 (e) H 5 (f) H 6

Table 3. Experimental results of tensile testing for unidirectional hybrid composites

Weight Fraction	Width (mm) (± 0.5)	Thickness (mm) (± 0.3)	Max. Breaking Load (N)	Max. Tensile Strength (MPa)	Tensile Modulus (GPa)
B 30%	15	4	2700	44.9	2.000
B10%, G20%	15	4	6210	104	2.810
B15%, G15%	15	4	5450	90.9	2.560
B20%, G10%	15	4	5930	98.8	2.820
G 30%	15	4	11100	184	3.720

Tensile Testing

The specimens for tensile testing were prepared from composite plate by cutting as per ASTM D 3039 for unidirectional composites while for bidirectional composites ASTM D 638 was used. The specimens were cut using CNC vertical milling machine and shaped to the accurate size using emery paper.

Flexural Testing

The specimens for tensile testing were prepared from composite plate by cutting as per ASTM D 790. The flexural test specimens dimension is 127 x 12.7 x 3 mm. The rate of loading taken was 5 mm/min.

Figure 6. Effect of weight fraction on tensile strength for unidirectional hybrid composites

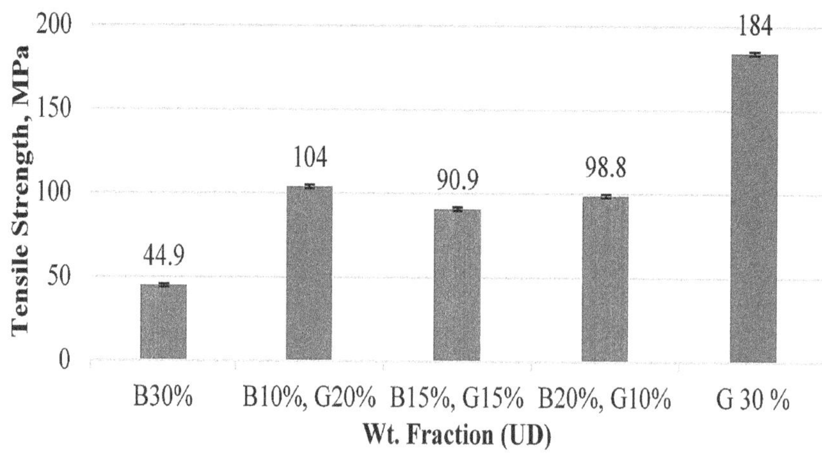

RESULTS AND DISCUSSION

Effect of Weight Fraction on Tensile strength

Table 3 shows the results of tensile testing for unidirectional hybrid composite with different weight fraction.

Figure 6 shows effect of weight fraction on tensile strength for unidirectional hybrid composites. Maximum tensile strength obtained is 184 MPa for G30% and the minimum tensile strength is 44.9 MPa for B30%. It is also seen that with decrease in the weight fraction of glass the tensile strength is decreasing.

Table 4 shows the results of tensile testing for bidirectional hybrid composite with different weight fraction.

Figure 7 shows the comparison of tensile strength for unidirectional and bidirectional hybrid composite. It is observed that unidirectional hybrid composites give better tensile strength than bidirectional composites.

Table 4. Experimental results of tensile testing for bidirectional hybrid composites

Weight Fraction	Width (mm) (± 0.5)	Thickness (mm) (± 0.3)	Max. Breaking Load (N)	Max. Tensile Strength (MPa)	Tensile Modulus (GPa)
B 30%	15	4	1720	18.1	0.700
B15%, G15%	15	4	5150	54.3	0.846
G 30%	15	4	6390	67.3	0.882

Figure 7. Comparison of tensile strength for unidirectional and bidirectional hybrid composite

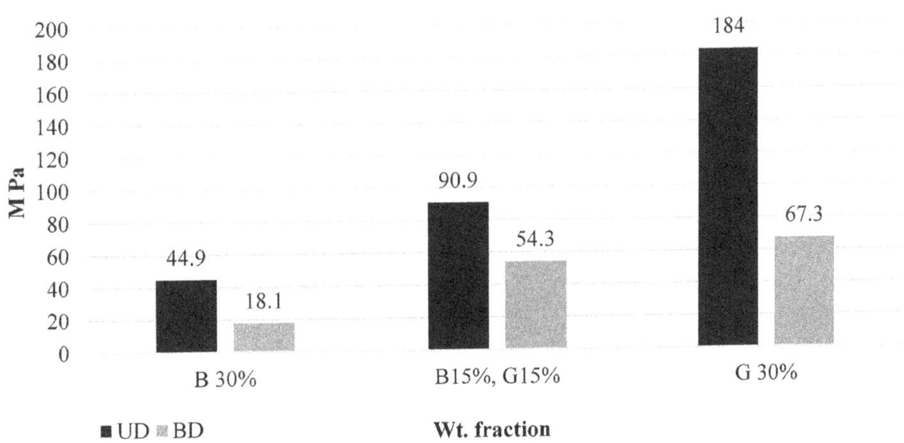

Figure 8. Effect of weight fraction on flexural strength for unidirectional hybrid composites

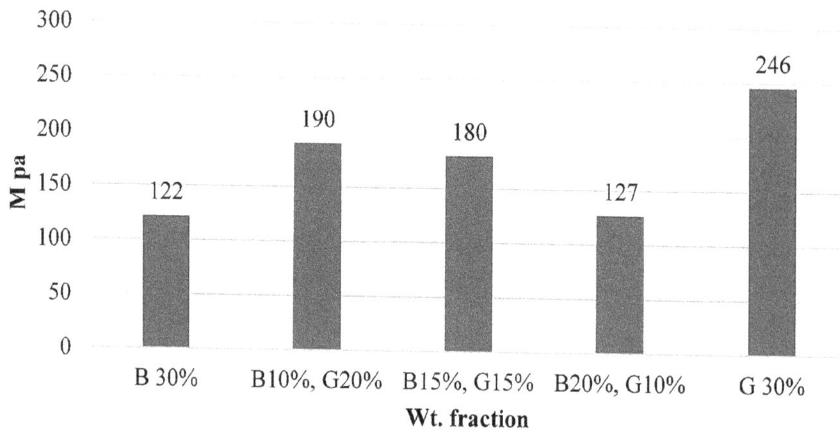

Table 5. Experimental results of flexural testing for unidirectional hybrid composite

Weight Fraction	Width (mm) (± 0.5)	Thickness (mm) (± 0.3)	Span Length (mm) (± 0.5)	Max. Fracture Load (N)	Max. Flexural Strength (MPa)	Flexural Modulus (GPa)
B 30%	12.70	3.2	50	193	122	4.540
B10%, G20%	12.70	3.2	50	328	190	4.780
B15%, G15%	12.70	3.2	50	294	180	4.380
B20%, G10%	12.70	3.2	50	212	127	4.830
G 30%	12.70	3.2	50	369	246	6.590

Effect of Weight Fraction on Flexural Strength

Table 5 shows the results of flexural testing for unidirectional hybrid composite with different weight fraction.

Figure 8 shows effect of weight fraction on flexural strength for unidirectional hybrid composites. Maximum flexural strength obtained is 246 MPa for G30% and the minimum flexural strength is 122 MPa for B30%. Figure 8 also shows that with decrease in the weight fraction of glass the flexural strength is increasing for hybrid composite.

Table 6 shows the results of flexural testing for BD hybrid composite with different weight fraction.

Table 6. Experimental results of flexural testing for bidirectional hybrid composite

Weight Fraction	Width (mm) (± 0.5)	Thickness (mm) (± 0.3)	Span Length (mm) (± 0.5)	Max. Fracture Load (N)	Max. Flexural Strength (MPa)	Flexural Modulus (GPa)
B 30%	12.70	3.2	50	219	111	5960
B15%, G15%	12.70	3.2	50	324	170	4870
G30%	12.70	3.2	50	427	213	6360

Figure 9 shows the comparison of flexural strength for unidirectional and bidirectional hybrid composite. It is observed that the difference in flexural strength is negligible for unidirectional hybrid composites and bidirectional composites.

Effect of Stacking Sequence on Tensile Strength

Table 7 shows the results of tensile testing of hybrid composites with different stacking sequence. Figure 10 shows effect of stacking sequence on tensile strength. It is observed that the H2 (GBGBGBGB) having maximum tensile strength and H1 (GGGGBBBB) having minimum tensile strength compared to other stacking sequence.

Figure 9. Comparison of flexural strength for unidirectional and bidirectional hybrid composite

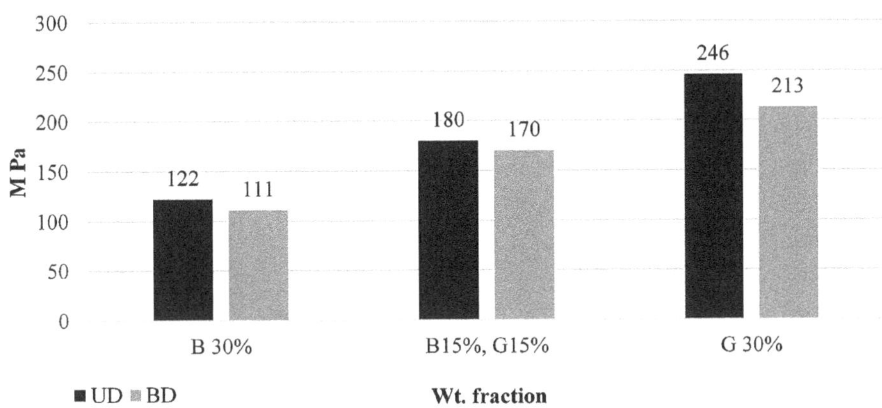

Table 7. Results of tensile testing of hybrid composite with various stacking sequence

Specimen No.	Width (mm) (± 0.5)	Thickness (mm) (± 0.3)	Max. Breaking Load (N)	Max. Tensile Strength (MPa)	Tensile Modulus (GPa)
H 1	15	4	16300	173	1.460
H 2	15	4	19200	202	1.450
H 3	15	4	16500	174	1.430
H 4	15	4	18800	198	1.550
H 5	15	4	18900	199	1.610
H 6	15	4	16800	177	1.490

Table 8. Results of flexural testing of hybrid composite with various stacking sequence

Specimen No.	Width (mm) (± 0.5)	Thickness (mm) (± 0.3)	Span Length (mm) (± 0.5)	Max. Fracture Load (N)	Max. Flexural Strength (MPa)	Flexural Modulus (GPa)
H 1	12.70	3.2	50	1150	661	20100
H 2	12.70	3.2	50	1290	742	33300
H 3	12.70	3.2	50	1380	796	43500
H 4	12.70	3.2	50	671	387	10000
H 5	12.70	3.2	50	940	542	18200
H 6	12.70	3.2	50	1060	611	24400

Figure 10. Effect of stacking sequence on tensile strength

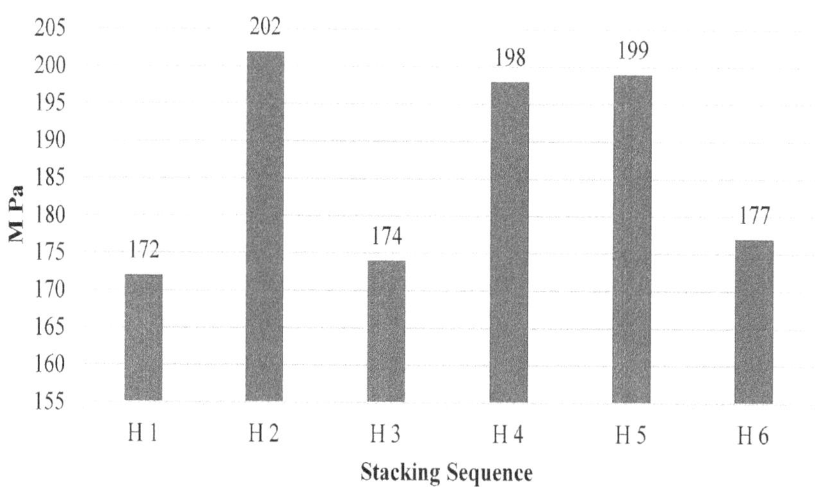

Figure 11. Effect of stacking sequence on flexural strength

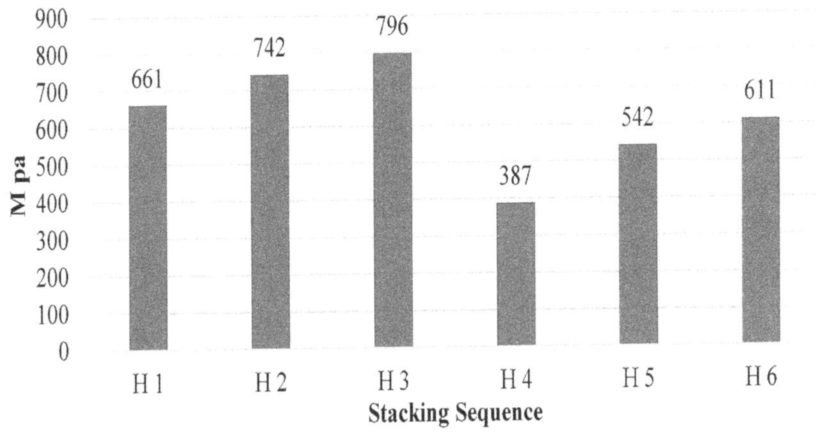

Effect of Stacking Sequence on Flexural Strength

Table 8 shows the results of flexural testing of hybrid composites with different stacking sequence. Figure 11 shows effect of stacking sequence on flexural strength. It is observed that the H3 (GGBBBBGG) having maximum flexural strength and H4 (BBGGGGBB) having minimum flexural strength compared to other stacking sequence.

CONCLUSION

In this study mechanical properties (tensile and flexural) of bamboo-glass fiber hybrid composites have been described. Successful fabrication of bamboo–glass fibre polyester hybrid composites has been done by the hand lay-up technique. It is seen that unidirectional hybrid composites gives better mechanical properties than bidirectional composites. It is also seen that with decrease in the weight fraction of glass the tensile strength is decreasing while the flexural strength increases. The stacking sequence was affected differently response to the various types of loadings (tensile and flexural) of the hybrid composite. It is observed that to maximize flexural behaviour, there must be the glass fibres on the top and bottom surfaces, but keeping some glass fibres at the mid-plane was also shown to be important. To optimize the desired properties, controlling stacking sequence is more important than hybridization. These hybrid composite materials may serve variety of applications. Few examples are as follow: Building and construction industry, Storage Devices, Furniture, Electric Devices and Transportations.

REFERENCES

Ahmed, K. S., Vijayarangan, S., & Rajput, C. (2006). Mechanical behavior of isothalic polyester-based untreated woven jute and glass fabric hybrid composites. *Journal of Reinforced Plastics and Composites*, *25*(15), 1549–1569. doi:10.1177/0731684406066747

Amico, S. C., Angrizani, C. C., & Drummond, M. L. (2008). Influence of the stacking sequence on the mechanical properties of glass/sisal hybrid composites. *Journal of Reinforced Plastics and Composites*.

Athijayamani, A., Thiruchitrambalam, M., Manikandan, V., & Pazhanivel, B. (2010). Mechanical properties of natural fibers reinforced polyester hybrid composite. *International Journal of Plastics Technology*, *14*(1), 104–116. doi:10.1007/s12588-009-0016-0

Bhargavi, D. S., Suresh, Y., Naresh, C., & Upendra, N. (2014). An Investigation of Mechanical Properties of Hybrid Polymer Composites. *International Journal of Engine Research*, *3*, 116–123.

Callister, W. D. Jr. (2006). *Materials science and engineering: an introduction* (7th ed.). New York: John Wiley & Sons.

Dabade, B. M., Reddy, G. R., Rajesham, S., & Kiran, C. U. (2006). Effect of fiber length and fiber weight ratio on tensile properties of sun hemp and palmyra fiber reinforced polyester composites. *Journal of Reinforced Plastics and Composites*, *25*(16), 1733–1738. doi:10.1177/0731684406068418

Girisha, C., Sanjeevamurthy, S., Gunti, R., & Manu, S. (2012). Mechanical performance of natural fiber reinforced epoxy hybrid composites. *International Journal of Engineering Research and Applications*, *2*(5), 615–619.

Gupta, K. M. (2009). Experimental investigation of the Behavior of dual fiber hybrid composite under different stacking sequence. In *17th International Conference on Composite Materials (ICCM-17)*.

Idicula, M., Neelakantan, N. R., Oommen, Z., Joseph, K., & Thomas, S. (2005). A study of the mechanical properties of randomly oriented short banana and sisal hybrid fiber reinforced polyester composites. *Journal of Applied Polymer Science*, *96*(5), 1699–1709. doi:10.1002/app.21636

Jarukumjorn, K., & Suppakarn, N. (2009). Effect of glass fiber hybridization on properties of sisal fiber–polypropylene composites. *Composites. Part B, Engineering*, *40*(7), 623–627. doi:10.1016/j.compositesb.2009.04.007

Jawaid, M., Khalil, H. A., & Bakar, A. A. (2011). Woven hybrid composites: Tensile and flexural properties of oil palm-woven jute fibres based epoxy composites. *Materials Science and Engineering A*, *528*(15), 5190–5195. doi:10.1016/j.msea.2011.03.047

Jayabal, S., Natarajan, U., & Sathiyamurthy, S. (2011). Effect of glass hybridization and staking sequence on mechanical behaviour of interply coir–glass hybrid laminate. *Bulletin of Materials Science*, *34*(2), 293–298. doi:10.1007/s12034-011-0081-9

John, K., & Naidu, S. V. (2004). Effect of fiber content and fiber treatment on flexural properties of sisal fiber/glass fiber hybrid composites. *Journal of Reinforced Plastics and Composites*, *23*(15), 1601–1605. doi:10.1177/0731684404039799

John, K., & Naidu, S. V. (2004). Sisal fiber/glass fiber hybrid composites: The impact and compressive properties. *Journal of Reinforced Plastics and Composites*, *23*(12), 1253–1258.

John, K., & Naidu, S. V. (2004). Tensile properties of unsaturated polyester-based sisal fiber–glass fiber hybrid composites. *Journal of Reinforced Plastics and Composites*, *23*(17), 1815–1819. doi:10.1177/0731684404041147

Khalil, H. A., Hanida, S., Kang, C. W., & Fuaad, N. N. (2007). Agro-hybrid composite: The effects on mechanical and physical properties of oil palm fiber (EFB)/ glass hybrid reinforced polyester composites. *Journal of Reinforced Plastics and Composites*, *26*(2), 203–218. doi:10.1177/0731684407070027

Khalil, H. A., Kang, C. W., & Adawi, T. O. (2008). The effect of different laminations on mechanical and physical properties of hybrid composites. *Journal of Reinforced Plastics and Composites*.

Ku, H., Wang, H., Pattarachaiyakoop, N., & Trada, M. (2011). A review on the tensile properties of natural fiber reinforced polymer composites. *Composites. Part B, Engineering*, *42*(4), 856–873. doi:10.1016/j.compositesb.2011.01.010

Laly, P. A., Thomas, S., & George, J. (2012).Tensile and impact properties of bananafibre/glass fibre hybrid polyester composites. ICCM 2012-Eupore.

Madhusudan, S., Rao, D. S. S., & Madhukiran, J. (2013). Fabrication and testing of natural fiber reinforced hybrid composites banana/pineapple. *International Journal of Modern Engineering and Research*, *3*, 2239–2243.

Malkapuram, R., Kumar, V., & Negi, Y. S. (2008). Recent development in natural fiber reinforced polypropylene composites. *Journal of Reinforced Plastics and Composites*.

Mishra, S., Mohanty, A. K., Drzal, L. T., Misra, M., Parija, S., Nayak, S. K., & Tripathy, S. S. (2003). Studies on mechanical performance of biofibre/glass reinforced polyester hybrid composites. *Composites Science and Technology*, *63*(10), 1377–1385. doi:10.1016/S0266-3538(03)00084-8

Mohanta, N., & Acharya, S. K. (2014). Investigation of mechanical properties of luffa cylindrica fibre reinforced epoxy hybrid composite. *International Journal of Engineering Science and Technology*, *7*(1), 1–10. doi:10.4314/ijest.v7i1.1

Mohanty, A. K., Misra, M., & Drzal, L. T. (Eds.). (2005). *Natural fibers, biopolymers, and biocomposites*. CRC Press. doi:10.1201/9780203508206

Nunna, S., Chandra, P. R., Shrivastava, S., & Jalan, A. K. (2012). A review on mechanical behavior of natural fiber based hybrid composites. *Journal of Reinforced Plastics and Composites*, *31*(11), 759–769. doi:10.1177/0731684412444325

Pervaiz, M., & Sain, M. M. (2003). Carbon storage potential in natural fiber composites. *Resources, Conservation and Recycling*, *39*(4), 325–340. doi:10.1016/S0921-3449(02)00173-8

Pickering, K. (Ed.). (2008). *Properties and performance of natural-fibre composites*. Elsevier.

Randjbaran, E., Zahari, R., Majid, D. L., Jalil, N. A. A., Vaghei, R., & Ahmadi, R. (2014). Experimental Study of the Influence of Stacking Order of the Fibrous Layers on Laminated Hybrid Composite Plates Subjected to Compression Loading. *Journal of Science and Engineering*, *4*(1), 1-8.

Saheb, D. N., & Jog, J. P. (1999). Natural fiber polymer composites: A review. *Advances in Polymer Technology*, *18*(4), 351–363. doi:10.1002/(SICI)1098-2329(199924)18:4<351::AID-ADV6>3.0.CO;2-X

Sreekala, M. S., George, J., Kumaran, M. G., & Thomas, S. (2002). The mechanical performance of hybrid phenol-formaldehyde-based composites reinforced with glass and oil palm fibres. *Composites Science and Technology*, *62*(3), 339–353. doi:10.1016/S0266-3538(01)00219-6

Thomas, S., Idicula, M., & Joseph, K. (2009). Mechanical performance of short banana/sisal hybrid fibre reinforced polyester composites. *Journal of Reinforced Plastics and Composites*.

Vara Prasad, V. (2011). Tensile and flexural properties of Glass fiber-reinforced polyester hybrid composites. *International Journal of Materials and Biomaterials Applications*, *1*(1), 14–16.

Velmurugan, R., & Manikandan, V. (2007). Mechanical properties of palmyra/glass fiber hybrid composites. *Composites. Part A, Applied Science and Manufacturing*, *38*(10), 2216–2226. doi:10.1016/j.compositesa.2007.06.006

Venkateshwaran, N., Elayaperumal, A., & Sathiya, G. K. (2012). Prediction of tensile properties of hybrid-natural fiber composites. *Composites. Part B, Engineering*, *43*(2), 793–796. doi:10.1016/j.compositesb.2011.08.023

Yahaya, R., Sapuan, S. M., Jawaid, M., Leman, Z., & Zainudin, E. S. (2014). Mechanical performance of woven kenaf-Kevlar hybrid composites. *Journal of Reinforced Plastics and Composites*, *33*(24), 2242–2254. doi:10.1177/0731684414559864

Yahaya, R., Sapuan, S. M., Jawaid, M., Leman, Z., & Zainudin, E. S. (2015). *Effect of fibre orientations on the mechanical properties of kenaf–aramid hybrid composites for spall-liner application*. Defence Technology.

Chapter 5
Estimation of Mechanical and Tribological Properties of Epoxy-Based Green Composites

Supriyo Roy
Birla Institute of Technology, India

J. Paulo Davim
University of Aveiro, Portugal

Sumit Bhowmik
National Institute of Technology, India

Kaushik Kumar
Birla Institute of Technology, India

ABSTRACT

Composites based on natural fibre reinforcement have generated wide research and engineering interest in the last few decades due to their small density, high specific strength, low cost, light weight, recyclability and biodegradability and has earned a special category of 'green composite'. Here, in our proposed research, wood dust reinforced epoxy composite was processed with different % filler weight primarily. For this, natural filler based epoxy composite from wood dust is developed and its mechanical behaviour, including Tensile, Flexural, Density etc., under various testing conditions and % of filler weight were studied. These samples were simultaneously tested for abrasive wear and friction coefficient measurement. Microstructure of the composites was studied to analyze the distribution of the filler in the epoxy matrix change using scanning electron microscopy.

DOI: 10.4018/978-1-5225-0424-5.ch005

INTRODUCTION

Composites constitutes of two main constituent materials: matrix and reinforcement. The matrix material surrounds and supports the reinforcement materials by maintaining their relative positions. The reinforcements impart their special mechanical and physical properties to enhance the matrix properties. A synergism produces material properties unavailable from the individual constituent materials, while the wide variety of matrix and strengthening materials allows the designer of the product or structure to choose an optimum combination. The reinforcing material can be of the form of particle (usually called filler) or fibre. In a particle reinforced composites, the reinforcement can be with large particle or small particle. It is not strictly the physical dimensions of the particles by which the materials are classified; rather it is the mechanism of reinforcement. In a small particle reinforced material the mechanism is on a molecular level whereas in case of large particle it is at macro level. The particles may be dispersed into or precipitated from the matrix; hence the properties can be varied with the amount and type of dispersion.

In the recent years, rising concern towards environmental problem and the need for more multifarious polymer-based materials has led to increasing interest about polymer composites reinforced with Green reinforcement, i.e., materials derived from natural resources like trees etc. The natural composites, or green composites, have shown a growth of interest because of their recyclability, biodegradability and abundant availability (Thakur et al., 2012).

Specific properties of natural composite such as light weight, low cost, renewable in nature, high specific strength and modulus have extended their usage and allow a considerable reduction in the use of non-biodegradable polymers and non-renewable resources (Bhowmick et al., 2012). Also in automobile industry the use of wood based natural composites enhance the mechanical strength and acoustic performance, reduce material weight and fuel consumption, improve biodegradability and production cost for the auto interior parts (Herrera-Franco and Valadez-González, 2004). They are also much less abrasive than inorganic-mineral counterparts to process machinery, less dangerous for the production employees in case of inhalation, easy to be scorched, and leads to final composites with lower specific weight (in comparison to mineral-filled counterparts) and allows obtaining interesting properties in terms of thermal and acoustic insulation. Also natural composite can be easily disposed at the end of their life cycle by compositing or by recovery of their calorific value in a furnace which is not possible in synthetic composite with reinforcement such as glass.

The woven banana fibre reinforced epoxy composites showed a very stable mechanical behaviour under different loading and speed condition (Sapuan et al., 2006). The epoxy based pissava fibres composite has rich silicon content on the surface and large dispersion in their mechanical properties (Nascimento et al., 2012). The flexural modulus of coconut fiber polypropylene matrix composite can be improved by using adequate fiber granulometry and extruder screw speed and that of agave fibre reinforced epoxy composite were significantly high due to alkali treatment of the fibre (Ishizaki et al., 2008; Mylsamy and Rajendran, 2011).

Almost all of the commonly available natural plant fibres are being used for reinforcement in combination with non-biodegradable matrix materials such as unsaturated polyester, epoxy resin, polyethylene and polypropylene (Ashori, 2008). Among these, epoxy resins are very versatile in nature. They are one of the most important classes of thermosetting polymers which are widely used as matrices for composite materials and as structural adhesives. They are amorphous, highly cross-linked polymers and this structure results in these materials possessing various desirable properties such as greater tensile strength and modulus, uncomplicated processing, fine thermal and chemical resistance, and dimensional stability (Song et al., 2000).

For the natural fibre composite, most of the researches were focusing on experimental test of mechanical properties of natural composites. The correlation between mechanical properties and characteristic parameters, e.g., the composition of composite and operating conditions is of prime importance for designing proper composites in order to satisfy various functional requirements.

Most widely known natural-organic fillers are wood flour. Wood flour can be easily and cheaply obtained from saw mill wastes and can be used after proper sieving (Sandberg et al., 2013). Besides wood derivatives, other natural-organic fillers have begun to find application as well. Examples among these are cellulose, sisal, kenaf, starch, banana, pissava, coconut coir, and agave. Sisal, and kenaf are relatively similar and are basically long fibers extracted from the bust of the plants, they can be used as fillers by proper cutting into long or short fibers.

Many researchers have studied the effect of various fillers on mechanical properties of polymer matrix (Ali and Elleithy, 2011, Chang et al., 2011, Chen et al., 2012, Hassan et al., 2012, Liu et al., 2012, Ma et al., 2012, Wang et al., 2012, Zeng et al., 2012, Asyadi et al., 2013, Kaseem et al., 2013, Krishnan et al., 2013, Ahmed and El-Sabbagh, 2014, Balakrishna et al., 2014). Most of the researchers have varied filler content from 0-30 wt% to enhance mechanical properties of polymer composites. The flexural strength and tensile strength of date wood palm flour based polyethylene composite was decreased by increasing the filler content while the flexural modulus was increased (Mirmehdi et al., 2014). The internal recycling of poly (vinyl chloride)-based composites was influenced by wood fibre filler, resulting with increase in number of cycle, the flexural strength is also increased (Augier et al., 2007). General trend noticed from the survey is that the tensile modulus and flexural modulus increase with increase in filler content but tensile strength and flexural strength increase up to a certain filler content but beyond that strength start to decrease. High volume of filler content also leads to agglomeration thus reduction in tensile properties.

Polymeric composite have emerged as one of the most significant materials in weight sensitive applications like automotive parts and aerospace industries. With this, Tribological behaviours like wear resistance and friction coefficients are come into serious attention. Wear resistance is one of the important properties as the parts made up with composite material may experience sliding motion. Wear resistant affects the coefficient of friction (Axen et al., 1996). Schön (1996) has also developed a model for predicting the coefficient of friction by applied load at different types of contact. Gamelas, et al. (2014) carried Inverse gas chromatography and found that, Lewis basic properties of the material surface increases with reduction of the dispersive component of surface energy. Addition of Ultra-high-molecular-weight polyethylene (UHMWPE) particles reduced wear rate of composites as well as coefficient of friction. Addition of UHMWPE particles improves abrasive behaviour as well as sliding wear resistance of epoxy resin (Chand et al., 2007). Roselman and Tavor (1976) showed variation of coefficient of friction over normal loads between crossed carbon fibers. Eiss and Czichos (1986) carried out experiments and measured the coefficient of friction for rubber reinforced epoxy composite sliding against glass and steel surfaces and obtained a value between 0.9 and 1.15 where coefficient of friction obtained consistently lower on glass surface than on steel surface. Wan et al. (2006) tested friction and wear behaviour of three-dimensional braided carbon fiber / epoxy composites and found that, higher volume fraction of composite presented lower specific wear rate at steady wear state as well as in running period and the effect of volume fraction on the coefficient of friction was less significant and more complicated throughout wear process. Modifications of tribological behavior of polymer by addition of filler have been reported to be quite encouraging (Ariffin et al., 2008, Aurrekoetxea et al., 2008, Kocserha and Gömze, 2010, Rasheva et al., 2010, Song et al., 2010, Xie et al., 2010, Ben Difallah et al., 2012, Esthappan et al., 2012, Li and Chen, 2012, Chang et al., 2013, Praveen et al., 2013, Akinci et al., 2014, Chang et al., 2014, Selvam et al., 2014, You et al., 2014). Researchers have used filler content range from 0-30 wt% on polymer composites with different normal load and sliding velocity which slides against metallic counter faces. In addition to tribo-performance obtained due to the addition of fillers in polymer composites, there is also cost reduction of components due to less consumption of matrix material.

In the present work, natural filler (wood dust) based epoxy composite is developed and its mechanical behaviour, including Tensile, Flexural, Density etc., under various testing conditions and % of filler wt. were studied. These samples were also tested for abrasive wear and friction coefficient measurement by using a pin block-on-roller machine. Tests were undertaken considering filler content (with two sets of values for normal load and speed) as the design parameters. The microstructures of the composites were studied to analyze the distribution of the filler in epoxy matrix using scanning electron microscope.

MATERIALS AND COMPOSITE PREPARATION

Materials and Composite Fabrication

Matrix Material

Here, low temperature curing epoxy resin (Standard Epoxy AW 106) chemically belonging to the 'epoxide' family and corresponding hardener (HV 953U) were mixed in a specified ratio (as declared by the manufacturer) for preparation of the matrix.

Filler Material

Sundi Wood dust was taken as the filler material. The principal organic constituents of wood are cellulose, glucomannan, xylan and lignin.

Conditioning of the Wood Fiber

The wood filler is a hygroscopic material and hence cannot be dried effectively by conventional hot air dryers, because hot air dryers are dependent on ambient conditions. They are also relatively inefficient in reducing moisture contents. The filler requires a steady low dew-point dry air (under 32^0C) and constant drying temperature which guarantee a residual moisture content of 0.02% or less. Once this is achieved, the composites can exhibit their optimum mechanical performance. For this the wood fillers were dried in a dehumidifier (Figure 1) at 80^0C for 6 hrs.

Composite Fabrication

For preparation of composites, wood dust particles having a particular particle size (around 75μm) and density (0.779 gm/ cc), after cleaning and drying, was reinforced with epoxy resin (density 1.26 gm/cc). This mixture (epoxy filled with wood) was used to prepare the specimen. Casting was performed in vacuum casting chamber (Figure 2) so that air bubbles and usual cause of imperfections could be minimized. Vacuum casting unit has two chambers (Figure 3). Resin and corresponding hardener, along with wood dust filler was mixed in the upper chamber. The mix was cast into

Figure 1. Dehumidifier

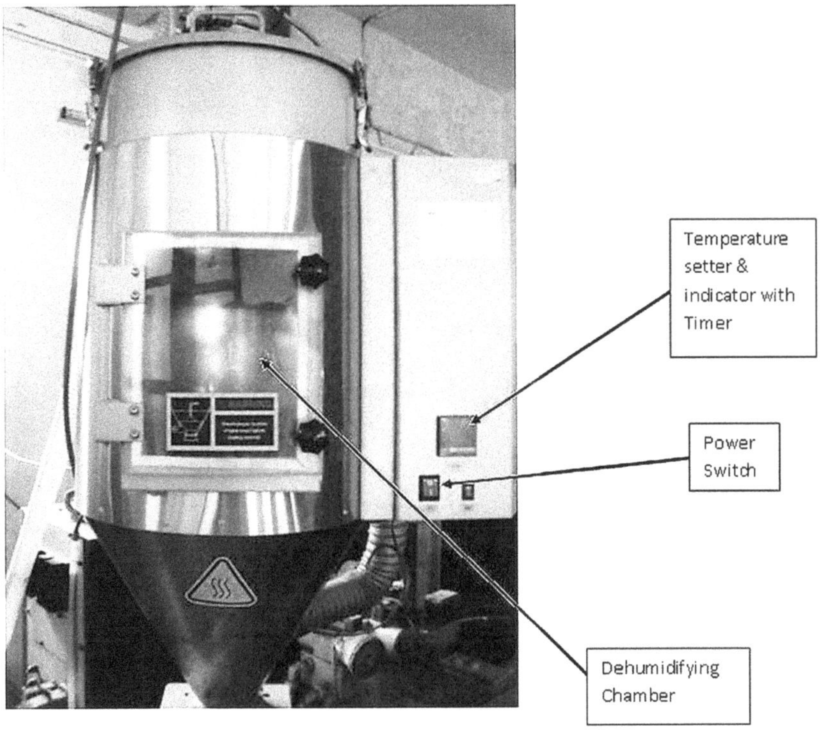

a rectangular glass mould coated with glass paper (for easy removal), placed in the lower chamber. The mould with composite was then kept in a Hot Air Oven (Figure 4) at a 60^0C for about 8 hrs so that complete curing could be obtained. After curing the specimen was removed from the mould and used for the experimental purpose.

EXPERIMENTAL DESIGN

Design of Experiments

Industrial physicists can no longer afford to conduct experiment in a trial-and-error manner, changing one factor at a time, the way Edison did in developing the light bulb. A far more effective method is to apply a computer-enhanced, systematic approach to experimentation, one that considers all factors simultaneously. That approach is called design of experiments (DOE), and corporations worldwide are adopting it as a cost-effective way to solve serious problems afflicting their operations. Design of experiment (DOE) is the process of planning a study to meet specific objectives. It is a powerful analysis tool which enables the designers for analyzing the influence

Figure 2. Vacuum casting chamber

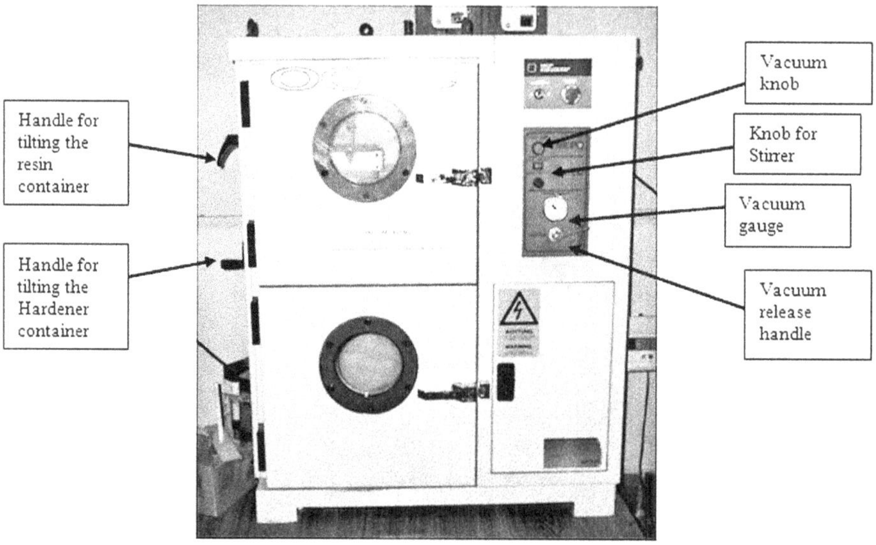

of individual and interactive effects of many factors on performance output in any design which helps to turn any standard design into a robust one. It also helps for the designers to know the sensitive parts and sensitive areas in design that cause problems in output and the designers are then able to fix the problems to get higher

Figure 3. Vacuum casting chamber (inside details)

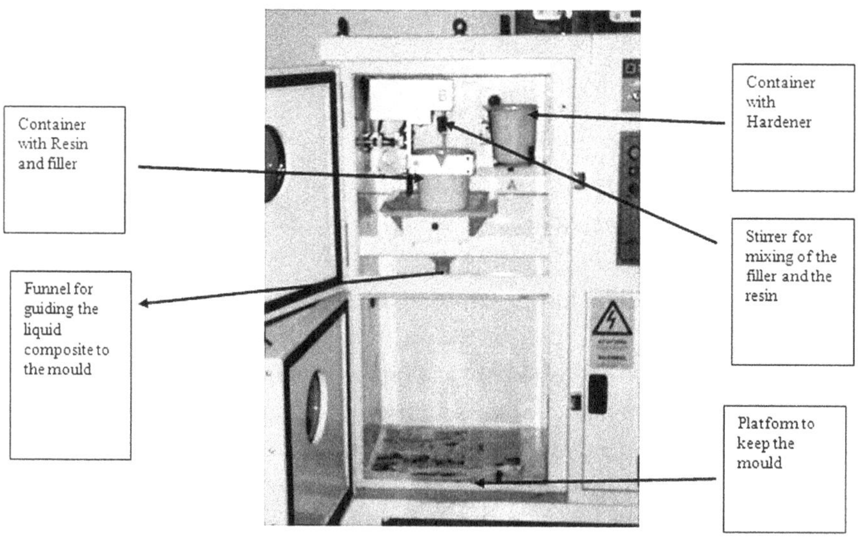

Figure 4. Hot air oven with inside details

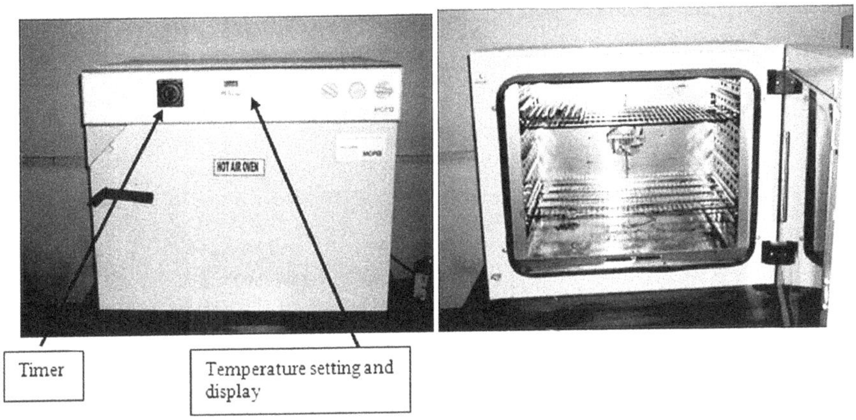

yield design prior going into production. With the help of DOE, the designers can no longer afford to experiment in a trial-and-error manner or changing one factor at a time. DOE is a systematic approach to experimentation that considers all individual and interaction factors simultaneously, and industries worldwide are adopting it as a cost-effective way to solve problems affecting their operations.

The concept of design of experiment has been in use since Fisher's (1935) work dealt with agricultural applications of statistical methods. Numerous applications of this approach are cited in the literature. The general model of a process or system is shown in Figure 5.

Figure 5. General model of a process/system

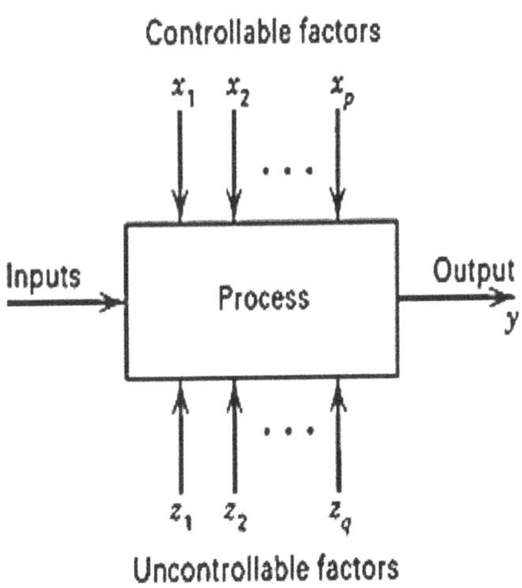

The following step provides the knowledge that need to be made for defining the design of experiment (DOE).

1. Identify the experimental unit.
2. Identify the types of variables.
3. Define the treatment structure.
4. Define the design structure.

Experimental Unit

The first step in DOE is to define the experimental unit. An experimental unit is a system or process can be visualized as a combination of machines, methods, people and other resources from which the data will be collected.

Types of Variables

Next step is to find the types of variables such as controllable factors which can be suitably changed to control the process response and uncontrollable variables are those variables that are known to exist, but conditions prevent them from being manipulated or it is very difficult to measure them.

Treatment Structure

The treatment structure consists of factors that the designer wants to study and about which the designer will make inferences. The designer should identify the controllable factors which are expected to show the effects on the response variables. The levels of factors should be clearly defined for the study.

Design Structure

The designer should allocate experimental units to treatments either randomly or randomly with constraints, as in blocked design (Montgomery, 1997) for most experimental runs.

DOE is a series or process of tests which considers how the experimental factors, both controlled and uncontrolled fit together into a model that will meet the specific objectives of the experiment and satisfy the practical constraints of time and money. If a designer wish to draw a meaningful conclusion from the observed data, statistical approach to DOE is required. So, basically DOE refers to the process of planning, designing and analysing the experiment through which valid conclusion can be drawn effectively and efficiently. DOE method has been applied in many functional areas which include product development, research, quality control, purchasing, engineering etc.

The advantages of DOE method are:

- It saves time and money and eliminates unnecessary work in experimentation.
- It solves complicated problems.
- It produces high quality products.
- It helps to determine the important variable that need to be controlled.
- It helps to measure interactions, which is very important.
- It helps to find out the tolerance level of the factors.
- It helps to optimizc product and process designs.
- It helps to study the influence of each factor on the performance.

The design of experiments technique is a very powerful statistical tool, used to modeling and analysis of experimental results on the response variables. The applying of DOE techniques can significantly reduce the time and money required for experimental investigations in the manufacturing industry.

Orthogonal Array

Orthogonal Arrays (OA) are generally used for the experimentation in Taguchi analysis. These are special matrices used as the design matrices in the fractional factorial design for the estimation of the effect of several factors in a highly efficient way. These designs are applicable even when the factors have more than two levels and for mixed level experiments where that factors do not have same number of levels. In general, when the number of process parameter increases, large number of experiments has to be carried out for factorial design. But using orthogonal array, carrying out small number of experiments in the specified range and studied the effects of process parameters can be studied. For a two level factors 8 (L8) experiments is needed for the experimentation. A three-level orthogonal arrays can be based on 9 (L9), 27 (L27) or 81 (L81) experimental points. For any pair of columns in the design matrix, all combinations of the factor levels appear equal number of times.

In this work Tensile and flexural tests of the composite were carried out under different operating conditions considering two parameters, viz., filler content and speed at seven and three levels respectively as listed in Table 1. In a conventional full factorial experiment design, it requires 7x3 = 21 runs to study two parameters at mixed levels.

Table 1. Design factors with their levels for mechanical tests

	Units	Levels						
Control Factors		1	2	3	4	5	6	7
Filler Content	wt.%	0	2.5	5	7.5	10	12.5	15
Speed	mm/min	1	1	1	1	1	1	1
		2	2	2	2	2	2	2
		3	3	3	3	3	3	3

Tribological study on the effect of filler content in composites, factors like normal load and sliding speed are also considered along with filler content. The experimentation was performed with two different sets of normal load and sliding speed and % ages of filler as shown in Table 2.

EXPERIMENTATION DETAILS

Density and Void Content

Density of polymer composites primarily depends upon the composition of the composite. It is usually defined as the ratio of mass to the unit volume of a substance. The theoretical densities of composite material were calculated by using weight additive principle and actual density of composite was determined experimentally by using simple water immersion method with the help of Mettler Toledo electronic balance. The difference indicated the void or air content (generally expressed in %age). Void content, in the fabricated composites, is an important physical property which can significantly affect the physical and mechanical properties of composite material. The knowledge of void content is important to estimate the quality of composites.

Table 2. Design factors with their levels for tribological tests

Design Factors	Unit	Levels						
		1	2	3	4	5	6	7
Filler content	wt %	0	2.5	5	7.5	10	12.5	15
Normal load	N	15/35	15/35	15/35	15/35	15/35	15/35	15/35
Speed	rpm	80/120	80/120	80/120	80/120	80/120	80/120	80/120

Tensile and Flexural Tests

Universal testing machine (UTM) was used to test the tensile and flexural behaviour (Tensile strength, tensile modulus, flexural strength and flexural modulus) of composite materials. The mechanical tests were carried out using Instron universal testing machine (Instron Ltd, UK) (Figure 6). The tensile tests were conducted according ASTM D 638. Rectangular specimens were cut from the prepared sample of size 100 mm long, 12 mm wide and 8 mm thickness. The tensile tests were performed with varying filler content and cross head speed. The flexural or 3 point bending tests were undertaken as per ASTM D 790 under the same condition as that of the tensile tests.

Hardness Test

Hardness is the resistance of a material to deformation, indentation or scratching. It can differentiate the grades of various polymers with hardness number and also reveals the dimensional stability of a composite. The indentation value has high importance for technical applications reflecting the resistance to deformation, which is a complex property and related to modulus, strength, elasticity, plasticity and dimensional stability of a material.

Figure 6. Instron universal testing machine

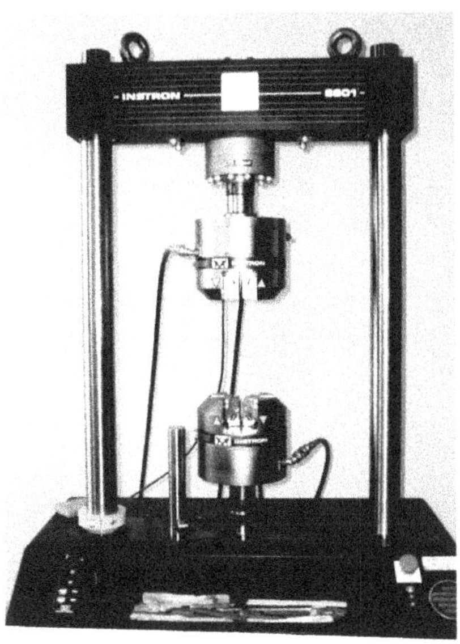

Hardness is generally classified into three categories with respect to the depth of indentation as follows:

1. Nano-hardness (d<1 μm),
2. Micro-hardness (d = 1-50 μm), and
3. Macro-hardness (d>50 μm).

The test methods commonly used for expressing relationship between hardness and the size of impression are Brinell, Vicker's and Rockwell hardness tests.

Vicker's Micro hardness tests can be used for ascertaining the hardness of composites (Ciccone-Nogueira et.al, 2007 and Poggio et. al, 2012). So in this study Vicker's hardness test setup was used to find out the micro-hardness value of polymer composites. Instrument used to carry out the experiment was a UHL micro hardness tester (Model - VMHT MOT, Sl. No. 1002001, Technische Mikroskopie) with a Vickers diamond indenter (Figure 7). The dwell time and speed of indentation were kept constant and a constant load was applied.

Tribological Test

Tribological tests of composites was performed on a block-on-roller multi-tribotester TR25 (Ducom, India) (Figure 8) under dry condition (ambient temperature is about 28°C with a relative humidity at 85%) with the constant time. The composite specimens of size 20 X 20 X 8 mm^3 was held stationary with the help of attachment provided

Figure 7. Vicker's micro-hardness tester

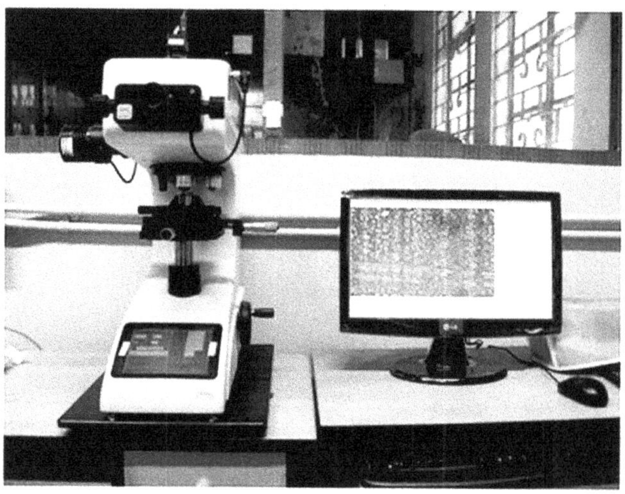

Figure 8. Experimental setup of multi-tribo tester

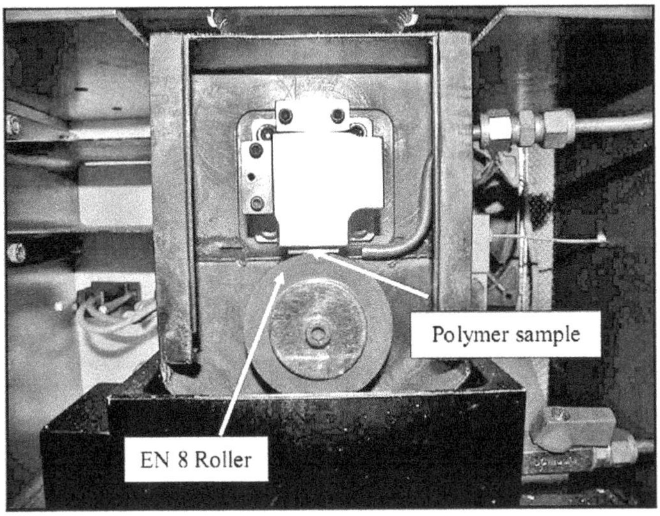

in the test rig and made to slide against the steel roller (diameter 50 mm, thickness 50 mm and material EN8 steel of hardness 55 HRc) which acts as a counterface. A loading lever was used to apply a normal load by placing dead weights on the loading pan on the top of the specimen. The experimental data of Coefficient of Friction (COF) were recorded on a computer attached to the testing apparatus. Weight loss of the composite are used to calculate the specific wear rate. Samples were weighed before and after the experiments in a Mettle Toddler electronic balance.

Morphological Study

It is important to study a composite material which is newly developed to ensure the structure; chemical composition, distribution etc. are exactly similar as desired. The microstructure characterization of a developed material helps one to quickly understand the reasons for variation of its properties, physical behavior, distribution of fillers into the matrix and performance under certain investigation, hence these techniques are in demand to study and characterize materials. These techniques also help the designer in designing new materials with newer property combinations.

Microstructure characterization of a material is mostly studied by using Scanning Electron Microscopy (SEM) images and it is very useful in order to observe the microstructure in detail. The surface of a specimen to be examined is generally coated with a thin coating of conductive material and is scanned with an electron beam. Reflected beam of electrons is collected and then displayed at the same scanning

Figure 9. Scanning electron microscope

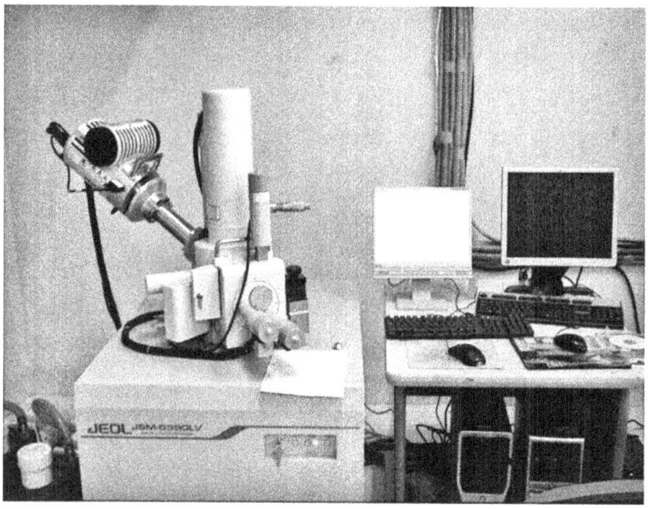

rate on a cathode ray tube monitor. The image displayed on the screen represents the surface features of the specimen are photographed for further examination. Magnification in SEM ranges from about 10 to 50000 times be possible with this technique.

Scanning electron microscopy examinations were carried out for the composite material to observe the distribution of the filler in the matrix. The composite surfaces were coated with thin platinum film on the worn out surface by sputtering to get a conducting layer and was observed in a JEOL (model JSM 6390LV, Japan) microscope (Figure 9).

RESULTS AND DISCUSSION

Density and Void Content

Density and void content is one of the important physical properties of material especially polymer based composite materials for estimation of quality of composites. Experimental density is measured by using simple water immersion technique and theoretical density is determined with the help of weight additive principle. Experimental and theoretical density along with their void content for different filler concentration of Wood-Epoxy composites is shown in Table 3. It is seen that the experimental densities is lower than the theoretical density; is due to presence of voids in samples. It may also be observed that the density decreases with increase in filler content, it is due to the fact that the density of the filler is lower than that

Table 3. Density and void

Wood Filler Content (%)	Density (g/cc)		Void Fraction (%)
	Experimental	Theoretical	
0	1.2551	1.2600	0.3889
2.5	1.2347	1.2480	1.0637
5	1.2215	1.2360	1.1691
7.5	1.2073	1.2239	1.3583
10	1.1941	1.2119	1.4688
12.5	1.1762	1.1999	1.9731
15	1.1635	1.1879	2.0471

of the matrix and hence as the % age filler increases the matrix content decreases and so does the density. Moreover % void of the sample increases suddenly from 0% filler to 2.5% filler, due to the voids at the interface of the filler and the matrix of the composite. In case of 12.5% and 15% filler content, again void content increases suddenly. This may be due to the fact that above 10% the agglomeration starts due to engulfing of the matrix. It can be verified by the SEM images in the morphological study. The acceptable range of void content for many applications is usually 5% and for the present study the void content lies within the acceptable range.

Tensile Test

The tensile test was carried out according to ASTM D 638-03 using universal testing machine with grip capacity of 100 kN. The testing was performed at ambient temperature of 24 °C and relative humidity of 53%. Seven different % filler wt. specimens (as 0%, 2.5%, 5%, 7.5%, 10%, 12.5% and 15%) were tested at three different speeds of 1 mm/min, 2 mm/min and 3 mm/min. The standard specimen was mounted by its ends into the holding grips of the testing instrument. The machine is designed to elongate the specimen at a uniform rate and using extensometer the instantaneous applied load and the resulting elongation are measured continuously and cumulatively.

The variations of maximum load versus filler content and tensile stress at maximum load versus filler content are shown in Figure 10 (a) and (b) respectively. The variation of tensile modulus and tensile strain at various speed for different filler content are shown in Figure 11 (a) and (b) respectively.

From the results, it can be observed that as the filler weight % increases, maximum load, tensile modulus, tensile stress and strain value increases and becomes maximum at 10% filler content by wt. and then these properties decreases.

Figure 10. Variation of (a) maximum load and (b) tensile stress with % filler wt. at various speed

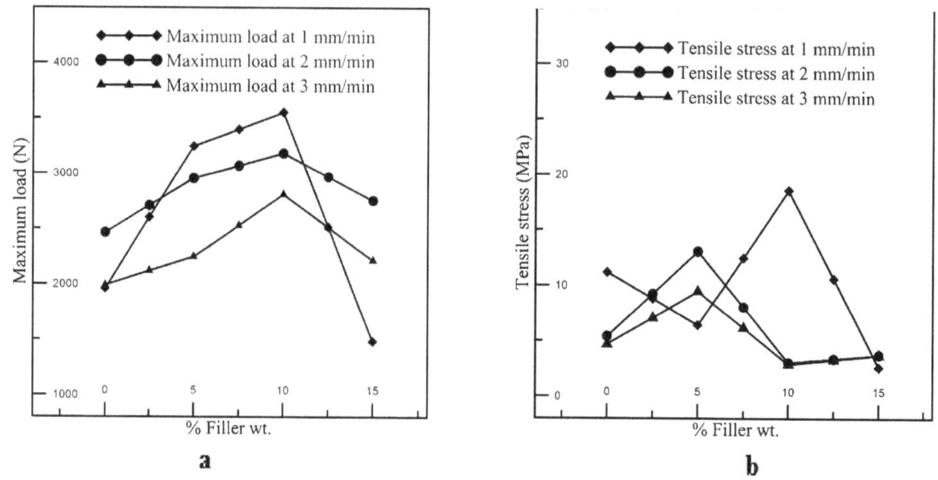

Maximum and minimum values of load with speed of 1 mm/min are found at 10% and 15% filler wt. respectively. Their respective values are 3550.36 N and 1482.08 N. The maximum and minimum values of tensile stresses with speed of 1 mm/min are 28.29 MPa and 10.83 MPa at 10% and 15% filler wt. respectively. The maximum and minimum strain values are 19.10 and 2.69 respectively at 10% and 15% filler wt. with speed of 1 mm/min respectively. For 2 mm/min speed the maximum and

Figure 11. Variation of (a) tensile modulus and (b) tensile strain with % filler wt. at various speed

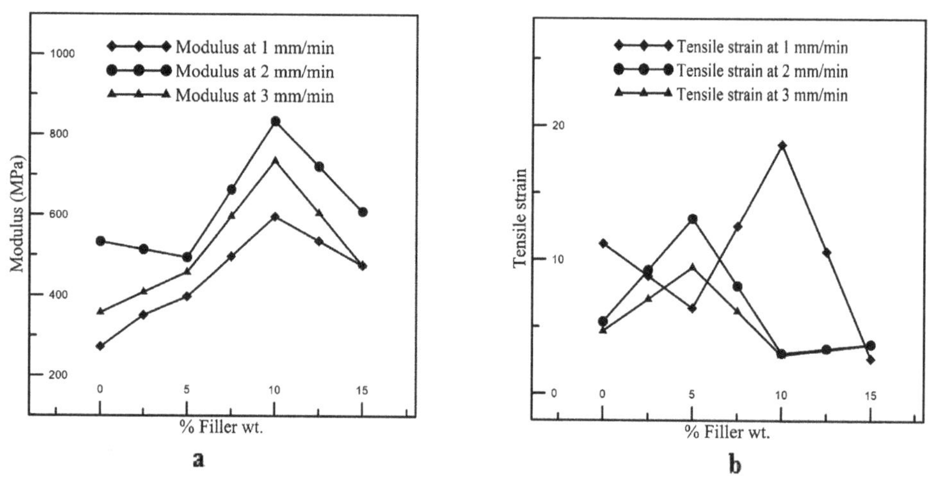

minimum modulus values are found at 10% filler wt. and the maximum and minimum modulus values are found 0% filler wt. at 1 mm/min speed and the values are 19.10 MPa and 2.69 MPa respectively. The maximum value of the stress is found at higher % age of filler with low speed. This is due to the fact that the fillers get time to reorient themselves and results in higher value of the stress whereas in case of higher speed the maximum value of stress is reached at a lower % age of filler.

Flexural Test

Three point flexural tests of composites are carried out using Universal testing machine as per ASTM D 790-03 at 24 °C and 53% relative humidity. The specimens of each filler contents were tested at three different crosshead speeds of 1 mm/min, 2 mm/min and 3 mm/min. The sample of rectangular cross section rests on two supports and is loaded by means of a loading nose midway between the anchors. The loading nose and anchors have cylindrical surface. Flexural modulus is calculated from the slope of the initial portion of load deflection curve.

Flexural test is carried out for seven different % filler wt. at three different speeds. Variation of flexural stress, modulus and strain vs. filler content at different speed are shown in Figure 12 (a), (b) and (c) respectively.

From the results, it can be inferred that as the filler wt. % increases, flexural modulus, flexural stress and strain value increases and becomes maximum at 10% filler wt. and then properties start decreasing.

Maximum and minimum values of flexural stress are 47.65 MPa and 18.24 MPa respectively for 10% and 15% filler wt. with 1mm/min speed. The maximum and minimum flexural strain values are 2.61 and 18.58 respectively for 10% and 15% filler wt. with 1 mm/min speed. The maximum and minimum values of flexural

Figure 12. Variation of (a) flexural stress, (b) flexural strain and (c) flexural modulus with % filler wt. at various speed

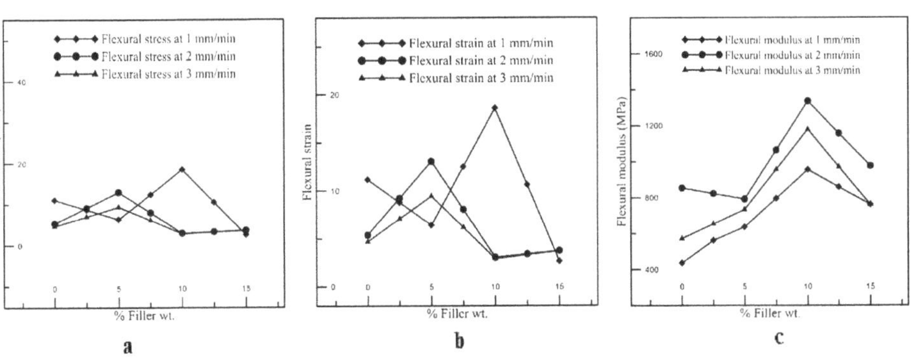

Figure 13. Micro-hardness of wood epoxy composite

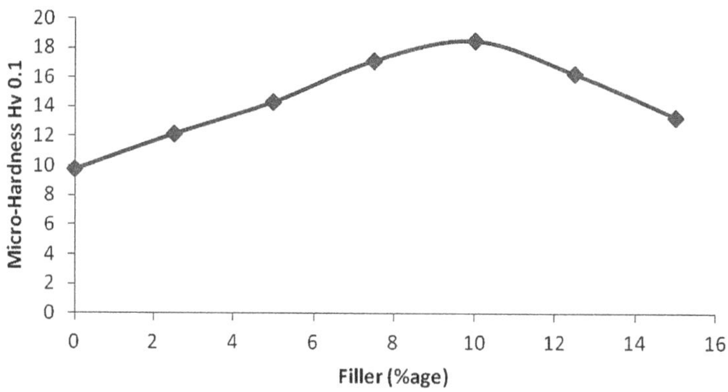

modulus are 1335.20 MPa and 435.70 MPa respectively for 10% and 0% filler wt. and speed of 2 mm/min and 1 mm/min respectively. As discussed in the case of tensile properties regarding the maximum stresses and its relationship with speed of testing, the same is also true for the flexural properties.

Hardness Test

The fabricated composites of different proportions are characterized for hardness using Vicker's micro-hardness tester. The hardness of the material is an important criterion which is the resistance of a material to deformation. Figure 13 shows the influence of Wood-Epoxy composites on micro-hardness. It reveals that the hardness of the composites increases from 9.5 $Hv_{0.1}$ to a maximum of 18.5 $Hv_{0.1}$ and then reduces to 13.33 $Hv_{0.1}$ with increasing filler content. This may be due to the rigidity of filler and increased modulus withstands the depth of penetration, hence micro-hardness increases with increase in filler content up to 10%. The subsequent decrease can be due to the agglomeration as has also been observed for the other mechanical properties.

Tribological Test

Tribological properties of friction coefficient and specific wear rate are studied by varying the design parameters viz., filler content, normal load and speed. The experimental procedure for the measurement of friction coefficient and specific wear rate were conducted as discussed earlier. Figure 14 (a) and (b) shows the variation of friction coefficient and specific wear rate on varying wood filler content of the composite at two different specific load and speed. It is noticed that the graph shows non-monotonic in nature i.e. it decreases upto a certain wt% and then it starts to increase.

Figure 14. Effect of filler content on (a) friction coefficient and (b) specific wear rate of epoxy based wood composite (with error bars)

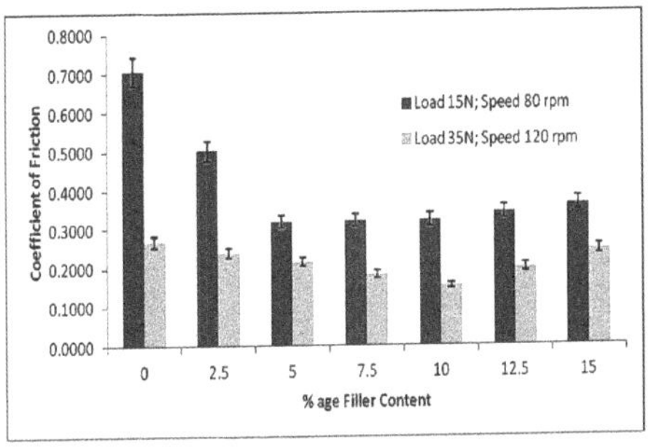

a

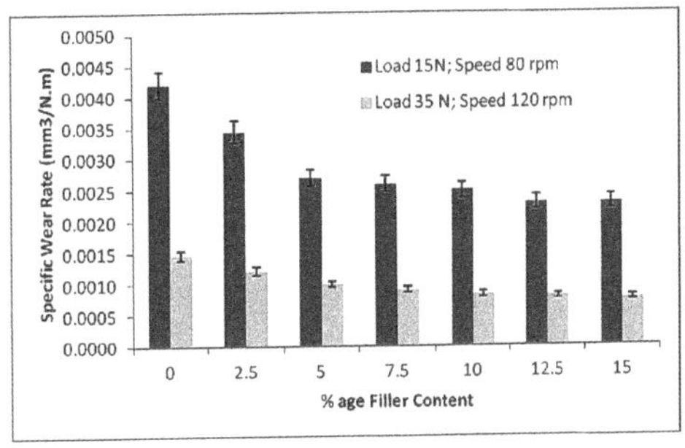

b

The addition of wood filler into Epoxy reduced the friction coefficient and wear rate with increase in normal load and sliding speed. This may be due to reduced contact surface between composite and counter face; hence decrease in friction coefficient is noticed. On the other hand, worn surface could be polished very rapidly at high sliding speed which leads to reduced wear rate. With increase in filler content, the wear rate decreases due to particle filled polymer matrix withstands the deformation and particle cannot be easily detached from the composite. This results show that the composite enhanced the friction coefficient and specific wear rate at certain filler content, normal load and sliding speed. It also shows that increased friction coefficient did not necessarily correspond to the increased wear rate.

Figure 15. SEM image of the composite with 0% filler

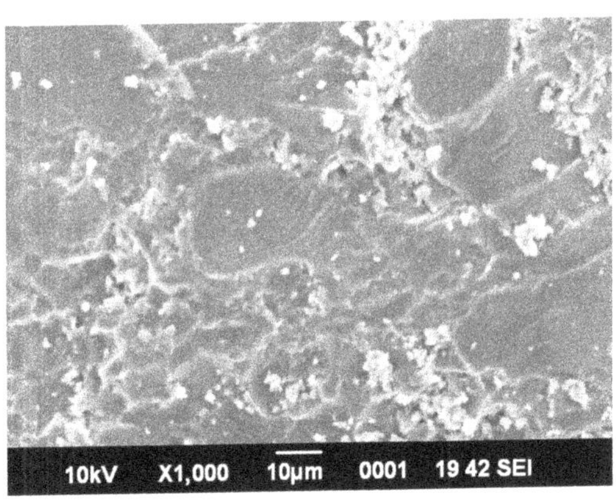

Figure 16. SEM images of the composite with 2.5% filler

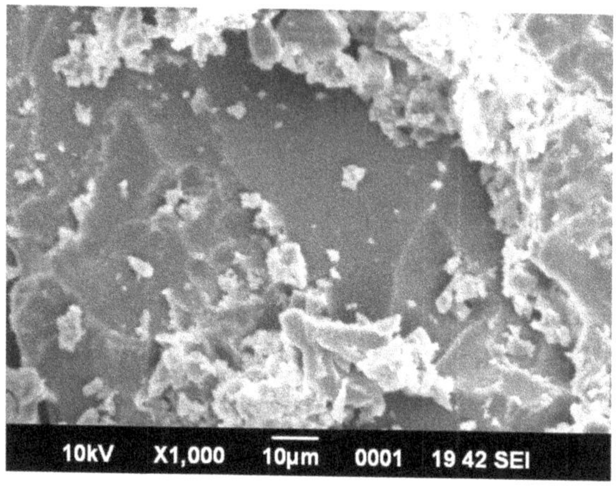

Figure 17. SEM images of the composite with 5% filler

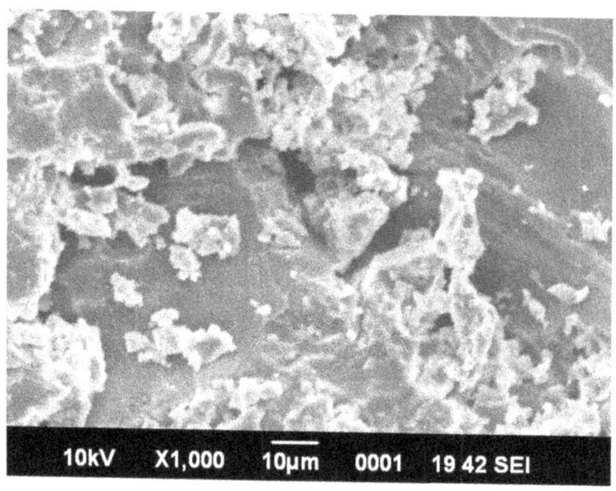

Morphological Study

Scanning electron microscopy examinations are carried out for the composite surfaces coated with thin platinum film on the worn out surface by sputtering to get a conducting layer on a JEOL (model JSM 6390LV, Japan) microscope to observe the filler distribution in the matrix. Figure 15 shows the SEM picture for the composite with 0% filler. The particles are debris resulting due to cutting of the sample. Figure 16 shows the SEM micrograph for the composite with 2.5% filler in which the fillers can be seen scattered. Figure 17 and Figure 18 shows the SEM images for the composite with 5% & 7.5% filler in which it can be observed that the filler concentration increases and hence results in increase in the mechanical properties.

Figure 18. SEM images of the composite with 7.5% filler

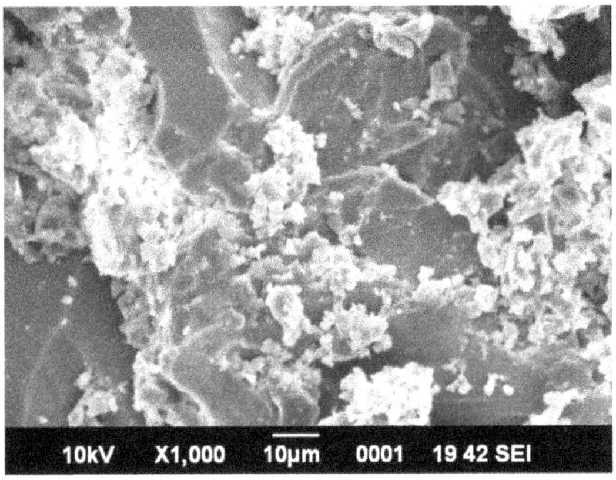

Figure 19. SEM images of the composite with 10% filler

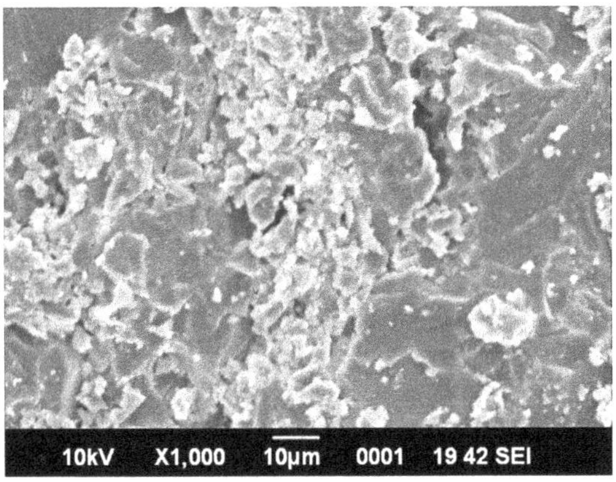

Figure 19 shows the SEM photograph for the composite with 10% filler in which it can be clearly seen that the fillers have started forming clusters. Figure 20 shows the SEM image for the composite with 12.5% & 15% filler in which it can be seen that the clusters have started engulfing the epoxy hence the curing is improper and mechanical properties falls. The trend increases with 15% filler concentration.

CONCLUSION

In this research, an attempt has been made to study the Mechanical and Tribological behaviour of wood reinforced epoxy composite reinforced with different filler content. Void content results supported the objective that wood dust possessed

Figure 20. SEM images of the composite with 12.5% and 15% filler

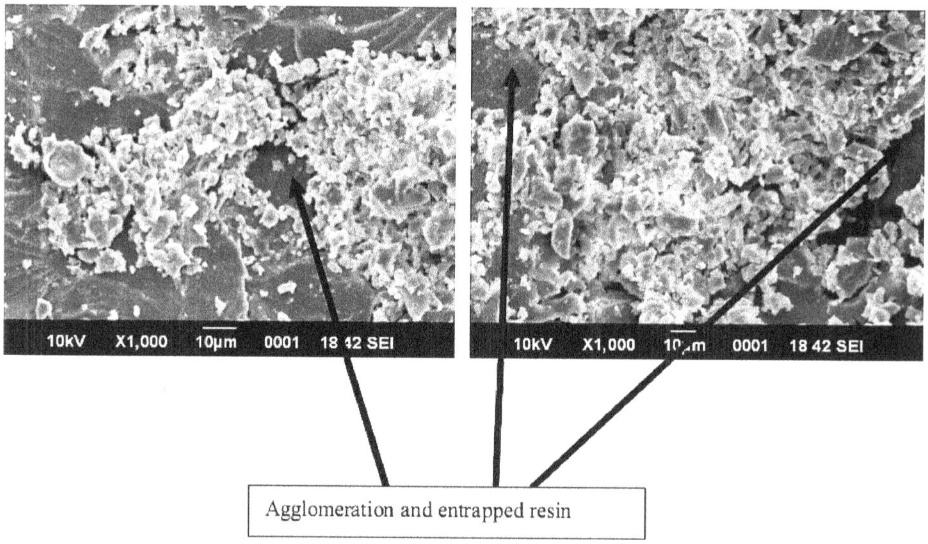

good filler characteristics and the processing technique adopted was strong enough to provide better composites. As desired, hardness of the composite increased with addition of filler into matrix and so did the strength (Tensile as well as Flexural). Best mechanical properties (maximum load, tensile stress and strain, and flexural stress and strain) were observed for 10% filler weight at a crosshead testing speed of 1mm/min and 2mm/min. Tribological results showed that, composite enhanced the friction coefficient and specific wear rate up to certain filler content. It is obvious as normal load and sliding speed influenced the output more than filler content. Tests clearly indicated that increased friction coefficient did not necessarily correspond to the increased wear rate and vice versa. The results for the same can be implementing on similar class of fillers i.e. green fillers with other thermosetting matrix but the effect of the controlling factors has to be evaluated accordingly as properties of natural filler based composites depends on the lignin and cellulose content to a great extent.

REFERENCES

Ahmed, N. M., & El-Sabbagh, S. H. (2014). The influence of kaolin and calcined kaolin on SBR composite properties. *Polymer Composites*, *35*(3), 570–580. doi:10.1002/pc.22697

Akinci, A., Sen, S., & Sen, U. (2014). Friction and wear behavior of zirconium oxide reinforced PMMA composites. *Composites. Part B, Engineering*, *56*, 42–47. doi:10.1016/j.compositesb.2013.08.015

Ali, I., & Elleithy, R. (2011). Toughness of HDPE / CaCO. *Composite Structures*, *48*(1), 49–58.

Ariffin, A., Mansor, A. S., & Jikan, S. S. (2008). Mechanical, morphological, and thermal properties of polypropylene / kaolin composite. Part I. The effects of surface-treated kaolin and processing enhancement. *Journal of Applied Polymer Science*, *108*(6), 3901–3916. doi:10.1002/app.27950

Ashori, A. (2008). Wood-plastic composites as promising green-composites for automotive industries. *Bioresource Technology*, *99*(11), 4661–4667. doi:10.1016/j.biortech.2007.09.043 PMID:18068352

Asyadi, F., Jawaid, M., Hassan, A., & Wahit, M. U. (2013). Mechanical Properties of Mica-Filled Polycarbonate / Poly (Acrylonitrile-Butadiene-Styrene) Composites. *Polymer-Plastics Technology and Engineering*, *52*(7), 727–736. doi:10.1080/03602559.2012.762672

Augier, L., Sperone, G., Vaca-Garcia, C., & Borredon, M. E. (2007). Influence of the wood fibre filler on the internal recycling of poly (vinyl chloride)-based composites. *Polymer Degradation & Stability*, *92*(7), 1169–1176. doi:10.1016/j.polymdegradstab.2007.04.010

Aurrekoetxea, J., Sarrionandia, M., & Gomez, X. (2008). Effects of microstructure on wear behaviour of wood reinforced polypropylene composite. *Wear*, *265*(5), 606–611. doi:10.1016/j.wear.2007.12.013

Axen, N., Hutchings, I. M., & Jacobson, S. (1996). A model for the friction of multiphase materials in abrasion. *Tribological International*, *29*(6), 467–475. doi:10.1016/0301-679X(95)00104-C

Balakrishna, N. S., Ismail, H., & Othman, N. (2014). Polypropylene / Rattan Powder / Kaolin Hybrid Composites: Processing, Mechanical and Thermal Properties. *Polymer-Plastics Technology and Engineering*, *53*(5), 451–458. doi:10.1080/03602559.2013.844833

Ben Difallah, B., Kharrat, M., Dammak, M., & Monteil, G. (2012). Mechanical and tribological response of ABS polymer matrix filled with graphite powder. *Materials & Design*, *34*, 782–787. doi:10.1016/j.matdes.2011.07.001

Bhowmick, M., Mukhopadhyay, S., & Alagirusamy, R. (2012). Mechanical properties of natural fibre reinforced composites. *Textile Progress*, *44*(2), 85–140. doi:10.1080/00405167.2012.676800

Chand, N., Dwivedi, U. K., & Sharma, M. K. (2007). Development and tribological behaviour of UHMWPE filled epoxy gradient composites. *Wear*, *262*(1-2), 184–190. doi:10.1016/j.wear.2006.04.012

Chang, B. P., Akil, H. M., Nasir, R. B. M., Bandara, I. M. C. C. D., & Rajakpase, S. (2014). The effect of ZnO nanoparticles on the mechanical, tribological and antibacterial properties of ultra-high molecular weight polyethylene. *Journal of Reinforced Plastics and Composites*, 0731684413509426.

Chang, B. P., Akil, H. M., Nasir, R. M., & Nurdijati, S. (2011). Mechanical and antibacterial properties of treated and untreated zinc oxide filled UHMWPE composites. *Journal of Thermoplastic Composite Materials*, *24*(5), 653–667. doi:10.1177/0892705711399848

Chang, B. P., Md Akil, H., & Bt. Md Nasir, R. (2013). Comparative study of micro- and nano-ZnO reinforced UHMWPE composites under dry sliding wear. *Wear*, *297*(1), 1120–1127. doi:10.1016/j.wear.2012.11.083

Chen, J., Feng, Y., Chen, N., Li, B., Chen, F., & Zhang, X. (2013). Fabrication and Properties of Silver-Doped Acrylonitrile Butadiene Styrene (ABS) Composites. *Polymer-Plastics Technology and Engineering*, *52*(11), 1081–1088. doi:10.1080/03602559.2013.769574

Ciccone-Nogueira, J. C., Borsatto, M. C., Souza-Zaroni, W. C., Ramos, R. P., & Palma-Dibb, R. G. (2007). Microhardness of composite resins at different depths varying the post-irradiation time. *Journal of Applied Oral Science*, *15*(4), 305–309. doi:10.1590/S1678-77572007000400012 PMID:19089149

Eiss, N. S. Jr, & Czichos, H. (1986). Tribological studies on rubber-modified epoxies: Influence of material properties and operating conditions. *Wear*, *111*(4), 347–361. doi:10.1016/0043-1648(86)90132-8

Esthappan, S. K., Kuttappan, S. K., & Joseph, R. (2012). Thermal and mechanical properties of polypropylene/titanium dioxide nanocomposite fibers. *Materials & Design*, *37*, 537–542. doi:10.1016/j.matdes.2012.01.038

Fisher, R. A. (1935). *The Design of Experiments* (1st ed.). Edinburgh, UK: Oliver and Boyd.

Gamelas, J. A. F., Ferraz, E., & Rocha, F. (2014). An insight into the surface properties of alcined kaolinitic clays: The grinding effect. Colloids and Surfaces A: *Physicochem. Eng. Aspects*, *455*, 49–57. doi:10.1016/j.colsurfa.2014.04.038

Hassan, A., Akbari, A., Hing, N. K., & Ratnam, C. T. (2012). Mechanical and thermal properties of ABS / PVC composites: Effect of particles size and surface treatment of ground calcium carbonate. *Polymer-Plastics Technology and Engineering*, *51*(5), 473–479. doi:10.1080/03602559.2011.651242

Herrera-Franco, P. J., & Valadez-Gonza'lez, A. (2004). Mechanical properties of continuous natural fibre-reinforced polymer composites. *Composites. Part A, Applied Science and Manufacturing*, *35*(3), 339–345. doi:10.1016/j.compositesa.2003.09.012

Ishizaki, M. H., Visconte, L. L., Furtado, C. R., Oliveira, M. G., & Leblanc, J. L. (2008). Characterization of Green Coconut Fiber Composites. *International Journal of Polymer Analysis and Characterization*, *13*(5), 319–330. doi:10.1080/10236660802351409

Kaseem, M., Hamad, K., & Deri, F. (2013). On-line rheological measurements and mechanical properties of acrylonitrile-butadiene-styrene / corn starch composite. *International Journal of Polymeric Materials and Polymeric Biomaterials*, *62*(5), 260–264. doi:10.1080/00914037.2011.641698

Kocserha, I., & Gömze, L. A. (2010). Friction properties of clay compounds. *Applied Clay Science*, *48*(3), 425–430. doi:10.1016/j.clay.2010.01.017

Kranthi, G., & Satapathy, A. (2010). Evaluation and prediction of wear response of pine wood dust filled epoxy composites using neural computation. *Computational Materials Science*, *49*(3), 609–614. doi:10.1016/j.commatsci.2010.06.001

Krishnan, A. K., George, T. S., Anjana, R., & George, K. E. (2013). Effect of modified kaolin clays on the mechanical properties of polypropylene / polystyrene blends. *Journal of Applied Polymer Science*, *127*(2), 1409–1415. doi:10.1002/app.38043

Li, Z. H., & Chen, T. (2012). The mechanical and tribological properties of MAH-grafted HDPE-filled clay / PEI composites. *Polymer-Plastics Technology and Engineering*, *51*(8), 776–779. doi:10.1080/03602559.2012.663848

Liu, H., Dong, L., Xie, H., & Xiong, C. (2012). Novel-modified kaolin for enhancing the mechanical and thermal properties of poly (vinyl chloride). *Polymer Engineering and Science*, *52*(10), 2071–2077. doi:10.1002/pen.23161

Ma, L., He, H., Jiang, C., Zhou, L., Luo, Y., & Jia, D. (2012). Effect of alkali treatment on structure and mechanical properties of acrylonitrile-butadiene-styrene / bamboo fiber composites. *Journal of Macromolecular Science, Part B: Physics*, *51*(11), 2232–2244. doi:10.1080/00222348.2012.669688

Mirmehdi, S. M., Zeinaly, F., & Dabbagh, F. (2014). Date palm wood flour as filler of linear low-density polyethylene. *Composites. Part B, Engineering*, *56*, 137–141. doi:10.1016/j.compositesb.2013.08.008

Montgomery, D. C. (1997). *Design and analysis of experiments* (Vol. 7). New York: Wiley.

Mylsamy, K., & Rajendran, I. (2011). The mechanical properties, deformation and thermo mechanical properties of alkali treated and untreated Agave continuous fibre reinforced epoxy composites. *Materials & Design*, *32*(5), 3076–3084. doi:10.1016/j.matdes.2010.12.051

Nascimento, D. C., Ferreira, A. S., Monteiro, S. N., Aquino, R. C., & Kestur, S. G. (2012). Studies on the characterization of piassava fibers and their epoxy composites. *Composites. Part A, Applied Science and Manufacturing*, *43*(3), 353–362. doi:10.1016/j.compositesa.2011.12.004

Poggio, C., Lombardini, M., Gaviati, S., & Chiesa, M. (2012). Microhardness of composite resins at different depths varying the post-irradiation time. *Journal of Conservative Dentistry*, *15*(3), 237–241. doi:10.4103/0972-0707.97946 PMID:22876009

Praveen, R. N., Muruli, D., & Nayak, S. S. (2013). Sliding wear behavior of Acrylonitrile Butadiene Styrene (ABS) thermoplastic composites. *International Journal of Mechanical Engineering and Robotics Research*, *2*(4), 361–372.

Rasheva, Z., Zhang, G., & Burkhart, T. (2010). A correlation between the tribological and mechanical properties of short carbon fibers reinforced PEEK materials with different fiber orientations. *Tribology International*, *43*(8), 1430–1437. doi:10.1016/j.triboint.2010.01.020

Roselman, I. C., & Tabor, D. (1976). The friction of carbon fibres. *Journal of Physics. D, Applied Physics*, *9*(17), 2517–2532. doi:10.1088/0022-3727/9/17/012

Sandberg, D., Haller, P., & Navi, P. (2013). Thermo-hydro and thermo-hydro-mechanical wood processing: An opportunity for future environmentally friendly wood products. *Wood Material Science and Engineering*, *8*(1), 64–88. doi:10.1080/17480272.2012.751935

Sapuan, S. M., Leenie, A., Harimi, M., & Beng, Y. K. (2006). Mechanical properties of woven banana fibre reinforced epoxy composites. *Materials & Design*, *27*(8), 689–693. doi:10.1016/j.matdes.2004.12.016

Schön, J. (1996). Coefficient of friction of composite delamination surfaces. *Wear*, *237*(1), 77–89. doi:10.1016/S0043-1648(99)00315-4

Selvam, B., Marimuthu, P., Narayanasamy, R., Anandakrishnan, V., Tun, K. S., Gupta, M., & Kamaraj, M. (2014). Dry sliding wear behaviour of zinc oxide reinforced magnesium matrix nano-composites. *Materials & Design*, *58*, 475–481.

Song, H. J., Zhang, Z. Z., Men, X. H., & Luo, Z. Z. (2010). A study of the tribological behavior of nano-ZnO-filled polyurethane composite coatings. *Wear*, *269*(1), 79–85. doi:10.1016/j.wear.2010.03.011

Song, X., Zheng, S., Huang, J., Zhu, P., & Guo, Q. (2000). Miscibility, morphology and fracture toughness of tetrafunctional epoxy resin/poly (styrene-co-acrylonitrile) blends. *Journal of Materials Science*, *35*(22), 5613–5619. doi:10.1023/A:1004824628535

Thakur, V. K., Singha, A. S., & Thakur, M. K. (2012). Green Composites from Natural Fibers: Mechanical and chemical aging properties. *International Journal of Polymer Analysis and. Characterization*, *17*, 401–407.

Wan, Y. Z., Chen, G. C., Raman, S., Xin, J. Y., Li, Q. Y., Huang, H., & Luo, H. L. et al. (2006). Friction and wear behavior of three-dimensional braided carbon fiber/epoxy composites under dry sliding conditions. *Wear*, *260*(9-10), 933–941. doi:10.1016/j.wear.2005.06.007

Wang, L. L., Zhang, L. Q., & Tian, M. (2012). Mechanical and tribological properties of acrylonitrile–butadiene rubber filled with graphite and carbon black. *Materials & Design*, *39*, 450–457. doi:10.1016/j.matdes.2012.02.051

Xie, G. Y., Zhuang, G. S., Sui, G. X., & Yang, R. (2010). Tribological behavior of PEEK / PTFE composites reinforced with potassium titanate whiskers. *Wear*, *268*(3), 424–430. doi:10.1016/j.wear.2009.08.032

You, Y. L., Li, D. X., Si, G. J., Lv, R. Y., Deng, X. (2014). Improvement in the tribological properties of polyamide 6: Talc, glass fiber, graphite, and ultrahigh-molecular-weight polyethylene. *Journal of Thermoplastic Composite Materials.*

Zeng, A., Zheng, Y., Guo, Y., Qiu, S., & Cheng, L. (2012). Effect of tetra-needle-shaped zinc oxide whisker (T-ZnOw) on mechanical properties and crystallization behavior of isotactic polypropylene. *Materials & Design*, *34*, 691–698. doi:10.1016/j.matdes.2011.05.028

Chapter 6
Fabrication and Processing of Pineapple Leaf Fiber Reinforced Composites

S. H. Sheikh Md. Fadzullah
Universiti Teknikal Malaysia Melaka, Malaysia

Zaleha Mustafa
Universiti Teknikal Malaysia Melaka, Malaysia

ABSTRACT

There is an increasing interest worldwide in the use of Pineapple Leaf Fibers (PALF) as reinforcements in polymer composites, since this type of natural fiber exhibit attractive features such as superior mechanical, physical and thermal properties, thus offer potential uses in a spectrum of applications. PALF contains high cellulose content (between 70-82%) and high crystallinity. However, being hydrophilic, it posed a compatibility issue particularly in a hydrophobic polymeric matrix system. Thus, their shortcoming need to be addressed to ensure good interfacial bonding at the fibers/matrix interphase before their full potential can be harnessed. This chapter summarized some of the important aspects relating to PALF and its reinforced composites, particularly the main characteristics of the fiber, extraction and pre-treatment process of the fibers. Following this, discussions on the available fabrication processes for both short and continuous long PALF reinforced composites are presented.

DOI: 10.4018/978-1-5225-0424-5.ch006

INTRODUCTION

This chapter is written to provide an insight about some of the past and recent advances in research and development of the natural fiber reinforced composites, with the focus on pineapple leaf fiber (PALF) and its reinforced composites. The main contents of this chapter are extraction, characterization, modification and fabrication techniques available in the literature as well the properties of various types of PALF reinforced composites. In addition, this chapter also highlights on the limitations and challenges in establishing reliable product from the proposed techniques.

BACKGROUND

To-date, with the growing concern to save and protect the environment, as well as with the noble notion to support sustainable manufacturing or "green" manufacturing, there is an increasing demand in the quest for finding alternative natural resources materials in developing renewable environmental-friendly composites, also known as *biocomposites*, as a substitute to the non-renewable synthetic petroleum-based composites. These efforts aid in minimizing global problems in dealing with the carbon footprint, global warming as well as waste management as a consequence of mass production of synthetic polymer composites worldwide. Such materials can be obtained from either renewable agricultural resources or waste or fully or partially degradable, hence features environmentally sustainable characteristics (Mishra, Mohanty, Drzal, Misra & Hinrichsen, 2004; Mitra, 2014; Smitthipong, Tantatherdtam & Chollakup, 2015).

Other reasons for the overwhelming attention on the natural resources materials are due to large scale agricultural production annually in the world market. As an example, pineapple is the third most important tropical fruit after banana and citrus. In addition, recent works from the last twenty years have shown that these materials can potentially be considered as candidate materials for both structural and non-structural applications, offering desirable or excellent mechanical properties, by tailoring the polymer and or fiber geometry as well as their architectures (Mishra, Mohanty, Drzal, Misra & Hinrichsen, 2004; Chollakup, Tantatherdtam, Ujjin and Sriroth, 2011; Faruk et al., 2012; Danladi and Shu'aib, 2014).

According to Summerscales and Grove (2014), there are three basic resources of natural fibers, which are animal, mineral and plant. The plant-based natural fibers can be further categorized into several basic divisions; these being bast, which is from the stem such as flex, hemp, jute and kenaf, grasses such as bamboo or wheat straw, leaf such as abaca, sisal or pineapple, seeds such as cotton or coir or wood fibers. The structure of plant fibers can be further described to exhibit a hierarchical structure with three main components;

Figure 1. Main classes of natural fiber
(Adapted from Munirah, Rahmat & Hassan, 2007)

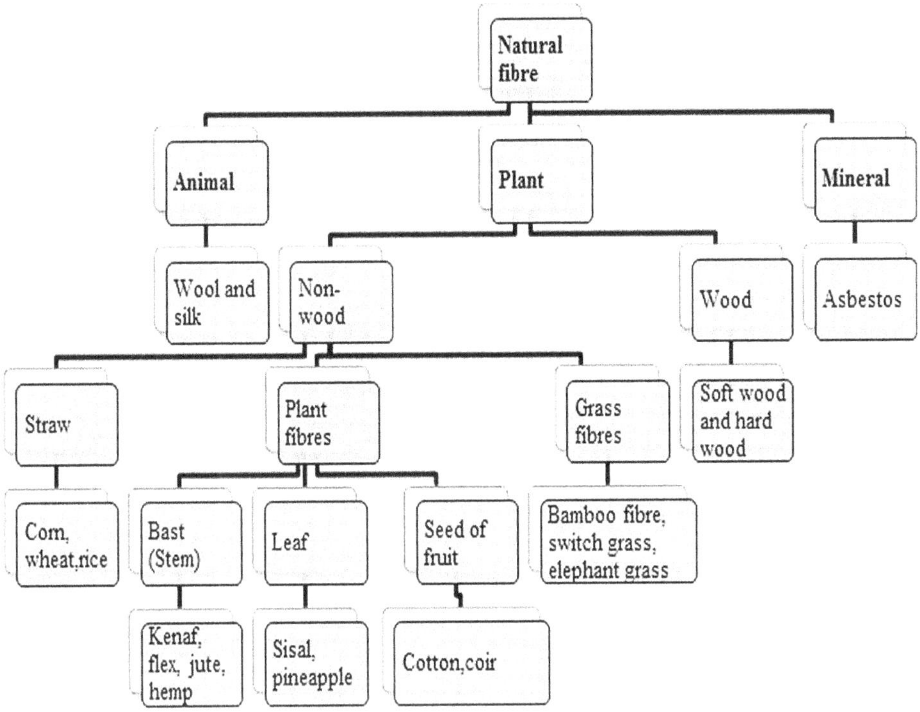

1. **At the Molecular Level:** Cellulose (structural fibers), hemicellulose (the matrix), lignin (accumulate as the plant ages) and pectin (binder that acts as an adhesive at interfaces);
2. Individual cells with a hollow core; and
3. **Cellular Arrays:** Fiber bundles or technical fibers.

The main classes of natural fibers are illustrated in Figure 1. Yu (2015) described cellulose as "the main content of such natural fibers, which is a linear polymer, or long chain molecule, combining several 100 anhydroglucose units", and regarded as the major component of reinforcement fibers.

The properties of a composite are dictated by the intrinsic properties of the constituents, with the main aspect being fiber architecture and fiber matrix interface. In addition, in a composite system, the reinforcing efficiency of natural fibers very much depends on their physical, chemical and mechanical properties. Sapuan et al. (2011) highlighted that some of the major drawbacks of natural plant fibers are "fiber non-uniformity, variation in properties, low degradation temperature, low microbial resistance and susceptibility to rotting". Moreover, they added that fiber extraction and processing techniques also strongly influence the final quality of the fiber and its cost and yield. Another important issue to overcome when using natural fibers with polymer matrices is poor fiber-matrix interfacial adhesion, which could lead to inferior mechanical and other properties of the composites.

PINEAPPLE LEAF FIBRE (PALF)

As mentioned earlier, in the agricultural sector, pineapple (*Ananas comosus*) is reported to be the third most important tropical fruit in the world after banana and citrus. For example, in Malaysia, three species locally grown are the *Queen*, also known as *Moris Gajah*, *Smooth Cayenne*, also known as *Sarawak* pineapple, and *Spanish* or *Josapine*, which is found to be the most appropriate species for PALF extraction in terms of fiber quantity, ease of extraction, fiber fineness, mechanical and thermal properties (Sapuan et. al, 2011). In most cases, the pineapple leaves from the plantations are being wasted as they are cut after the fruits harvested before being either composted or burnt, causing environmental pollution (Sapuan et al, 2011). Traditionally, this material has also been predominantly used as textile materials, i.e. threads and textile fabrics such as dresses, table linens, bags and mat in Philippines and textile materials in Indonesia and Thailand (Sapuan et al., 2011; Kengkhetkit & Amornsakchai, 2014).

Some of the main features of PALF as reinforcement materials in composite systems are:

1. Low density,
2. Low cost,
3. Nonabrasive,
4. Low energy consumption,
5. High specific properties,
6. Biodegradability, and
7. Generation of rural or agricultural economy (Mishra et al., 2004).

In addition, these materials have been considered as the reinforcement material for various types of polymer composite materials, such as polyester, polyethylene, polypropylene, biodegradable polymer such as PLA, as well as thermoplastic starch (TPS) since it contains high cellulose content (approximately 70-82%) and exhibit high degree of crystallinity (Mishra et al., 2004; Chollakup, Tantatherdtam, Ujjin and Sriroth, 2011; Smitthipong, Tantatherdtam and Chollakup, 2015; Danladi and Shu'aib, 2014; Kengkhetkit and Amornsakchai, 2014).The PALF fibers have also been used in fabricating hybrid composites such as kenaf/pineapple leaf reinforced high density polyethylene (Aji et al., 2012).

PALF is described as multi-cellular and lignocelluloses materials, which is extracted from the leaves of the *Ananas cosomus* (belong to *Bromeliaceae* family) via retting process. In general, the leaves are sword-shaped which arise from a stem with an overall dimensions ranging from a length between 0.9 m to 1.5 m, with a width between 2.54 m and 5.1 m respectively. In terms of colors of the leaves, they could come in purely green or with spots of red, yellow or even ivory. Among the good features as a source of fiber is the length of fiber strand, ranging between 7.5 cm to 10 cm long (Danladi and Shu'aib, 2014).

It is essentially important to understand the basic characteristics of the pineapple leaf fiber (PALF) before discussing further into the fabrication and processing aspects relating to its reinforced composites. In general, pineapple leaves have nominal dimensions of length ranging between 70 – 90 cm, while the width is between 5 to 7.5 cm (Sapuan et. al, 2011). Examples of the sword-shaped, dark green pineapple leaves are given in Figure 3 (a), whilst in (b), the extracted PALF bundles from leaves are given. In addition, strong, white and silky luster PALF is shown in Figure 3 (c). However, it must be noted that these type of fibers exhibit a variation in properties based on several factors such as the species, geographical regions, age, location in each plant as well as weather conditions.

Chemical Compositions

The chemical compositions of pineapple leaf fiber as characterized by selected researchers are tabulated in Table 1, showing traces of cellulose, hemicellulose, hollocellulose, alpha cellulose, lignin, pectin, fat and wax, ash and extractive as possible chemical composition.

Physical, Mechanical and Thermal Properties of PALF

Some of the main physical, thermal and mechanical properties of pineapple leaf fiber are tabulated in Table 2. In addition, a summary of selected physical and mechanical properties of bast and leaf fibers are given in Table 3, which provides information about the fiber length (commercially available), length of spinnable fiber, linear density, tenacity and the extension of break.

Table 1. Chemical compositions of pineapple leaf fiber

Chemical Composition (%)	Source		
	Yu (2015)	**Mishra et.al (2004)**	**Munirah, Rahmat & Hassan, 2007[1]**
Cellulose	56 - 62	70 - 82	-
Hemicellulose	16 -19	5-12	9.45
Hollocellulose	-	-	87.56
Alpha cellulose	-	-	78.11
Lignin	9 - 13.0	5 - 12	4.78
Pectin	2.0 - 2.5	-	-
Fat and wax	4.7	-	-
Ash	2.0 - 3.0	-	-

Note:
[1]*Josephine* pineapple

Table 2. Properties of pineapple leaf fiber (PALF)

Properties	Source				
	George et al., 1995	**George et al.,1998**	**Luo and Netravalli*, 1999**	**George et al., 2001**	**Leão et al., 2015***
Physical Properties					
Cell length (mm)	-	-	-	-	-
Diameter (μm)	-				5.0-30.0
Moisture content (%)	-	-	-	-	-
Vicat softening point (°C)	-	104.0	-	-	-
Density (gcm^{-3})	1.526	1.526	-		1.44
Mechanical Properties					
Tensile strength (MPa)	170.0	170.0	-	-	170
Ultimate tensile strength (MPa)	-	-	-	413-1627	-
Young's Modulus (MPa)	6.26	6.26	-	-	6.26
Elongation at break (%)	1.6	3.0	-	0.8 - 1	1.6
Flexural Modulus	-	-	-	0.24 - 0.40	-

Table 3. Mechanical properties of bast and leaf fibers

Type of Plant Fiber	Length of Commercial Fiber (mm)	Length of Spinnable Fiber (mm)	Linear Density (tex)	Tenacity (cN/dtex)	Extension at Break (%)
Jute	750 -1500	60 -150	1.5 - 4.5	1.0 - 3.5	1.8 - 2.5
Kenaf	750 -1500	80 -150	1.9 - 6.0	3.0	1.6 - 2.3
Flax	700 - 900	60 -150 (wet spun)	1.2 - 2.5	3.5 - 5.5	2.0 - 5.0
Ramie	800 - 1300	70 - 100	0.3 - 0.6	4.0 - 6.0	2.0 - 6.0
Hemp	2500 (long hemp)	50 -150	0.3 - 2.2	3.5 - 5.8	2.0 - 5.0
Pineapple	600 - 1000	30 - 80	1.5 - 2.3	3.5	2 - 4.8
Sisal	800 -1200	500 - 1000	12 - 20	2.0 - 5.8	1.8 - 3.5

(Adapted from Yu, 2015)

EXTRACTION OF PINEAPPLE LEAF FIBRE (PALF)

It is essentially important to understand the extraction of pineapple leaf fibers before discussing further into the fabrication processes involved in producing the reinforced composites. Traditional extracting techniques of the PALF are by using hand scraping, decortications and retting (Danladi and Shu'aib, 2014).

It is more common to find that the PALF leaves are extracted from the plant by hand, which involve stripping off of the fiber from the retted leaves, hence leading to several constraints; labor intensive and costly process in addition to affecting the properties of the fiber. Due to these problems, extraction by machine have been developed and used, although such processes is relatively much slower than hand scrapping but aid in production processes (Mwaikambo, 2006). However, mechanical processes may introduce damage to the natural fibers through breaking, scotching and hackling actions which resulted in reduction in tensile strength properties of the fibers in comparison to those of the elementary fibers produced manually.

Asim et al., (2015) described the extraction process of the pineapple leaves as a process that involves the scrapping, retting, followed by mechanical separation of the fibers prior to drying in air. During the scrapping process, the pineapple leaves are fed through a feed roller, passing through the second roller called scratching roller in which the upper leaf is scratched and the wax layer is removed. Lastly, the leaves will enter the area where the serrated roller is position for the leaves to be crushed and makes several breaks for the entry passage for the retting microbes. They further described that following the scarping process, the scratched pineapple leaves are immersed in a water tank containing substrate: liquor in 1:20 ratio and 5%

of urea or diammonium phosphate (DAP) for fast reaction during the retting process. Consequently, the fibers are segregated mechanically during rinsing process in pond water prior to drying in the air. In addition, there are two ways in which PALF can be extracted; these being ball mill and disc mall, with the wet ball milling offering greater number of elementary fiber although the speed of process is much slower. Such extraction methodology is claimed to be simpler and able to produce high fiber yield and smaller fiber relative to conventional methods.

An example of a mechanical extraction process is the work reported by Munirah, Rahmat & Hassan (2007) in which the waste pineapple leaves collected were collected from Sedenak and Pekan Nanas, Johore, Malaysia during the harvest process. The leaves were then pressed using a two-roll milling machine to remove circa that is about 90% of the water content. The fibers were then extracted manually from the semi-dried pineapple leaves using knife or sharp tool. Following these, the short fibers undergone cleaning process where they were washed thoroughly in 2% detergent solution at temperatures of approximately 70°C followed by rinsing in tap water. The purpose of this cleaning process is to remove foreign objects and impurities present in the fibers. This is followed by drying process in the oven at 70°C for 24 hours before undergoing further processing and characterization.

More recently, Yusof, Yahya and Adam (2015) have proposed a novel sustainable technology in extracting the pineapple leaf fibers as well as fabricating the reinforced composites. They have developed a customize decortication machine called Pineapple Leaf Fiber Machine 1 (PALF M1) in Universiti Tun Hussein Onn (UTHM), Batu Pahat, Johore, Malaysia to extract the pineapple leaf fiber in a more efficient and sustainable way. The special features of this machine are that it used blades instead of crusher technology in removing the waxy layer on the pineapple leaf. In addition, they have also considered specific number of blades used, sizes and certain angle of the two blades required to ensure that the leaf will not snap during the extraction process.

PRE-TREATMENT OF THE PINEAPPLE LEAF FIBRE

To-date, the development of natural fibers reinforced composites or biocomposites does not progress as rapidly as its competing synthetic glass fiber due to two main major problems; which are poor moisture resistance and low impact strength. Moreover, fabricating natural fiber reinforced composites may involve bringing together two materials that are incompatible, since natural fibers are *hydrophilic* while the matrix systems may be *hydrophobic*. Therefore, to overcome this compatibility issue, it is a common practice to introduce surface treatment of the fibers that can change the interfacial chemistry or physical properties of the fibers, in order to enhance the composite system performance. Surface treatments of the raw natural fibers may include acetylation, biological treatments, bleaching, grafting, *mercerization,* oxidation, plasma treatment and *scouring*.

As an example, scouring which is a solvent treatment process, as discussed previously in the extraction process of PALF aims to remove natural fats, waxes, proteins and other constituents such as dirt, oil and other impurities. In addition, it may also be necessary to consider the use of coupling agents such as organometallic silicates, titanates and zirconates to improve interfacial bonding (Summerscale and Grove, 2014).

Lopattananon et al. (2006) studied the effect of the fiber surface treatment on the performance of pineapple leaf fiber–natural rubber composites using Sodium hydroxide (NaOH) solutions (1, 3, 5, and 7% w/v) and benzoyl peroxide (BPO) (1, 3, and 5 wt. % of fiber) to treat the surfaces of PALFs. The chopped PALF fibers and natural rubber composite were fabricated using two roll mill at room temperature and care were taken to ensure the alignment and uniformity of fibers dispersion. They reported that significant improvement in the mechanical properties of the composite with optimum condition is given by usage of 5% NaOH and 1% BPO of 28 and 57% respectively in comparison to the untreated fiber.

George et al. (1998) have summarized how the chemical treatment work to improve the wettability of fiber/matrix interphase in PALF reinforced LDPE composite. Alkaline treatment such as NaOH, work by reducing the polarity of the PALF fibers thus prevents the penetration of the water molecule. The enhance bonding in silane treatment due to interaction of functional group of coupling agent with hydroxyl group of cellulose. The resulting –OH group will then bond with hydroxyl group of the cellulose and formed a long hydrophobic polymer chain of polymerized silane which in turn adhere to matrix through the van de Waals's bonding, producing strong interfacial bonding at the fiber/matrix interphase. In case of Isocyanate treatment (PMPPIC), the strong formation is resulting from covalent bonds between –OH group of cellulose and the –NCO groups from PMPPIC. These long chain molecules interact with LDPE through van de Waals's interaction creating a better adhesion fiber/matrix.

Often more combination of two types of treatment was employed to improve the adhesion of the fiber/matrix. Siregar et al. (2012) reported that by combining the alkaline treatment and poly (styrene-co-maleic anhydride) compatibilising agent in 50% PLAF reinforced high impact polystyrene (HIPS) gave a tremendous improvement on their tensile strength. Alkaline treatment removed the natural and artificial impurities which in turn increased wettability of the alkali treated fibers with the matrix. Mwaikambo and Ansell (2003) commented that by removing of the impurities created a rougher surface and help to form a better mechanical interlocking and increased the interfacial adhesion. It is also noted that by adding 2 wt. % poly(styrene-co-maleic anhydride) coupling agents, the composite strength increased by about 34 MPa, approximately by 48% in comparison to the untreated fibers. During fabrication, the chemical composition from the compatibilising agent interacts with the fibers and forming a bridge of chemical bonds between the fiber and matrix resulting in increased in their mechanical properties.

Lopattananon et al. (2006) studied the effect of the fibers surface treatment on the performance of pineapple leaf fiber–natural rubber composites using sodium hydroxide (NaOH) solutions (1, 3, 5, and 7% w/v) and benzoyl peroxide (BPO) (1, 3, and 5 wt. % of fiber) to treat the surfaces of PALFs. The chopped PALF fibers and natural rubber composite were fabricated using two roll mill at room temperature and care were taken to ensure the alignment and uniformity of fibers dispersion. They reported that significant improvement in the mechanical properties of the composite with optimum condition is given by usage of 5% NaOH and 1% BPO of 28 and 57% respectively in comparison to the untreated fiber.

Munirah et al. (2007) suggested that alkaline treatment not only removed the natural and artificial impurities of the fibers but at the same time destroyed the hydroxyl group in the cellulose make them more susceptible to react with functional group from the coupling agents resulting in better adhesion with the polymer matrix.

Mohanty et al. (1996, 2000) have studied the graft copolymerization of acrylonitrile (AN) onto chemically modified PALF. Ce (IV) in combination with N-acetyl glycine (NAG) has been used as an initiator in the temperature range 40–60°C. While Tripathy et al. (1999) investigated the Cu (II)–KIO_4 initiated graft copolymerization of MMA from defatted PALF. They compared various factors such as variation of time and temperature, concentration of Cu (II), KIO_4, and MMA, the amount of PALF, inorganic salts and organic solvents to the graft yield. Results from these studies showed that grafting substantially improve the thermal stability of the fibers.

Huda et al. (2008) studied the effect of modifying surfaces of the PALF using both alkaline treatment and silane treatment on the overall mechanical properties of the PALF reinforced PLA biocomposites. For the alkaline treatment, sodium hydroxide (5 wt. % w/v) was used in which the PALF were immersed in the solution for 2 hours at room temperature. Following this, the PALF fibers were washed with distilled water which contained a few drops of acetic acid and distilled water until the sodium hydroxide was removed. Consequently, the fiber underwent air drying for 2 days and lastly, the fibers were oven dried at 80°C for six hours before vacuum oven dried at the same temperature and duration of time. They claimed that through this treatment will roughened topography of the fiber surface hence improve the adhesion between the fiber and the polymer matrix.

The second surface treatment considered in the study by Huda et al. (2008) is using silane treatment. Here, 5 wt. % silane (3-APS) was hydrolyzed in a mixture of water and ethanol with the ratio of 40:60 w/w. Such chemical treatment is expected to initiate bonding between the silanol groups with the PALF fiber surface, which in simple words function as connector molecules between the matrix and the fiber. In addition, they also investigated the effect of using both alkaline and silane treatment on the mechanical properties of the biocomposites. From the experimental work, it has been reported that both silane-treated fiber reinforced composites and alkaline treated fiber reinforced composites offered superior mechanical properties compared to untreated fiber reinforced composites.

FABRICATION OF PINEAPPLE LEAF FIBRE (PALF) REINFORCED COMPOSITES

There are various approaches in which fabrication process of the PALF composites have been discussed; these include processing method based on matrix materials such as thermoplastic, thermoset, rubber as well as based on techniques such as melt-mixing, compression molding or solution mixing. In this section, however, available fabrication process of PALF reinforced composites are discussed based on the fiber architecture, particularly its geometry, which is classified as short and continuous long fibers, with the latter having length of more than 100 mm.

Short Fiber PALF Reinforced Composites

Short fiber PALF reinforced composites can be produced using various methods depending on the types of matrix materials. In general, thermoplastic matrix composites can be produce by adopting one or two techniques such as solvent casting such as prepregging, melt mixing (e.g. extrusion) and melt forming such as injection molding, compression molding, thermoforming and etc.

Extrusion process is often used for melt mixing the raw materials such as resin/matrix with the reinforcement being in the form of particulate or short fibers. During the extrusion process, raw materials (often supplied in pellet form) and reinforcement are fed into the hopper and the motor-driven screw will then extrude the material through the nozzle. Shape of the extruded materials can be controlled by carefully selecting the type of die required. The temperature of the extrusion head is specifically controlled and the temperature profile typically increases from the feed zone to the die. This is to allow sufficient melting time and prevent from thermal degradation of the mixer (Zhou et al., 2012). Major drawback of this method is it only limited to fibers with the length up to 100 mm.

Injection molding process is a well-known process for thermoplastic-based resin; normally use to produce plastic materials with no reinforcement. However, this method can be adopted to produce short fibers composites when stiffness is of important requirement. Generally, this process starts with the resin in the pellet form containing short fibers is fed into the hopper and passed through to the barrel of the machine. In return, the screw pushes the materials are further forward into the barrel and squeezing it to remove any entrapped gasses. The resin is often transformed into liquid form due to friction and shear induced by the screw heats before it reaches the injection chamber. Mechanical shearing cause splitting of the fibers and reduced their length. Following this, the materials are pushed forward by the screw and injected into the mold. The mold is then cooled to solidify the molded component before it is being injected out.

Another method that is quite popular to form the final shape of the composites is using compression molding. The process begins with placing the mixture containing matrix and reinforcement materials onto the bottom half of preheated cavity mold. This process also enables production of laminated short fiber to be carried out. The mold then is closed and top half is lower and the pressure is increase to the pre-set level. With increasing pressure, the materials start to melt and flow into the cavity. The mold then is then cool down and the product is removed from the mold.

Often, two-step methods are adopted to produce composites. The first stage is mixing method which can be done through melt mixing using an internal mixer or rotor mill or via solution mixing where the matrix is dissolve in solvent. This is followed by forming method such as injection molding or compression molding to form the composites at the desired shape.

George et al. (1995) has successfully produced short pineapple-leaf-fiber-(PALF)-reinforced low-density polyethylene (LDPE) composites using melt-mixing and solution-mixing methods. Prior to mixing process, short fibers with different length of 2, 6, and 10 mm were cleaned with water and dried in oven for 24 hours at 70°C. The process of melt mixing was carried out using an internal mixer at different rotor speed, mixing time and chamber temperature. Meanwhile, for the solution-mixing method, the LDPE were dissolved in toluene to form viscous slurry. The short fibers were then added to the slurry and stirred to ensure a good coating and were later dried at room temperature to evaporate off the solvent. Both of the composites were extruded at 120°C. The profile were directly extruded into the mold for random fibers orientation composite, while for controlled fiber orientation composite, the extruded profile were carefully aligning in a mold followed by compression molding at 120°C. They also reported that for the melt mixing method, the optimum fabrication conditions were achieved at mixing time of 6 minutes with the rotor speed of 60 rpm, and mixing temperature of 130°C. Morphology analysis revealed that a higher degree of damage such as splitting and peeling to the fibers occurred at higher rotor speed resulting in declining of the mechanical properties. Composites (f similar fiber loading) produced by solvent mixing exhibited better mechanical properties in comparison to those produced by melt mixing process.

They argued that such observation is possibly due to better dispersion of the fibers in the matrix, low fibers damage and better adhesion were achieved with solvent mixing. They also reported that only 5% of 6 mm of fibers length were retained using melt mixing in comparison to the 70% produced by solution mixing. Reductions in fiber length are the direct result of damage occurred during mixing which lead to lower mechanical properties.

In another study, Liu et al. (2005) considered two processing methods to produce soy-based bioplastic and pineapple leaf fiber composites, using twin-screw extrusion followed by injection molding, with the focus on assessing the effect of different fiber loading and polyester amide grafted glycidyl methacrylate (PEA-g-GMA) as compatibilizer on the properties of pineapple leaf fiber reinforced soy based biocomposites. Morphological study using ESEM revealed two important observations:

1. Morphological change in the PALF fiber bundles as an effect of the processing which changes the fiber diameter hence resulting in an improved fiber dispersion in matrix and the improved fiber reinforcement efficiency, and
2. Dispersion of the fibers is poorer at high fiber loading (30 wt. %), however this is not in the case when compatibilizer was used. They suggested that such observations are possibly due to interactions between epoxy groups in PEA-g-GMA and hydroxyl groups in the pineapple leaf.

Similar techniques were used by Threepopnatkul et al. (2009) to produced PALF reinforced composite with fibers length of 2 mm. Prior to the fabrication process, the fibers surface were treated using 1% w/w silane (weight percentage compared to the fiber) in which was dissolved for hydrolysis in methanol solution with pH was adjusted to 3.5 using acetic acid. During the surface treatment, the fibers were soaked in the solution for 6 hours followed by air dried at room temperature. Both untreated and silane-treated PALFs were mixed with PC using diisononyl phthalate 3% w/w (weight percentage compared to PC) internal mixer, at temperature of 230 °C with rotor speed of 10 rpm for 3–6 min. In this study, four different fiber concentration (0, 5, 10 and 20 wt. %) were used. The mixtures were then injected using injection molding to produce the specimen for mechanical testing.

In another study, Kengkhetkit and Amornsakchai (2014) produced PALF reinforced polypropylene composite using two steps process. Initially, the matrix and fibers were mixed using two-roll mill for 15 min by keeping temperatures of the front and back rollers at 185 °C and 175 °C, respectively. The mixture then was stretched to form unidirectional prepreg. Later, the prepreg was then compression molded at a temperature of 195 °C under a pressure of 3.5 MPa for 5 minutes, followed by cooling under the pressure of 7.0 MPa for 3 minutes to form sheets with thickness of 1 and 3 mm respectively.

Short fiber reinforced rubber composite have exceptional mechanical properties due to unique combination of fiber rigidity and rubber elasticity. More recently, Prukkaewkanjana et al. (2015) developed rubber composites that exhibit a very high modulus at low elongation, high elongation at break and strong. In this study, they produced hybrid composites by combining short fine pineapple leaf and carbon black. The mixture with fibers aligned to the machined direction was produced using a laboratory two-roll mill compounder machine. The amount of PALF was fixed at 10 parts (by weight) per hundred of rubber (phr) in combination of various loading carbon black (0 to 30 phr). The compound later vulcanized using compression molding at temperature of 150 °C, under pressure of 1500 psi using a 1mm thick mold. The tensile properties were measured in both longitudinal and transverse to the fiber axis directions. It was reported that there is an increase in the slope in early part of the stress–strain curve with addition of carbon black, which simultaneously increase the strain at break. At carbon black contents of 20 phr and above, the stress–strain relation displayed an improvement at high elongations, providing greater ultimate strength. They also concluded that the composite behavior at low strains is governed by the PALF and at high strains by the carbon black. These high performance PALF-carbon black reinforced NBR composites offers great potential for engineering applications.

Often, when non-woven mat or long fibers are involved, due the limitation of the fibers length, the fabrication process is simplified by using only compression molding. Chollakup, Tantatherdtam, Ujjin and Sriroth, (2011) have study the effects of PALF length and content in LDPE and PP matrices. The PALF reinforced composites were produced by using hot compression molding technique. The LDPE and PP matrix were pressed a temperatures of 180°C and 240°C respectively, with a constant pressure of 12.5 MPa for 3 minutes and then cooled down to room temperature to produce 0.25 mm polymeric sheet film. The PALF in the form of nonwoven fibers mat with fibers length of 40 mm were then sandwiched in between the thin films and compressed at melting temperatures of 160°C and 185°C respectively to produce 0.5 mm composites. The composites performance was compared with similar composites containing orientated long fibers produced by similar method. The tensile test data analysis showed that the strength of the composites is increase with increasing fibers content up to 25%. In relation to fiber length, the strength of the composite (with fiber loading of 15 wt. %) containing long fibers is higher by more than 4 fold. SEM micrographs show evidence of good dispersion and adhesion achieved in long fibers better than short fibers using this fabrication techniques indicating successful fabrication methods in producing short and long fibers reinforced composites.

The interest to produce bio-based polymers made from degradable sources such as starch, poly (lactic acid) (PLA), poly (butylene succinate) (PBS) are gaining increasing attention among researchers due to sustainability and environmental issues. As an example, PLA is a versatile bioplastic; however it has limited application because of its brittleness.

To overcome these problems, Huda et al. (2008) investigated the effect of the chemical treatments on the interfacial surfaces and mechanical properties of the laminated composites. PALF reinforced PLA composites were produced using compression molding process. Prior to fabrication of the composites, the PALF fibers (with fiber length of 18 – 24 mm) were chemically treated either using alkaline treatment using 5 wt.% NaOH or silane treatment using 5 wt.% silane using 3- Amino propyl triethanoxy silane (3-APS). In addition, the PLA pellets were air dried at 80°C under vacuum for 10 hours. Following this, the PLA pellets were compression molded using a picture frame mold with dimensions of 250 mm x 200 mm to produce 1-mm thick thin PLA films. Here, the temperature used was at 190°C and a pressure of approximately 260 psi for 10 minutes of contact time, and the pressure was increased up to a maximum of 624 psi at 100°C. The PALF composites of various fibers loading were fabricated using a film-stacking procedure using compression molding machine, to produce samples with dimensions of 180 mm x 140 mm with nominal thickness of 3 mm prior to material characterization tests. Thermal analysis via thermogravimetric (TGA) analysis of the pre-treated PALF fiber showed that both chemical treatments have improved thermal performance of the fibers, which together with an improvement in the fiber-matrix adhesion of the composite system led to enhanced mechanical properties of the composites. In addition, from mechanical testing of the composites, both the silane-treated PALF reinforced composites and alkali-treated PALF fiber reinforced composites exhibit superior mechanical properties in comparison to those of the untreated PALF fiber reinforced composites. These findings also suggest that manufacturing laminated composites using surface modified or unmodified PALF reinforcement for the PLA matrix with superior mechanical and thermal properties can be successfully achieved using the proposed techniques.

Recently, Smitthipong, Tantatherdtam and Chollakup (2015) manufactured the PALF reinforced TPS/PLA composites via blending the component using single screw extrude. Prior to blending, the TPS was prepared by mixing of cassava starch, glycerol and water as a plasticizer with ratio of 65 wt.% cassava starch, 26 wt.% glycerol, and 9 wt.% water manually. PALF short fibers of 1-mm length were then added at 2–10 wt. % of the total TPS. The blend was extruded and pelletized. Consequently, the PLAF/TPS was blended with PLA (0, 20, 40, 60 and 80 wt. %) using single screw extruder at 180°C and 180 r/min. The extruded profiles were then hot compression molded at 95°C and 180°C, respectively, with a pressure load of 15 MPa. Viscosity, water solubility as well as tensile properties of the composite were measure. The results showed that using short pineapple fibers in thermoplastic starch (TPS) and PLA blends improved the mechano-static properties as well as water resistance of the composites.

Another problem in using biopolymer is their inferior mechanical properties in comparison to their petroleum-based counterpart. Luo and Netravali (1995, 1999) reported that a low value of interfacial shear stress was observed in PALF/poly (hydroxybutyrate-co-valerate) (PHBV) composite. The compressed molded the sandwiching of three layers of PALF fibers in between a four layers of PHBV film at temperature and pressure of 180°C and 140 MPa respectively. Interestingly, by adding 20 wt. % unidirectional PALF substantially increased the tensile and flexural strength of the composite by 75% and 32%, respectively. Additional of another 10 wt. % fibers content lead to significant increase of tensile and flexural strengths by 100% and 60%, respectively.

Hybrid biocomposites are fabricated to take advantage of individual fibers or overcome each other's short coming. Hybrid biocomposites are combinations of two or more different types of fibers in a single matrix (Sapuan et al, 2011). Idicula et al. (2006) investigated the thermo-physical properties PALF/ glass fiber-reinforced polyester composites. Initially both PALF and glass fibers were cut at a uniform length of 30 mm. The specimen was then prepared by combining the PALF and glass fibers at a volume ratio of 1:1 and impregnating the fiber with the polyester resin. The liquid resin was degassed before poured into close mold and kept under pressure for 12 hours. Post curing was carried out for another 48 hours at 30°C. Following this, the samples were cut to desired size. They reported that hybridization of natural fibers with glass allow a significantly improvement in heat transport ability of the composites.

Long Fiber PALF Reinforced Composite

To-date, there is still limited amount of published work on the continuous long PALF fiber reinforced composites.

Arib et al. (2006) manufactured continuous PALF fiber reinforced polypropylene composites using compression molding technique by laying down one layer of PALF fibers between two layers of thin polypropylene films in parallel array onto a stainless steel mold platen using a carver press machine with a temperature and pressure control. Here, heating was started immediately and a pressure of 12.4 MPa was introduced when the temperature reached 180°C, with a soaking time of 5 minutes. Following this, the stainless steel mold platen was removed quickly from the press for cooling process to take place, with an aid of a cooling fan to expedite the cooling process. Consequently, 3-mm thick composite laminates produced from this process were characterized mechanically via tensile testing as per ASTM-D638 and three-point bending testing as per ASTM- D790, as well as surface analyzed using a scanning electron microscope (SEM). It has been observed that even with small fiber loading (approximately 2.7 by volume fraction) in the composite system showed enhancement in both the tensile and flexural properties in comparison to that of the plain polymer. However, further work on improving the fiber-matrix bond strength and achieving homogeneous fiber distribution are required to significantly improve the overall composite performance as potential substitute to synthetic polymer composites such as glass fiber.

Chollakup, Tantatherdtam, Ujjin and Sriroth (2011) studied the effects of fiber length and fiber content on the PALF reinforced thermoplastic composites. In their study, both the short and long PALF fibers, with fiber final length being 40 mm and 140 mm respectively were considered as reinforcement for LDPE and PP thermoplastic materials. Fiber mats with different directions, in which the fibers were aligned along the leaf length direction were produced. The long PALF fiber was arranged separately in parallel arrays which covered a 100 mm x 100 mm stainless steel plate, whereas the short PALF fibers were randomly arranged by means of a mini card machine to form a non-woven mat. For the plain thermoplastic polymer matrices, both the LDPE and the PP thin films with nominal thickness of 3 mm were prepared by a heat press machine at temperatures of 180° and 240° respectively, with an applied pressure of 12.5 MPa for 3 minutes. To minimize manufacturing defects such as voids and air bubbles, the specimens were repeatedly depressurized and pressurized at each melting temperature for 3 minutes. Following these, slow cooling to room temperature at a pressure of 12.5 MPa for duration of 30 minutes, to produce thin films with nominal thickness of 0.25 mm. Composite laminates with thickness of 0.5 mm were produced by laying up one layer of the fiber mat between two layer of either the LDPE or the PP sheets at the melting temperature of 160°C and 185°C respectively. It was observed that the long PALF fibers reinforced composites exhibit superior tensile strength in comparison to those of the short fiber reinforced composites. In addition, SEM analysis showed evidence of good dispersion of the long fiber in the composites.

More recently, Yaacob (2015) manufactured continuous long alkaline-treated PALF fiber reinforced PLA composites via compression molding process in which the pineapple leaf fibers (extracted from *Josapine* pineapple) with fiber weight fraction of 30 wt.% and fiber length of more than 100 mm were pre-treated using 5 wt.% sodium hydroxide. Here, care was taken while handling the PALF fibers to ensure significant improvement in the mechanical properties of the reinforced PLA composites. The treated PALF fiber bundles were air dried for 48 hours before stored in a desiccator prior to usage and brushed to ensure good fiber alignment prior to lamination process in which the PALF fibers were laid down in between two layers of thin PLA films with nominal thickness of 1 mm onto a stainless steel mold platen and compression molded with an applied pressure and temperature before slow cooling to room temperature and demolding process. The 3-mm thick PALF fiber reinforced PLA laminated composites were further characterized via Tensile Testing as per ASTM D3039, three-point bending test as per ASTM D790 and Charpy Impact test as per ASTM D6110 in addition to morphological study via scanning electron microscope (SEM). It has been reported that there is a significant improvement in the tensile properties, i.e. modulus of elasticity of the reinforced composites is approximately 60% higher than the plain PLA. In addition to that, the modulus of elasticity following three-point bending test showed superior performance of the PALF reinforced PLA composites, with an increase of approximately 32% in comparison to that of the plain PLA. An example of the cross-sectional view of the PALF reinforced PLA composites following Charpy Impact test is shown in Figure 2 whilst a summary of the main findings are tabulated in Table 4.

Figure 2. SEM micrograph showing cross-sectional area of the PALF fiber reinforced PLA biocomposites following tensile testing (500X)
(Adapted from Yaacob, 2015)

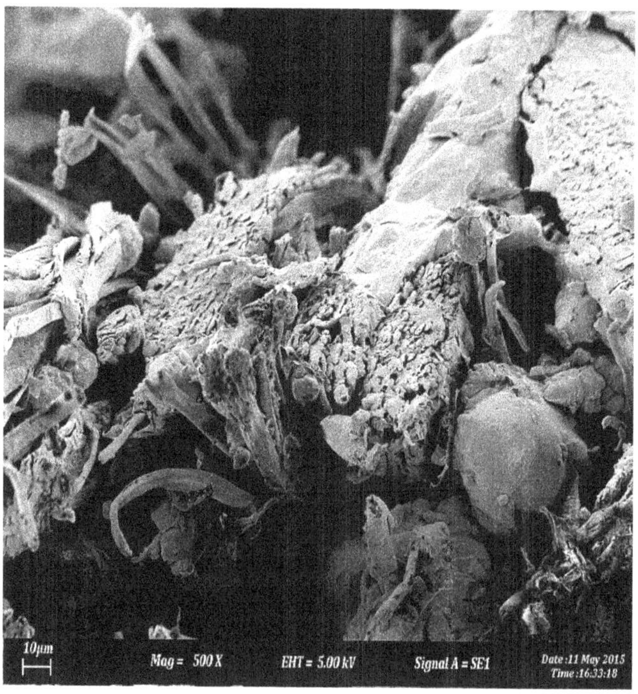

Table 4. Summary of main findings from experimental work

Parameter	Plain PLA	PALF-Reinforced Composites
	Tensile	
Strength (MPa)	34.85 ± 4.3	73.26 ±11.0
Modulus (MPa)	1641.12 ± 98.6	2735.36 ± 510.0
	Flexural	
Strength (MPa)	88.25 ± 10.7	44.40 ±15.4
Modulus (MPa)	1649.33 ± 550.2	3168.73 ± 1603.6
	Impact	
Energy absorption (J/cm^2)	0.35 ± 0.05	0.92 ± 0.13

(Adapted from Yaacob, 2015)

SOLUTIONS AND RECOMMENDATIONS

Past and present work in the study of pineapple leaf fibers and their reinforced composites have revealed attractive features of such natural fiber reinforced bio-composites for useful applications in widespread applications, which include both structural and non-structural applications as substitute materials to those of the non-renewable petroleum-based polymer composites available. However, some constraints that must be highlighted and to be dealt with include constant supply, quality control, reliable and replicable standards to establish PALF reinforced composites for commercial uses.

CONCLUSION

Pineapple leaf fibers (PALF) and its reinforced composites exhibit attractive features such as superior mechanical, physical and thermal properties, thus offer potential uses in a spectrum of applications. PALF contains high cellulose content (between 70-82%) and high crystallinity. However, the aforementioned shortcomings need to be addressed carefully to ensure good interfacial bonding at the fibers/matrix interphase before their full potential can be harnessed and utilize as a substitute to non-renewable petroleum based polymer composites. This chapter has summarized some of the important aspects relating to PALF and its reinforced composites, particularly focusing on some important aspects of the PALF itself, extraction and pre-treatment process of the fibers. Following this, discussions on the available fabrication processes for both short and continuous long PALF reinforced composites have been discussed at some length. Finally, brief discussions on solutions and recommendations to known fabrication problems and limitations have been proposed for future works in establishing sustainable manufacturing of the PALF reinforced composites for commercial uses.

FUTURE RESEARCH DIRECTIONS

Natural fibers can potentially be the replacement materials for the non-renewable synthetic fibers; however, to-date, there are still high variability in their properties. Thus, to overcome this shortcoming, future research shall be directed to look into issues relating to fiber extraction to produce better quality technical fibers with minimal variability, with reduced moisture absorption to ensure long term durability and reliability, optimization of fiber modification to enhance the fiber/matrix interfacial bonding, hybridization and optimization of composite processing to suit specific product requirement as well as a thorough study on degradation and life cycle assessment, expanding research on incorporating organic and in organic nano filler into natural fiber reinforced composite and other.

REFERENCES

Aji, I., Zainudin, E., Abdan, K., Sapuan, S., & Khairul, M. (2012). Mechanical properties and water absorption behavior of hybridized kenaf/pineapple leaf fiber-reinforced high-density polyethylene composite. *Journal of Composite Materials*, *47*(8), 979–990. doi:10.1177/0021998312444147

Arib, R. M. N., Sapuan, S. M., Ahmad, M. M. H. M., Paridah, M. T., & Zaman, H. K. (2006). Mechanical properties of pineapple leaf fiber reinforced polypropylene composites. *Materials & Design*, *27*(5), 391–396. doi:10.1016/j.matdes.2004.11.009

Asim, M., Abdan, K., Jawaid, M., Nasir, M., Dashtizadeh, Z., Ishak, M. R., & Hoque, M. E. (2015). A Review on Pineapple Leaves Fibre and Its Composites. *International Journal of Polymer Science*, 2015.

Chollakup, R., Tantatherdtam, R., Ujjin, S., & Sriroth, K. (2011). Pineapple leaf fiber reinforced thermoplastic composites: Effects of fiber length and fiber content on their characteristics. *Journal of Applied Polymer Science*, *119*(4), 1952–1960. doi:10.1002/app.32910

Danladi, A., & Shu'aib, J. (2014). Fabrication and Properties of Pineapple Fibre/High Density Polyethylene Composites. *American Journal of Materials Science*, *4*(3), 139–143.

Faruk, O., Bledzki, A. K., Fink, H. P., & Sain, M. (2012). Biocomposites reinforced with natural fibers: 2000–2010. *Progress in Polymer Science*, *37*(11), 1552–1596. doi:10.1016/j.progpolymsci.2012.04.003

George, J., Bhagawan, S. S., Prabhakaran, N., & Thomas, S. (1995). Short pine-apple-leaf-fiber-reinforced low-density polyethylene composites. *Journal of Applied Polymer Science*, *57*(7), 843–854. doi:10.1002/app.1995.070570708

George, J., Bhagawan, S. S., & Thomas, S. (1998). Effects of environment on the properties of low-density polyethylene composites reinforced with pineapple-leaf fiber. *Composites Science and Technology*, *58*(9), 1471–1485. doi:10.1016/S0266-3538(97)00161-9

George, J., Sreekala, M. S., & Thomas, S. (2001). A Review on Interface Modification and Characterization. *Polymer Engineering and Science*, *41*(9), 1471–1485. doi:10.1002/pen.10846

Huda, M. S., Drzal, L. T., Mohanty, A. K., & Misra, M. (2008). Effect of chemical modifications of the pineapple leaf fiber surfaces on the interfacial and mechanical properties of laminated biocomposites. *Composite Interfaces*, *15*(2-3), 169–191. doi:10.1163/156855408783810920

Idicula, M., Boudenne, A., Umadevi, L., Ibos, L., Candau, Y., & Thomas, S. (2006). Thermophysical properties of natural fibre reinforced polyester composites. *Composites Science and Technology*, *66*(15), 2719–2725. doi:10.1016/j.compscitech.2006.03.007

Kengkhetkit, N., & Amornsakchai, T. (2014). A new approach to "greening" plastic composites using pineapple leaf waste for performance and cost effectiveness. *Materials & Design*, *55*, 292–299. doi:10.1016/j.matdes.2013.10.005

Leão, A., Cherian, B., Narine, S., Souza, S., Sain, M., & Thomas, S. (2015). The use of pineapple leaf fibers (PALFs) as reinforcements in composites. Biofiber Reinforcements. *Composite Materials*, 211-235.

Liu, W., Misra, M., Askeland, P., Drzal, L. T., & Mohanty, A. K. (2005). 'Green' composites from soy based plastic and pineapple leaf fiber: Fabrication and properties evaluation. *Polymer*, *46*(8), 2710–2721. doi:10.1016/j.polymer.2005.01.027

Lopattananon, N., Panawarangkul, K., Sahakaro, K., & Ellis, B. (2006). Performance of pineapple leaf fiber–natural rubber composites: The effect of fiber surface treatments. *Journal of Applied Polymer Science*, *102*(2), 1974–1984. doi:10.1002/app.24584

Luo, S., & Netravali, A. N. (1999). Mechanical and thermal properties of environmental-friendly" green" composites made from pineapple leaf fibers and poly (hydroxybutyrate-co-valerate) resin. *Polymer Composites*, *20*(3), 367–378. doi:10.1002/pc.10363

Mishra, S., Mohanty, A. K., Drzal, L. T., Misra, M., & Hinrichsen, G. (2004). A review on pineapple leaf fibers, sisal fibers and their biocomposites. *Macromolecular Materials and Engineering*, *289*(11), 955–974. doi:10.1002/mame.200400132

Mitra, B. C. (2014). Environment Friendly composite materials: Biocomposites and Green composites. *Defence Science Journal*, *64*(3), 244–261. doi:10.14429/dsj.64.7323

Munirah, M., Rahmat, A. R., & Hassan, A. (2007). *Characterization and Treatments of Pineapple Leaf Fiber Thermoplastic Composite for Construction Application*. Universiti Teknologi Malaysia.

Mwaikambo, L. Y., & Ansell, M. P. (2003). Hemp fiber reinforced cashew nut shell liquid composites. *Composites Science and Technology*, *63*(9), 1297–1305. doi:10.1016/S0266-3538(03)00101-5

Prukkaewkanjana, K., Thanawan, S., & Amornsakchai, T. (2015). High Performance Hybrid Reinforcement of Nitrile Rubber using Short Pineapple Leaf Fiber and Carbon Black. *Polymer Testing*, *45*, 76–82. doi:10.1016/j.polymertesting.2015.05.004

Sapuan, S. M., Mohamed, A. R., Siregar, J. P., & Ishak, M. R. (2011). Pineapple Leaf Fibers and PALF-Reinforced Polymer Composites. In Cellulose Fibers: Bio- and Nano-Polymer Composites (pp. 325-343). Springer Berlin Heidelberg.

Siregar, J. P., Sapuan, S. M., Rahman, M. Z. A., & Zaman, H. M. D. K. (2012). Effects of Alkali Treatments on the Tensile Properties of Pineapple Leaf fibre Reinforced High Impact Polystyrene Composites. Editorial Board, 409.

Smitthipong, W., Tantatherdtam, R., & Chollakup, R. (2015). Effect of pineapple leaf fiber-reinforced thermoplastic starch/poly (lactic acid) green composite Mechanical, viscosity, and water resistance properties. *Journal of Thermoplastic Composite Materials*, *28*(5), 717–729. doi:10.1177/0892705713489701

Summerscales, J., & Grove, S. (2014). Manufacturing Methods for Natural Fiber Composites. *Natural Fiber Composites: Materials. Processes and Properties*, *7*, 176–215.

Threepopnatkul, P., Kaerkitcha, N., & Athipongarporn, N. (2009). Effect of surface treatment on performance of pineapple leaf fiber–polycarbonate composites. *Composites. Part B, Engineering*, *40*(7), 628–632. doi:10.1016/j.compositesb.2009.04.008

Yaacob, Q. (2015). *The effect of dispersion on the mechanical properties of PLA-based polymer composites*. (Unpublished undergraduate dissertation). Universiti Teknikal Malaysia Melaka.

Yu, C. (2015). *Natural Textile Fibers: Vegetable Fibers. In Textiles and Fashion* (pp. 29–56). Woodhead Publishing.

Yusof, Y., Yahya, S. A., & Adam, A. (2015). Novel Technology for Sustainable Pineapple Leaf Fibers Productions. *Procedia CIRP*, *26*, 756–760. doi:10.1016/j.procir.2014.07.160

Zhou, S. F., Reekie, L., Chan, H. P., Chow, Y. T., Chung, P. S., & Man Luk, K. (2012). Characterization and modelling of Bragg gratings written in polymer fiber for use as filters in the THz region. *Optics Express*, *20*(9), 9564–9571. doi:10.1364/OE.20.009564 PMID:22535048

KEY TERMS AND DEFINITIONS

Biocomposite: Composite material containing one or more biologically derived phases.

Cellulose: A polysaccharide $(C_6H_{10}O_5)x$ of glucose units that constitutes the chief part of the cell walls of plants, occurs naturally in such natural fibers such as pineapple leaf, kenaf, cotton and coir.

Grafting (Polymer): A polymer modification where mostly a different monomer is attached to an existing polymer through a covalent bonding attachment.

Hydrophilic: Having a strong affinity for water; tending to dissolve in, mix with, or be wetted by water.

Hydrophobic: Tending to repel or fail to mix with water.

Mercerization: A chemical treatment applied to cotton fibers or fabrics to permanently impart a greater affinity for dyes and various chemical finishes.

Scouring: An act to remove dirt, grease, etc., from or to cleanse or polish by hard rubbing, as with a rough or abrasive material.

Chapter 7
Green Composites and Their Properties: A Brief Introduction

Deepak Verma
Graphic Era Hill University, India

Xiaolei Zhang
Queen's University Belfast, UK

Prakash Chandra Gope
College of Technology, India

Siddharth Jain
University of Alberta, Canada

Rajneesh Dabral
Graphic Era Hill University, India

ABSTRACT

Green composites are important class of biocomposites widely explored due to their enhanced properties. The biodegradable polymeric material is reinforced with natural fibers to form a composite that is eco-friendly and environment sustainable. The green composites have potential to attract the traditional petroleum-based composites which are toxic and nonbiodegradable. The green composites eliminate the traditional materials such as steel and wood with biodegradable polymer composites. The degradable and environment-friendly green composites were prepared by various fabrication techniques. The various properties of different fiber composite were studied as reinforcement for fully biodegradable and environmental-friendly green composites.

DOI: 10.4018/978-1-5225-0424-5.ch007

INTRODUCTION

The first natural fibre based composites were appeared in 1908. Green composites are a specific part of bio-composites, in which bio-based polymer matrix is reinforced by natural fibers; they symbolize a developing area in polymer science. In a situation like increment in oil price, the use of green composites is helpful not only in making the environment better but also from an economical perspective. The newest development in the field of bio-composites is the substitution of oil-derived polymers with polymers from renewable resources as the matrix component. Such materials are termed as ''green composites''. Green composites are less in cost and are decomposable. There is a mounting need to develop bio- based products and other advanced skills that can decrease our dependence on fossil fuel. Bio-based resources include industrial goods, wood, wood wastes and remains. Bio composites which are eco-friendly are new materials they are not just a solution to growing environmental danger but also as a solution to the ambiguity of petroleum supply. Green composites, which are a mixture of natural fibers and bio plastics, have appeared as hopeful replacements to conventional glass fiber composites because they carry a widespread range of benefits, such as biodegradability, renewable and low density. In recent years, efforts have been made to lessen the use of expensive glass, carbon fibers and also reduce the car's weight by taking advantage of the lower density natural fibers. Natural fibres like sisal, jute, abaca and coir are been in use as reinforcement in composites. Most of the composites are built using polymers and synthetic fibers. These polymers are threat to the environment. As a result, bio based composites have drawn much attention. As compared to synthetic fibres, natural plant-based fibres have various advantages. These natural fibres are low in cost, high thermal insulation and biodegradability. At present, numbers of biodegradable and green composites have been advanced with enhanced mechanical properties using different natural fibers. The performance of green composites relies on the properties of the natural fibers used for reinforcement. Though, green composites have some drawbacks, but these natural fibers don't have any effect on the environment hence, are useful.

Mechanical Properties

Georgios Koronis et.al (2013) have explored about the use of numerous green composites in automobile sector. Green composites made with the help of resins and plants (using plant fibres) are already been in use in automobile. Some tests were performed on the green composites to check their various performances for different application. Table 1 shows results of various tests done on several reinforced bio-resins with different kinds of natural fibres. From the data in Table 1 most of the

Table 1. Mechanical properties of several green composites fibers and PP + GFR composites

S. No.		Elongation to Break (%)	Tensile Strength (MPa)	Young's Modulus (GPa)
1	Starch + 30% jute	2 ± 0.2	26.3 ± 0.55	2.5 ± 0.23
2	PLA + 30% ramie	4.8 ± 0.2	66.8 ± 1.7	n.s
3	PLA + 30% jute	1.8 ± 0	81.9 ± 2.9	9.6 ± 0.36
4	PLA + 25% hemp	n.s	62 ± 2	7.2 ± 0.3
5	PHBV + 30% jute	0.8 ± 0	35.2 ± 1.3	7 ± 0.26
6	PLLA + 30% flax	2.3 ± 0.2	98 ± 12	9.5 ± 0.5
7	PHB + 30% flax	7 ± 1.5	40 ± 2.5	4.7 ± 0.3
8	PLA + 30% flax	1 ± 0.2	53 ± 3.1	8.3 ± 0.6
9	PP + 30% flax	2.7 ± 1.5	29.1 ± 4.2	5 ± 0.4
10	PP + 30% jute	1.4 ± 0.1	47.9 ± 2.7	5.8 ± 0.47
11	PP + 30% fiberglass	3.01 ± 0.22	82.8 ± 4.0	4.62 ± 0.11

Adapted from Georgios Koronis et.al (2013)

green composites are made up of Poly Lactic Acid (PLA), Poly-L-Lactide (PLLA) and other natural fibres. Increment in Flax performance when reinforced with PLLA as compare to PP-fibreglass and highest tensile strength was shown by jute fibres as compared among other natural fibres but reinforcement of jute fibre also reported with the low mechanical properties on comparing to abaca fibres. The lower in performance in the mechanical properties could be due to the different manufacturing method used. Several physical and mechanical properties are presented in Table 1. For the structural performance of automobile panels and the automobile design specific stiffness and strength happens to be important indicators, since these have special distinction in application. The different values of tensile strength and young's modulus have been attributed to the different harvesting seasons.

E-glass is clearly better in terms of specific strength on the other hand is lower in term of specific stiffness comparing with kenaf, hemp and ramie. Therefore more factors should be considered to choose the ideal material.

Tensile Properties

The studies of advance composites have been done by Anil N. Netravali et.al. (2007) Table 2 presents the tensile properties of the advanced composites. The Kevlar show the outstanding performance than E-glass fibre and LC cellulose reinforced

Table 2. Tensile properties of SPC resin based advanced composites

Composite Type	Fracture Stress (MPa)	Fracture Strain (%)	Young's Modulus (GPa)
Kevlar — (#1)	921 (4.3)	3.71 (5.6)	30.4 (3.9)
Kevlar — (#2)	1002 (5.4)	3.97 (6.2)	31.7 (3.2)
Kevlar — (#3)	1086 (4.1)	4.02 (6.8)	34.0 (4.3)
Kevlar — (#4)	984 (5.6)	4.02 (5.7)	27.9 (4.7)
Glass — (#1)	332 (6.6)	4.29 (7.9)	14.5 (5.8)
Glass — (#2)	385 (7.1)	4.03 (7.5)	14.9 (6.4)
Glass — (#3)	379 (8.0)	4.32 (8.2)	16.4 (6.1)
Glass — (#4)	383 (5.3)	4.58 (6.9)	15.8(5.9)
LC Cellulose — (#1)	588 (6.3)	9.98 (7.4)	12.2 (7.7)
LC Cellulose — (#2)	565 (4.8)	9.59 (6.5)	13.1 (5.2)
LC Cellulose — (#3)	638 (5.5)	10.07 (5.9)	12.5 (6.1)
LC Cellulose — (#4)	583 (6.4)	9.99 (7.1)	11.9 (5.8)

Adapted from Anil N. Netravali et.al. (2007)

composite having fracture stresses in the series of 1 GPa (for Kevlar) and modulus of 30 GPa (for Kevlar). The lower performance of E-glass fibre and LC cellulose may be due to the hand layup technique used to build the composite.

The LC cellulose fibres reinforced due to the high fracture strain have higher energy absorbing performance during fracture. While for E-glass fibre reinforced composites due to poor bonding with the resins and their penetration into the fibre showed lower fracture stress and higher modulus than LC cellulose fibre reinforced composites which in turn affects the tensile properties of the composite. These composites have been developed by hot pressing method therefore there is more chance of weaker interfacial bonding which affects the flexural properties. The interfacial bonding can be improved by adding resin before hot pressing.

Calistor Nyambo et.al (2010) have studied the consequence of changing jute fibre content in polypropylene on tensile strength. Author observed higher gathering of fibre, due to poor adhesion within them which ultimately led to the decrement in tensile strength, which was found the same when percentage of wheat straw increased.

The increase in stiffness was observed when wheat straw quantity varied by 10-40%. The decrease in tensile strength has also been found while using fibres alone or combining other fibres. Stress-strain curves of the composites are presented in Figure 1. The stress-stress shows the reduction in % elongation and toughness at break due to high quantity of rigid fibre fillers.

Figure 1. Stress-strain curves for PLA with corn stover, wheat straw, soy stalks, and hybrid fibers
Adapted from Calistor Nyambo et.al (2010)

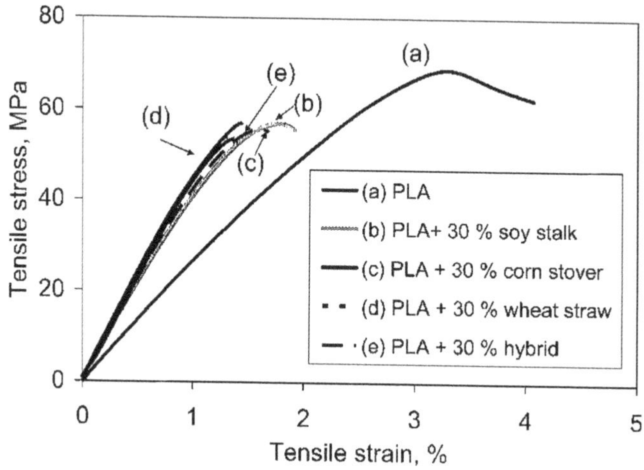

Koichi Goda et.al (2006) observed the effect of load application during mercerization of ramie fibres. Ramie fibres consist of hemicellulose and lignin experiences infringement when treated with alkali which also removed the waxy substances which in turn led to the reduction in weight of the fibre due to the removal of binding substances.

The untreated fibre of ramie, in which applying load decreases the small fibrillar angle but as treated fibre undergoes tension, they show orientation of small fibrils in the direction of tension. When excess tensile stress is applied some of the bonds break from the amorphous region. Hence micro fibrillar maintains maximum stress when the small fibrillar angle is less.

M. S. Huda, et.al, (2005) studied the properties of PLA and PLA based composite, while comparing them the improvement in the modulus of PLA was observed.

From Table 3 the increased in modulus has been observed but not any improvement was observed when fibre TC1004 added to PLA. While the addition of R 0083 improved tensile strength but as the increment in the fibre content is done a decrease in tensile strength is observed which has been attributed to the fibre collection. As the % of fibre content increased, the increase in modulus and decrease in tensile strength was observed. This result has been due to weak bond between cellulose fibres and PLA. The addition of fibre content lead to agglomeration which created stress concentration region hence lowered the impact strength; usage of coupling agent may help in improving impact strength.

Table 3. Tensile properties of the PLA and PLA based composites

Polymer/Cellulose (wt.%)	Tensile Strength (MPa)	Modulus of Elasticity (GPa)	Improvement (Modulus) (%)
PLA/Cellulose (100/0)	62.9 ± 4.9	2.7 ± 0.4	–
PLA/TC 1004(70/30)	47.3 ± 2.5	6.3 ± 0.4	132
PLA/R 0083(70/30)	67.4 ± 3.5	6.1 ± 0.4	124
PLA/R 0084 (70/30)	72.5 ± 2.0	6.7 ± 0.6	148
PLA/R 0083 (60/40)	55.0 ± 4.2	6.4 ± 0.8	136
PLA/R 0084 (60/40)	58.6 ± 5.8	6.6 ± 0.9	144

Adapted from M. S. Huda, et.al, (2005)

Pengju Pan, et.al (2007) studied the tensile properties of Neat PLLA (Poly-L-Lactide) and PLLA/KF (Kenaf Fibre) composites which show increase in tensile strength for neat PLLA as compared with other composite. Figure 2 shows the stress-strain curve for neat PLLA and various composite of PLLA in which higher tensile strength is shown by pure PLLA(47.7MPa) than PLLA composites, while other composites of PLLA showed brittleness in nature since KF is rigid than pure PLLA hence more filler led to the decrement in the amount of polymer for elongation. The KF has led to the increase in modulus of PLLA composites while decrease in tensile strength as compared to the pure PLLA which has been attributed to the weaker adhesion between KF and PLLA.

Figure 2. Stress–strain curves of neat PLLA and PLLA/ KF composites: (a) neat PLLA, (b) PLLA/KF (90/10), (c) PLLA/KF (80/20), and (d) PLLA/KF (70/30)
Adapted from Pengju Pan et. al (2007)

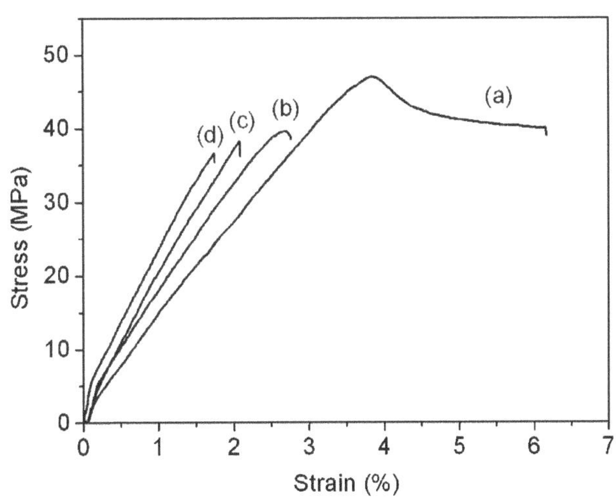

Tensile Properties of Green Composites Fabricated Using the Newly Proposed Methods

Alexandre Gomes et.al (2007) have developed and studied the effect of alkali treatment on tensile of curaua fibre green composites made-up using Pre-Forming, Direct Method and Prepreg Sheet methods. Prepreg sheet (PS) composites show the higher tensile strength followed by Pre forming (PF) composite. Lowest tensile strength was shown by direct method composites (DM). DM composites show deprived fibre arrangement which may have occurred due to the shrinkage of resin, but the shrinkage of resin did not disturb the fibre arrangement of PF and PS composites since pre-forms of resin were absolutely dried in the case of PF composites and silvers were slightly stretched during the building of PS composites. The improvement in the tensile strength of PF was also attributed to their longitudinal arrangement of the fibres. Prepreg sheet composite also showed a rise in the Young's modulus (36 GPa), which is similar to that of some Glass fibre reinforced composites which had young's modulus around (38 GPa). Figure 3 demonstrates stress–strain figure of the composites.

PS composite has highest slope in stress-strain diagram followed by PF and DM. Moreover, PF and PS composites show a linear graph. The increase in the angle of slope in the case of PS was credited to the higher fibre loading.

Characterization of interfacial and mechanical properties of "green composite" used with soy protein isolate was studied by Netravali et al. The stronger bond between hydrogen bond and protein molecules were expected due to the presence

Figure 3. Stress–strain diagrams of curaua fibre green composites
Adapted from Alexandre Gomes et.al (2007)

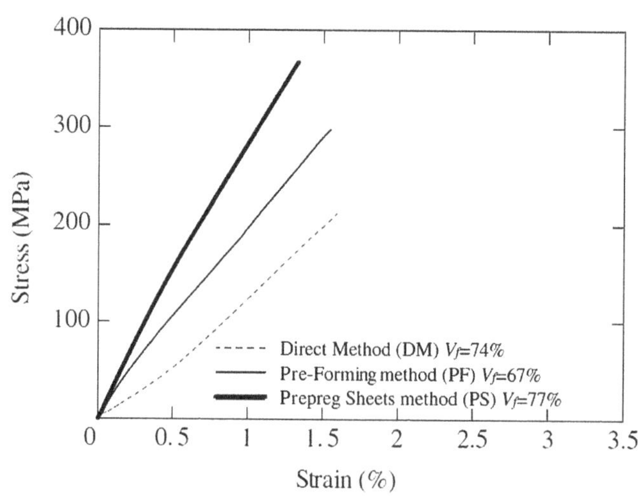

Figure 4. Tensile strength vs. strain plots of ramie fiber/SPC composites
Adapted from Anil N. Netravali et al. (2002)

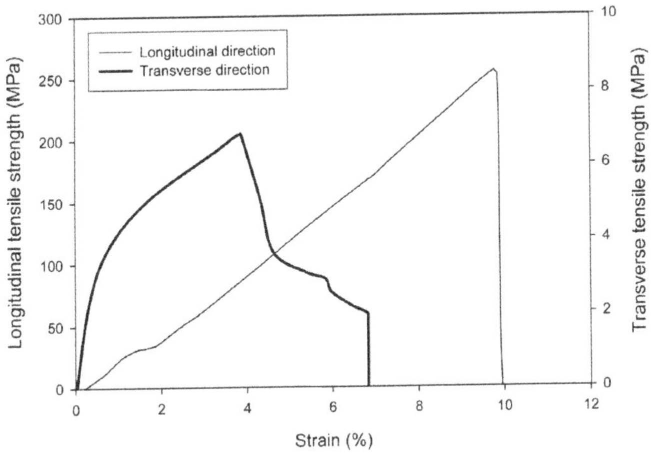

of 3 OH group in glycerol. Since it acts a plasticizer hence increases the fracture strain (20%-50%) on increasing the content of glycerol while the same reason was attributed to the decrease in fracture stress (15.5MPa-5.3MPa) and young's modulus (505MPa-40MPa).

Sunghyun Nam and Anil N. Netravali (2002) studied about fully biodegradable and environment friendly green composite made using ramie fibres and soy protein concentrate. Glycerin was used earlier to reduce brittleness of the material by adding it to SPC (soy protein concentrate) resins and to avoid cracking of the material. Pure SPC resin showed brittleness nature without glycerine, but increase in the content of glycerine, greatly reduced both tensile and young's modulus with an increase in fracture strain. Since glycerine has 3 hydroxyl groups, hence attracts water and leads to glycerine to act as a plasticizer. As the percentage of glycerine content increased above 40%, the significant drop in tensile strength was seen. The collection of glycerine molecules have been attributed to this cause which has reduced the plasticization of the material. In this study author has selected 30% glycerine for ramie fibre/SPC resin composites, because less glycerine (>30%) leads to the shrinkage during curing. The author performed the test for tensile strength, both longitudinally and transversely shown in Figure 4.

Unreinforced SPC resin was also tested for the same. Tensile strength for longitudinally measured composite was found 271MPa while for transversely and unreinforced fibre it was 7.4MPa and 6.9Mpa respectively.

The tensile strength-strain plots of the green composites in both longitudinal and transverse directions showed linearity for longitudinal direction while non-linear for transverse due to matrix failure or interfacial debonding. The interfacial shear strength values at 30% Glycerin were found to be 23 MPa for un-plasticized SPC and 15 MPa for SPC resin containing 30% glycerine.

Effect of Fibre Length

Characterization of interfacial and mechanical properties of "green" composites with soy protein isolate (SPI) and ramie fibre were studied by Preeti Lodha, Anil N. Netravali, (2002) and found fracture stress and Young's modulus to be lower for composite than SPI polymer of length 5mm having fibre wt.10% in Figure 5 while composite having less fibre content did not influence the fracture stresses. Instead, the fibres with low aspect ratio acted like particles, low stress bearing points. Addition of short fibres at low fibre content led to the inadequacies to reduce the strength of the composite. As the fibre content increased, the increase in fracture stress was observed. Similar behaviour is shown in case of young's modulus in Figure 5. Longer fibres will have less fibre-end and hence there would be less flaws or low stress bearing points. The composite fracture stress and modulus increase and fracture strain decreases on increasing the fibre length due to more fibres are available per unit area. However, long length fibres at 30% fibre weight content showed delamination hence anticipated strength values for these composites were not reached. These delaminated surfaces show poor wetting of the fibres.

Figure 5. Effect of fiber content on the Young's modulus of ramie fiber reinforced SPI-polymer composite
Adapted from Netravali et al. (2002)

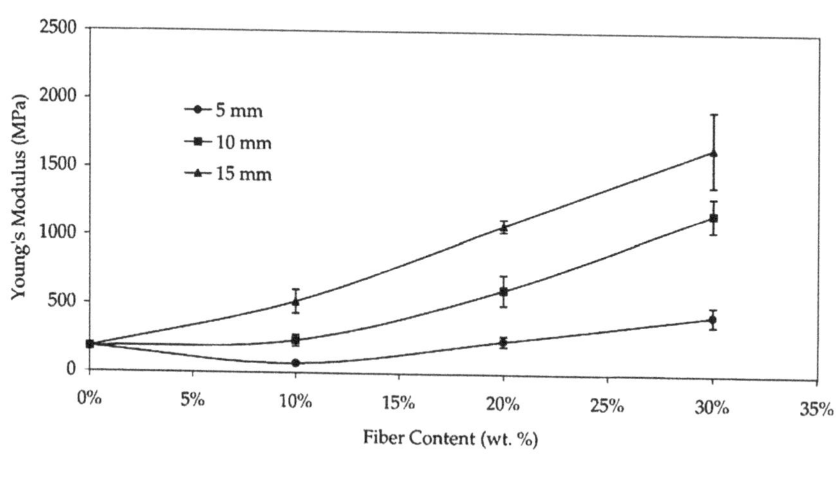

Composites containing 10% Fiber wt. failed from the region where composite had lower fiber content which corresponds to the fracture stress values. Some deviations were observed due to lack of randomness, affecting tensile properties of the composites. The fibre length distribution also adds towards lower fracture stress. However, these flaws can be avoided by better processing methods.

Effect of Curing Temperature

Sunghyun Nam and Anil N. Netravali studied the effect of curing temperature on tensile strength. The effect of curing temperature on the SPC resin tensile properties are presented in Figure 6.

Glycerin was used earlier to prepare SPC resin since glycerine acts as a plasticizer. The alkaline pH helps soy protein molecule to dissolve in water. In this with the alkaline treatment, normal heating was also used to take away the natural characteristics of soy protein resulting in unfolding of the protein molecules. Tensile strength (7MPa) and young's modulus (90MPa) increased as the temperature increased to 120°C and decreased afterward. Samples containing 30% glycerine were cured for 2 hours at 80, 100, 120 and 140 °C Amino acid in soy proteins forms cross links on heating. The combination of two cysteine forms disulphide crosslink. These cross linkage help in upholding the 3-dimensional structure of the protein molecules. Author observed inter and intra-molecular cross linkage between soy proteins due to thermal treatment which ultimately increased tensile strength and modulus of the

Figure 6. Effect of curing temperature on the tensile properties of SPC resin
Reproduced with permission from Springer.

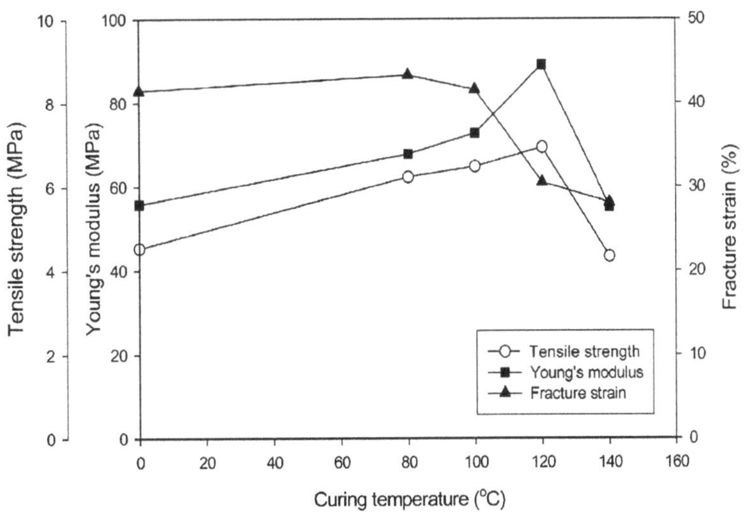

Table 4. Flexural properties of SPC resin based 'advanced' composites

Composite	Flexural Stress (MPa)	Flexural Strain (%)	Flexural Modulus (GPa)
Kevlar — (#1)	319 (6.3)*	0.78 (10.2)	79.1 (8.9)
Kevlar — (#2)	379 (5.2)	0.74 (8.4)	88.5 (7.0)
Kevlar — (#3)	387 (8.4)	0.74 (7.5)	89.6 (3.8)
Kevlar — (#4)	357 (6.8)	0.77 (11.3)	83.5 (9.2)
Glass — (#1)	126 (11.9)	1.34 (13.7)	20.4 (10.2)
Glass — (#2)	148 (9.6)	1.45 (11.5)	23.7 (8.1)
Glass — (#3)	155 (8.6)	1.37 (12.1)	24.2 (7.9)
Glass — (#4)	151 (10.2)	1.52 (9.8)	22.4 (7.5)
LC Cellulose — (#1)	202 (8.2)	2.51 (7.8)	17.6 (3.8)
LC Cellulose — (#2)	228 (5.1)	2.08 (7.1)	24.3 (5.7)
LC Cellulose — (#3)	248 (8.7)	2.03 (9.2)	27.8 (5.9)
LC Cellulose — (#4)	240 (9.4)	2.02 (12.7)	26.9 (6.8)

Reproduced with permission from Springer.

SPC resin. Curing at 140°C, led to the degradation of the resin and led to the decrease in the mechanical properties. Resin coloured changed to dark brown. While fracture strains were found higher in the series of 28% to 43% for pre-cured and cured SPC. The hike in the value of fracture strains of SPI were observed in the range of 120% to 215%, but with the increase in curing temperature sudden decrement in the fracture strain was also reported. Hence curing at 120°C for 2 hours is suggested for optimum results.

Anil N. Netravali, Xiaosong Huang & Kazuhiro Mizuta have studied the flexural properties of the advanced composites fabricated using the three fibres and 4 resins. Their values are presented in the Table 4. Author concluded from the Table 4 that Kevlar shows higher flexural stress and modulus properties than E-glass and LC cellulose fibre. Comparable values can be seen for E-glass and LC cellulose fibre in terms of flexural modulus while flexural stress values are higher than E glass fibre, for lower flexural strength and modulus insufficient penetration and weaker interfacial bonding since composites were made by hot pressing methods.

The Kevlar and LC cellulose fibre reinforced composites had lower fibre towed due to hydroxyl and amide groups on the LC cellulose and Kevlar fibres respectively. Based on the flexural stress and strain values, LC cellulose composite absorbed maximum energy. The presence of clay in resin #3 in all type of composite mentioned in Table 7 showed the best flexural performance due to better penetration of resin within the composite while resin #1 of all composite revealed lower value for flexural stress as compared with other resins. Hence the properties of the composite depend upon fibre properties.

Hitoshi Takagi et.al (2008) have studied the change in the flexural properties of Nano fibre "green composite" when processed with different conditions. A plot between flexural stress and crosshead displacement for composites molded at different pressures are presented in Figure 7. It has been observed in the graph that with the increase in stress level and molding pressure raises the slope but with the further increase in molding pressure the slope decreases. The maximum value attained was 50 MPa.

The increase in flexural strength occurred with the increase in molding pressure for the control composite. Both vacuum treated and stirrer treated composite showed similar variation in flexural strength as for control one while composites treated with stirrer presented greater flexural strength.

Mitsuhiro Shibata et.al studied the mechanical properties and bio-degradability of Green composites based on Biodegradable polyester and Lyocell Fabric. Author did the impact test on PLA and PLA composites containing different number of ply as shown in Figure 8. The PLA composites with 8-ply showed higher impact strength of about 40.8 kJ/m^2 as compared to the PLA/Lyocell with 6 ply (28.2 kJ/m^2) and pure PLA (14.1 kJ/m^2).

The author observed Lyocell fibre composite carry higher impact strength than with the natural fibres as compared on the basis of same length fibre and methods and the longer elongation of the lyocell fabric improved the impact strength of the composites.

Figure 7. Typical stress–crosshead displacement curves for control composite materials
Adapted from Takagi et.al (2008)

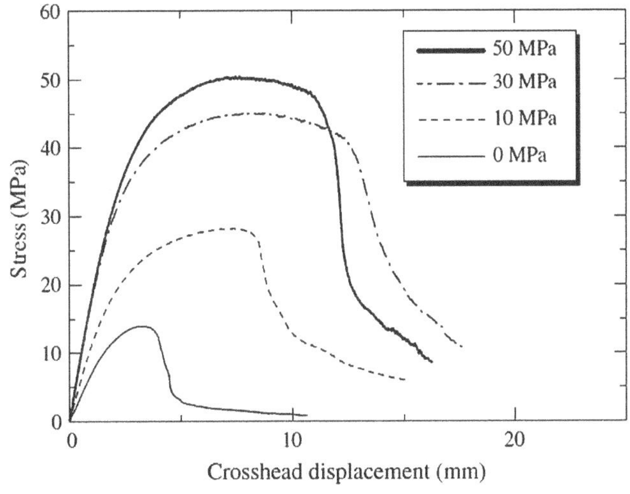

Figure 8. Izod impact strength of the PLA and multilayered PLA/lyocell composites: (a) PLA, (b) PLA/lyocell (6 ply), and (c) PLA/lyocell (8 ply)
Adapted from Mitsuhiro Shibata et.al (2004)

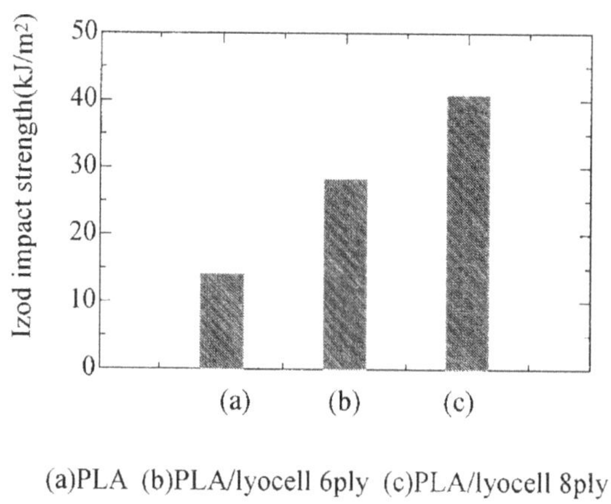

Morphology

Calistor Nyambo et.al studied the morphology of the tensile fractures with the help of scanning electron microscopy (SEM). The Figure 9 (a-h) shows the indication of voids and tow of fibres, which may have backed the reduction in tensile, flexural and impact strength. Moreover, clear sign of fibre debonding can be observed, due to weak grip in the corn stalk composites and fibre rupture in the wheat straw composites (Figure 9f).

This is reasonable because no surface alteration of the fibres was done to improve interfacial bonding. The main cause of failure was attributed to the weak adhesion between fibres and matrix. Surface modifications of fibres can be completed by alkaline action or the use of compatibilzer to improve bonding. Here, the voids are observed in the image on the surface of tensile fracture might have been created, as an effect of fibre tow during testing.

Hitoshi Takagi et. al studied the SEM images (Figure 10) and found invisibility of resin rich and resin poor surfaces in case of stirrer treated fibers, while vacuum treated composite had some resin rich and poor surfaces. The same was observed for the control composites, having no major difference in the density. The increase in the mechanical properties of the stirrer treated composite was credited to the stirrer treated fiber which looked more of straight and less twisted and uniform diffusion of Nano fibers.

Pengju Pan et.al examined the morphologies of KF and PLLA/KF composites with the help of SEM. The SEM micrographs exposed the rough surface of the fibres

Figure 9. Scanning electron microscopy images (SEM) at (a) high and (b) low magnification for PLA + 30% soy stalks, (c) high and (d) low magnification for PLA + corn stover, (e) high and (f) low magnification for PLA + wheat straw and low (g) and high magnification for PLA + 30% hybrid composite (i.e., 10% corn stover + 10% wheat straw + 10% soy stalks). The scale bars for high and low magnification are 50 and 200μm, as indicated on the scale bars
Adapted from Calistor Nyambo et.al (2010)

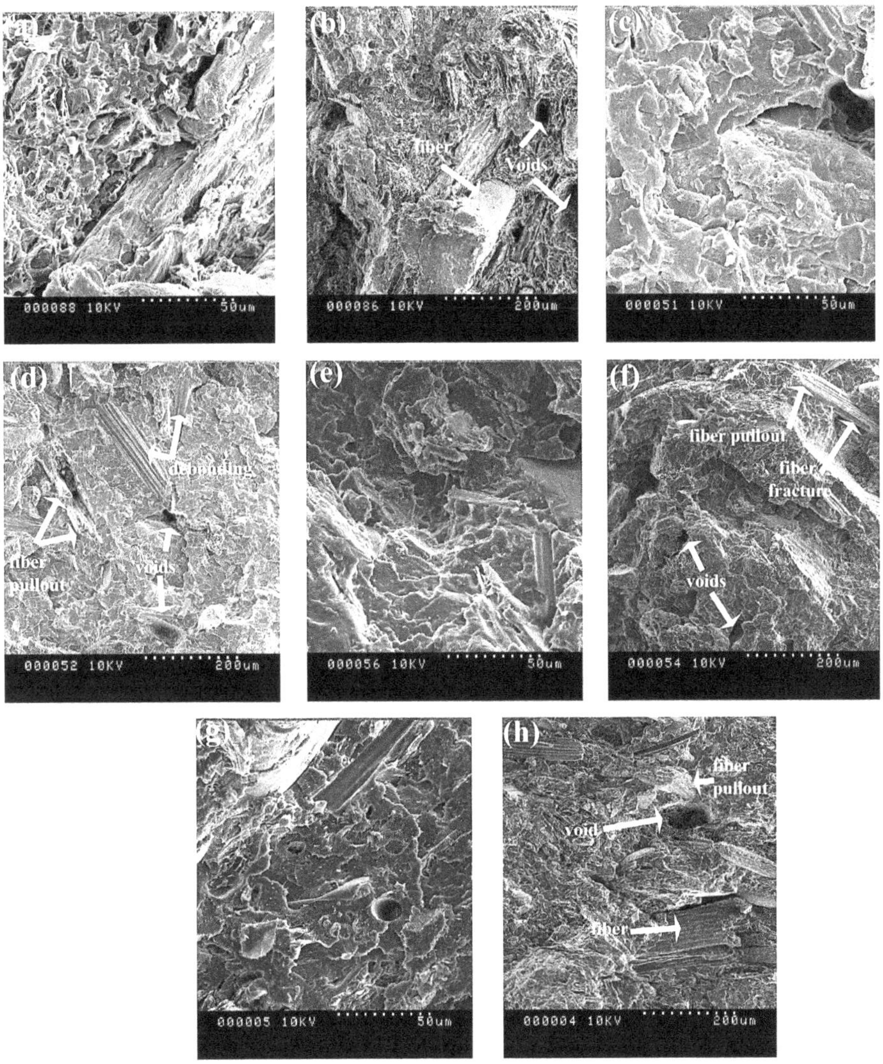

Figure 10. SEM photomicrographs of sectioned surfaces of composites molded at 50 MPa showing differences in the uniformity of nanofibers dispersion; (a) control composite, (b) vacuum-treated composite, and (c) stirrer-treated composite, circles and rectangles respectively indicate resin-rich and resin-poor areas
Adapted from Takagi et al. (2008)

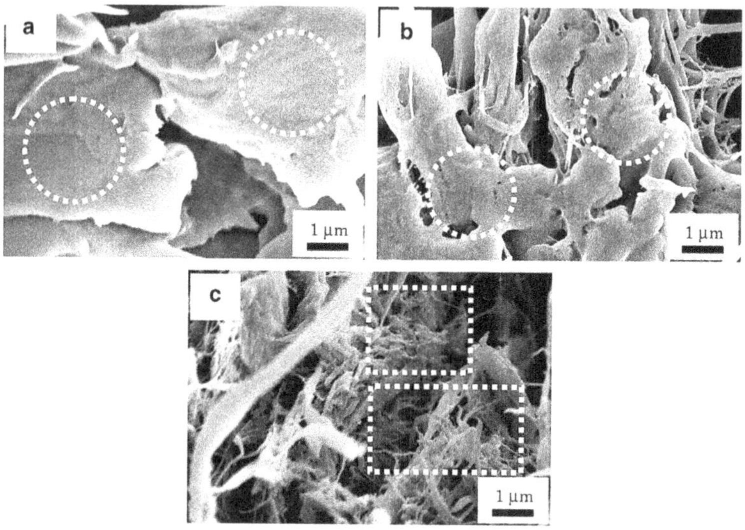

Figure 11. SEM micrographs of the fracture surface of (a) pure PLLA and the PLLA/ KF (70/30) composites: (b) 200, (c) 500 and (d) 500
Adapted from Pengju Pan et. al

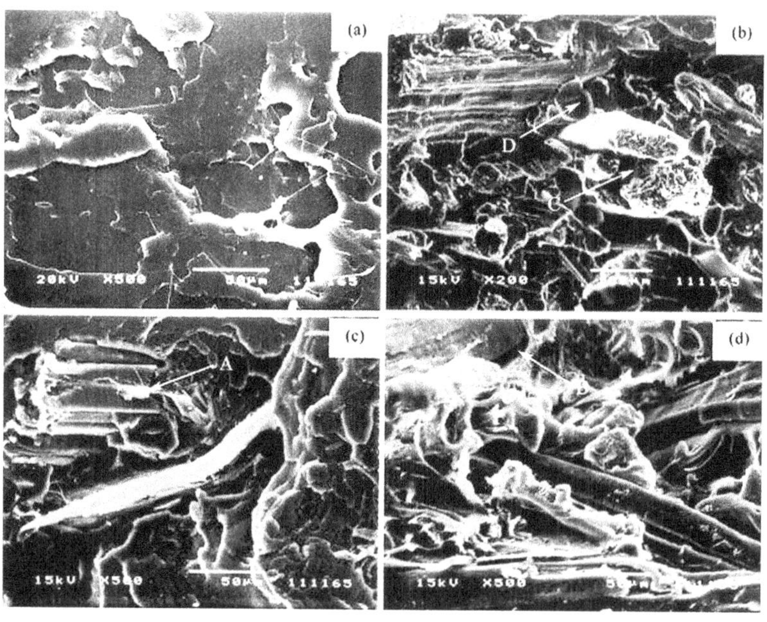

with the adhesion of smaller fibers. SEM micrographs (Figure 11) of the fracture surface of the pure PLLA and PLLA/KF (70/30) composite shows the uneven and coarse surface of pure PLLA, revealing brittle'' manner fracture. Most fibers were found attached firmly with the matrix pointed by "A". Thin layer of matrix on the surface has been attributed to this attachment of fibres with matrix, which provided better stress transfer between matrix and fibers.

Though, some fibres were found unglued from the matrix during tensile test as indicated by "B" in the Figure 11. In addition, accumulation of kenaf fibres and small amount of air bubbles as shown by "C" and "D" in the Figure 11 respectively were also observed. The mechanical properties of the composite were found affected by the dispersion of filler and the interfacial interaction. The dependent of crystallization rate on the interfacial adhesion and nucleating agent was also reported. Hence; the mechanical properties as well as the crystallization behaviour can be enhanced with the use of compatilizers or coupling agents.

CONCLUSION

The above studies show that the green composite becoming the backbone of materials in industries. Its biodegradable nature increases its interest in the research and development of biocomposites. The improved and advanced mechanical properties enable the use of biocomposites in place of traditional composite. The biodegradation of biocomposites are enhanced by natural fibers due to lignocellulosic materials, which can be decomposed in less time in the environment. Being biodegradable they have remarkable performance with respect to the specific weight.

REFERENCES

Goda, K., Sreekala, M. S., Gomes, A., Kaji, T., & Ohgi, J. (2006). Improvement of plant based natural fibers for toughening green composites- Effect of load application during mercerization of ramie fibers. *Composites. Part A, Applied Science and Manufacturing*, *37*(12), 2213–2220. doi:10.1016/j.compositesa.2005.12.014

Gomes, A., Matsuo, T., Goda, K., & Ohgi, J. (2007). Development and effect of alkali treatment on tensile properties of curaua fiber green composites. *Composites. Part A, Applied Science and Manufacturing*, *38*(8), 1811–1820. doi:10.1016/j.compositesa.2007.04.010

Huda, M. S., Mohanty, A. K., Drzal, L. T., Schut, E., & Mishra, M. (2005). Green composites from recycled cellulose and poly lactic acid: Physico-mechanical and Morphological Properties evaluation. *Journal of Materials Science*, *40*(16), 4221–4229. doi:10.1007/s10853-005-1998-4

Koronis, G., Silva, A., & Fontul, M. (2013). A review of adequate materials for automotive applications. *Composites. Part B, Engineering*, *44*(1), 120–127. doi:10.1016/j.compositesb.2012.07.004

Lodha, P., & Netravali, N. (2002). Characterization of interfacial and mechanical properties of green composite with soy protein isolate and ramie fiber. *Journal of Materials Science*, *37*(17), 3657–3665. doi:10.1023/A:1016557124372

Nam, S., & Netravali, A. N. (2002). Green Composites ii. Environment-friendly, Biodegradable composites using Ramie Fibers and Soy Protein Concentrate (SPC) Resin. *Fibers and Polymers*, *7*(4), 380–388. doi:10.1007/BF02875770

Netravali, A. N., Huang, X., & Mizuta, K. (2007). Advanced 'green' composites. *Advanced Composite Materials*, *16*(4), 269–282. doi:10.1163/156855107782325230

Nyambo, C., Mohanty, A. K., & Misra, M. (2010). Polylactide-Based Renewable Green Composites from Agricultural Residues and Their Hybrids. *Biomacromolecules*, *11*(6), 1654–166. doi:10.1021/bm1003114 PMID:20499931

Pan, P., Zhu, B., Kai, W., Serizawa, S., & Ji, M. (2007). Crystallization behavior and mechanical properties of bio-based green composites based on poly(L-lactide) and kenaf fiber. *Journal of Applied Polymer Science*, *105*(3), 1511–1520. doi:10.1002/app.26407

Shibata, M., Oyamada, S., Kobayashi, S., & Yaginuma, D. (2004). Mechanical properties and biodegradability of green composites based on biodegradable polyesters and lyocell fabric. *Journal of Applied Polymer Science*, *92*(6), 3857–3863. doi:10.1002/app.20405

Takagi, H., & Asano, A. (2008). Effects of processing conditions on flexural properties of cellulose Nano fiber reinforced "green" composites. *Composites. Part A, Applied Science and Manufacturing*, *39*(4), 685–689. doi:10.1016/j.compositesa.2007.08.019

Chapter 8
Rice Husk Reinforcement in Polymer Composites

Sanjay Sharma
Graphic Era Hill University, India

Deepak Verma
Graphic Era Hill University, India

ABSTRACT

Increasing concern about global warming and depleting petroleum reserves and the high cost of petroleum products had made scientists to focus more on the use of natural fibres such as rice husk, baggase, coconut husk, hemp, sisal, jute, flax, banana etc. Past decade has shown many efforts to develop composites to replace the Petroleum and other non-decaying material products. Reinforcement with natural fibre in composites has recently gained attention due to low cost, easy availability, low density, acceptable, strength full, stiffness, ease of separation, enhanced energy recovery, biodegradability and recyclable in nature. Natural fibre composites are suitable as wood substitutes in the construction sector. All these have excellent physical, thermal and mechanical properties and can be utilized more effectively in the development of composite materials. In this connection, an investigation has been carried using rice husk, a natural fibre abundantly available in India.

INTRODUCTION

The unique and diverse characteristics of composite materials have caused an increase in their utilization worldwide. Composites are a need in the evolution of engineering materials. Composites can be very specific in their properties with their

DOI: 10.4018/978-1-5225-0424-5.ch008

unique combination of materials it is possible to overcome many of the limitations of conventional materials for instance, brittleness and poor process ability of stiff and hard polymers. Industry has begun to recognize that the commercial applications of composites promise to offer much larger business opportunities in transportation sector due to the sheer size of the industry. Agricultural wastes can be used to prepare fibre reinforced polymer composites for commercial use and fortunately natural fibres are abundantly available in India. This also gives the chance to explore value-added application avenues of composites. India alone produces more than 400 million tonnes of agricultural waste annually. It has got a very large percentage of the total world production of coconut husk fibre, baggase fibre, rice husk, jute and stalk. These materials have a great potential to be used in composite preparation and also be helpful in conserving the environment as well as lifting the economy.

RICE

Rice or *Oryza Sativa* (as botanists prefer to call it) is not a tropical plant but is still associated with a wet, humid climate. It is generally believed that the domestication of rice began somewhere in the Asia. Rice is the world's second largest cereal crop and produces the largest amount of crop residues. Rice, rice husk and rice straw are the main products of rice cultivation and processing (Binod *et al.*, 2010). The average ratio of rice grain: rice husk: rice straw is 1:0.25:1.25 (Haefele *et al.*, 2011).

The outermost layer of the paddy grain/rice is the rice husk, also called rice hull. It is separated from the brown rice in rice milling. Burning rice husk produced rice husk ash (RHA), if the burning process is incomplete carbonized rice husk (CRH) is produced. Globally, approximately 600 million tons of paddy is produced each year. On average 20% of the rice paddy is husk, giving an annual total production of 120 million tones. Paddy, on an average, consists of about 72 percent of rice, 5-8 percent of bran, and 20-22 percent of husk. Of all the plant residues, the ash of rice husk contains the highest proportion of silica.

It is estimated that every ton of paddy produces about 0.20 tons of husk and every ton of husk produces about 0.18 to 0.20 tons of ash, depending on the variety, climatic conditions and geographical location.

PRODUCTION OF RICE OVER THE WORLD

The Table 1 shows the production of rice over the world particularly in China and India as there is huge production in comparison to rest of the countries.

Figure 1. Rice

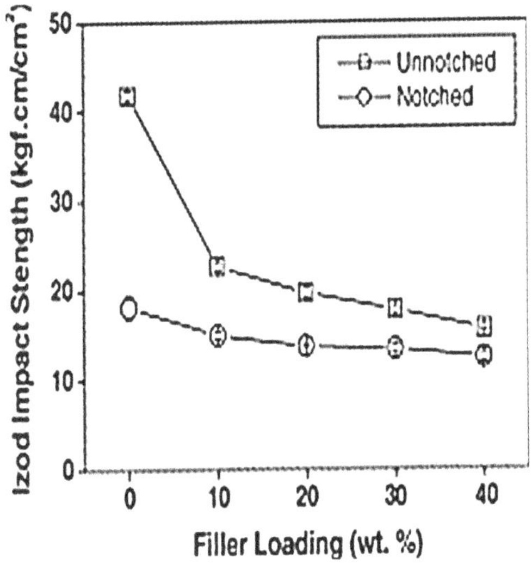

RICE HUSK

Rice husk is one of the most widely available agricultural wastes in many rice producing countries around the world. Farm income can be increased both directly and indirectly if economically profitable means of utilizing rice husk generated are utilized in industry. There are many reported uses of rice husk such as a fuel in brick kilns, in furnaces, in rice mills for parboiling process, in the raw material for

Table 1. Rice production in metric tons

Countries	Year 2000	Year 2010
Afghanistan	260,000	672,000
Bangladesh	37,627,500	49,35,5000
Brazil	11,089,800	11,308,900
China	189,814,600	197,212,010
Egypt	6,000,890	43,29,500
India	1,27,465,000	120,620,000
Indonesia	51,898,000	66,411,500
Japan	11,863,000	10,600,000
Malaysia	2,140,800	25,48,000
Madagascar	2,480,470	47,37,970

the production of xylitol, furfural, ethanol, acetic acid, lingo-sulphonic acids, as an cleaning or polishing agent in metal and machine industry, in the manufacturing of building materials, etc. Despite having so many well established uses of rice husk, little portion of rice husk produced is utilized in a meaningful way, remaining part is allowed to burned in open piles or dumped as a solid waste or it is used as a cattle feed. There are so many reasons associated with rice husk for not being utilized effectively, like,

1. Lack awareness of its potential to a farmers and industry persons,
2. Insufficient information about proper use,
3. Socio-economic problems,
4. Penetration of technology,
5. Lack of interest,
6. Lack of environmental concerns,
7. Inefficiency of information transfer, etc.

Solution to the problems associated with utilization of this solid waste needs to be worked out not only from the quality point of view but quantitatively as well, because quantity of rice husk produced is very large. But the most promising and profitable use of this biomass is its use for producing composites in efficient way besides this adopting efficient equipment and techniques can give very valuable by products.

Table 2 shows organic composition of rice husk in which Cellulose is in largest quantity and extractives is in lowest quantity. Cellulose is an important structural component of the primary cell wall of plants, due to this rice has good structural properties.

A typical analysis of rice husk is shown in Table 3. The content of each of them depends on rice variety, soil chemistry, climatic conditions, and even the geographic localization of the culture.

Table 2. Typical organic composition of rice husk

Composition	Percentage
Cellulose	31.12
Hemicelluloses	22.48
Lignin	22.34
Extractives	2.33

Table 3. Properties of rice husk along with their range and percentage

Property	Range
Bulk density(kg/m^3)	96-160
Ash	22-29%
Carbon	35%
Hydrogen	4-5%
Oxygen	31-35%
Nitrogen	0.23-0.32%
Sulphur	.04-0.08%
Moisture	8-9%
Hardness(mohr's scale)	5-6%

RICE HUSK REINFORCEMENT IN POLYMER COMPOSITES

Hyun-Joong Kim et al (2001), studied the feasibility of a raw material for manufacturing agricultural lignocellulosic fiber-thermoplastic polymer composite. Rice husk flours were analyzed by chemical composition and thermo gravimetric methods in nitrogen atmosphere. In the chemical composition analysis, rice husk flour was composed of, 5.0% moisture 21.6% lignin, 60.8% hemicellulose, and 12.6% ash. In thermo-gravimetric test, thermal decomposition of rice husk flour from room temperature to 350"C was similar to that of wood flour but rice husk flour showed approximately 60% of thermal loss in comparison with 20% of thermal loss in wood flour at the temperature of 350°C or higher. And rice husk flour with larger particle size was more considerably decomposed from 350 to 800°C because of its lower content of organic material. He also evaluated the activation energy of thermal decomposition using Flynn & Wall expression. As the thermal decomposition proceeded in rice husk flour, the activation energy of thermal decomposition appeared almost constant up to a= 0.25, but thereafter increased. Activation energy of thermal decomposition in wood flour, however, decreased steeply up to a= 0.3, but thereafter remained almost constant. From his work he concluded that rice husk flour can be used as a substitute for woof flour in manufacturing agriculture lingo-cellulose fibre-thermoplastic polymer composite in aspect of thermal decomposition.

B. Waswa Sabuni et al (2002), studied samples of rice husks from Mwea Rice Mills to confirm the pozzolanic properties of rice husk ash (RHA). The results of his study showed that there were good prospects of using RHA as pozzolana in combination with OPC or lime. RHA/lime binder in the ratio of 1:1 is suitable for mass concrete and costs about one-third of OPC. For mortar used in plastering 3:2 ratio mix of RHA and OPC gives a saving of 32%. The production process for RHA would also provide employment within the rice growing areas as well provide clean environment devoid of heaps of rice husk.

Hyun joong kim et al (2003) investigated the possibility of using rice husk flour as a partial substitute of the wood particles for manufacturing particleboards. In his experiment he used commercial wood particles and two types of rice husk flours (A type (30 urn), B type (300 urn)) .E1 and E2 class urea-formaldehyde resin was used as the composite binder, combined with 10 wt.% NH4CI solution as a hardener. Rice husk flour-wood particleboards was manufactured at a specific gravity of 0.7 with rice husk flour contents of 0, 5, 10, and 15 (wt. %). After examining the physical and mechanical properties of the composite he concluded that composites made from rice husk flours were of somewhat poorer quality than those made from wood; however, blending in small amounts of rice husk flour (e.g., 5% to 10% by weight) may have no significant impact on quality.

H.J. Kim et al (2004) studied the thermal degradation and thermal stability of rice husk flour (RHF) filled polypropylene and high density polyethylene (HDPE) composites in a nitrogen atmosphere using thermo gravimetric analysis. He found that the content of homo-cellulose and lignin in RHF is a slightly lower than that in WF (Wood Flour). Because of the higher cellulose content of WF, the thermal stability of WF is a slightly higher than that of RHF. Also the mass loss of the composites made of PP or HDPE to which RHF was added at different filler loadings and one heating rate (10°C min^{-1}). It was found that as the filler loading is increased, the thermal stability of the composites decreased and the ash content increased.

From Figure 2, it can be observed that the Thermo Gravimetric curves of both RHF and WF reveals two mass loss steps. The first mass loss is below 100°C. This is due to the slow evaporation of absorbed moisture. The second mass loss from (approx.) 150 to 500°C is because of the decomposition of the three major constituents such as cellulose, hemicellulose and lignin of the natural fillers. Results also shows that lignocellulose materials are chemically active and decompose between 150 and 500°C; similarly, in case of hemicellulose, it decomposes between 150 and 350°C, cellulose between 275 and 350°C and lignin between 250 and 500°C. At 800°C, the ash content of RHF is greater than that of WF.

Hyun-Jong Kim et al (2004), studied the possibility of using lignocellulosic materials as reinforcing fillers in the thermoplastic polymer composite, polypropylene as the matrix and rice-husk flour as the reinforcing filler were used to prepare a particle-reinforced composite. Generally, fiber-reinforced composites have higher tensile strength than particle-reinforced composites. Tensile strengths of the composites slightly decreased while the tensile modulus improved as the filler loading and crosshead speed increased, but the composites had an acceptable strength up to a filler loading of 40 wt.%. As the filler loading and crosshead speed increased,

Figure 2.

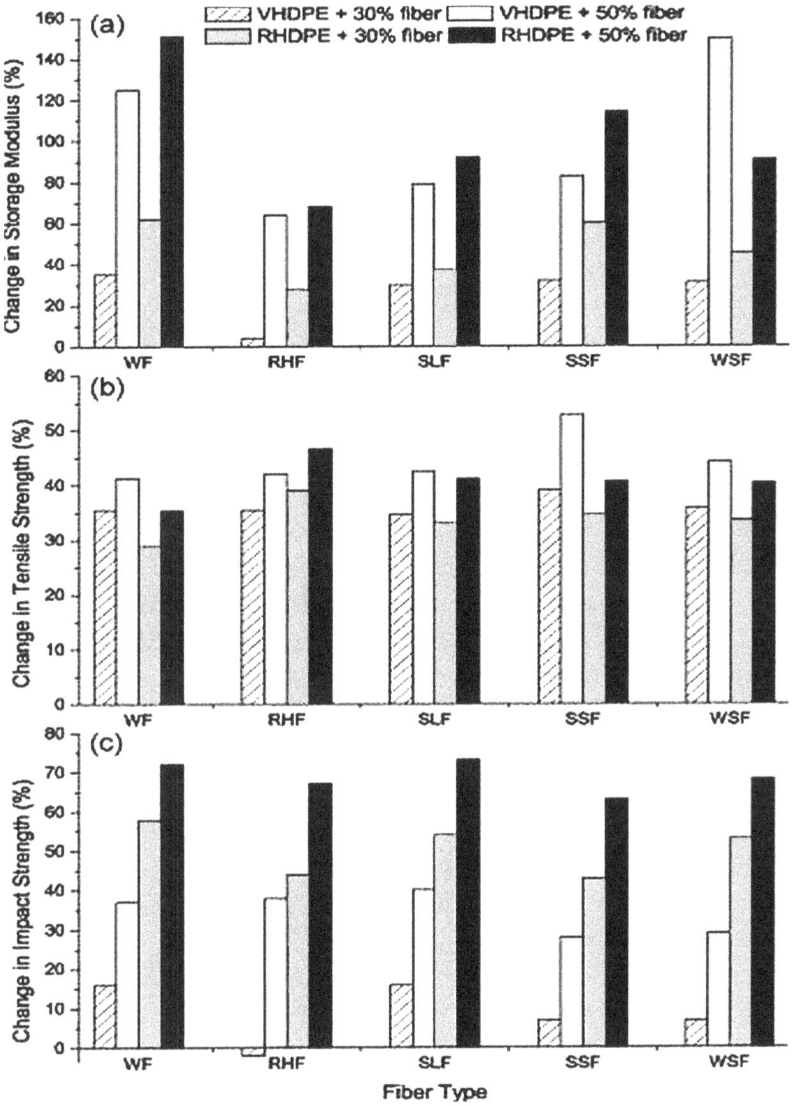

the tensile property became more brittle, and the poor interfacial bonding between filler and matrix polymer caused decreased tensile strength and Izod impact strength of the composites.

Han-seung yang et al (2004), studied a bio-composite made by combining melt-blended cellulosic material-thermoplastic polymer. Rice husk flour was used as reinforcing filler to verify its usability and to provide the physical, mechanical, thermal and morphological properties of the bio-composite. The results showed

Figure 3.

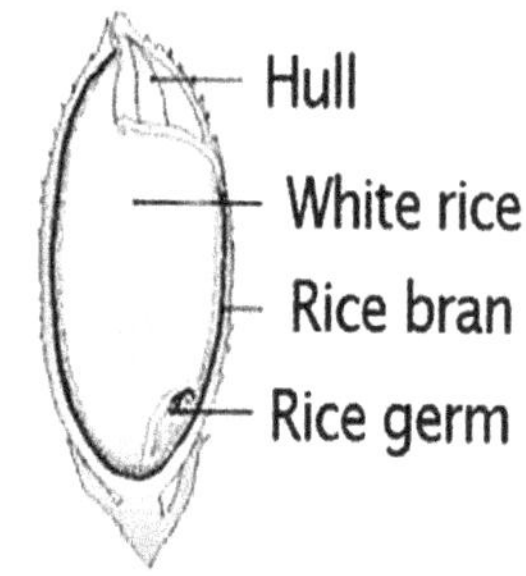

that the increasing of the mixing ratio of the rice husk flour increases composite thickness swelling and the water absorption but the tensile strength was somewhat decreased while the tensile modulus increased. The impact strength reduced with the increase in the mixing ratio of the rice husk flour and the wood flour, the unnotched impact strength was higher than the notched impact strength due to the difference of the mechanism fractured in the impact. Meanwhile, with the increase of the mixing ratio, the thermal expansion of the bio-composite was decreased and was remarkably low when the compatibilizing agent was added.

From Figure 3 it can be noticed that the izod impact strength (at room temperature) of composites decreased moderately by increasing the filler percentage the impact strength of the notched samples decreases by increasing the quantity of RHF about 20 wt.%. it is also observed that a poor interfacial bonding results microspaces between the fiber and matrix and also results many micro-cracks when impact occurs, which originates crack propagation and ultimately decreases the impact strength of the composites. Also the unnotched Izod impact strength of unfilled polypropylene composite showed notably high impact strength, which was acutely reduced at a filler loading of 10 wt.%.

Jutarat Prachayawarakorn et al (2005), studied the use of rice husk as a filler for polypropylene and its recycle ability. Rice husk (200 mesh and 40% by weight) and polypropylene were compounded in a twin screw extruder and injection molding technique was applied in order to obtain test specimens. It was found that tensile modulus was improved with increasing filler loadings and the specimen became brittle at higher crosshead speeds. Moreover, hygrothermal aging behavior of rice husk filled polypropylene was investigated. Rice husk roughly contains 35% cellulose, 35% hemicellulose, 20% lignin and 10% ash (94% silica), by dry weight basis. Tensile, flexural and impact properties including % water absorption of the rice husk-filled polypropylene was found to be slightly decreased as the result of the increase in melt flow index by the recycling process.

Sang-Yong Park et al (2005) evaluated the impacts of thermoplastics polymer which is a matrix polymer and the rice husk flour which was used as a reinforcing filler relative to the manufacturing temperature and time when bio-composites were manufactured. In order to evaluate the impacts on the rice husk flour relative to the manufacturing temperature, the rice husk flour was preserved for 10 minutes to 2 hour period at 220 degree Celsius temperature which was then added with the poly-propylene and LDPE to complete the manufacturing process of bio-composites and measure the corresponding mechanical properties. It was found that the temperature was considerably affected by the presence of rice husk flour that the temperature increased. Due to which manufacturing time also increased.

H.-J. Kim et al (2005), studied and found that the thermal stability was found to slightly decrease and the ash content increased as the lignocellulosic filler loading increased. The dispersion and interfacial adhesion between the lignocellulosic filler and thermoplastic polymer were important factors affecting the thermal stability of the composite system. A thermoplastic polymer, a polyolefin, was used as a matrix polymer to compound with the natural fillers for the purpose of manufacturing the bio-composite. Lignocellulosic materials are chemically active and decompose thermo chemically in the range of 150 to 500°C; hemicellulose mainly between 150 and 350°C, cellulose between 275 and 350°C and lignin between 250 and 500°C. As the filler loading increased, the thermal stability of the composites slightly decreased, whereas the final ash content increased. These results showed that the thermal stability of the composites decreased as the lignocellulosic filler content increased. In order to improve their compatibility and interfacial adhesion, the incorporation of a compatibilizing agent into the lignocellulosic material- thermoplastic polymer bio-composites is recommended.

Hyun-Joong Kim et al (2006), investigated the biodegradability of PBS and bio-flour, which is a poly(butylene succinate) (PBS) bio-composite filled with rice-husk flour (RHF) reinforcing, in natural and aerobic compost soil. The biodegradability of the bio-composites was enhanced with increasing bio-flour content because the bio-flour is easily attacked by microorganisms. These biodegradable materials can be completely degraded into natural ecosystems such as active sludge, natural soil, lake and marine. Accordingly, the biodegradability of biodegradable polymers corresponds to the ability to be chemically transformed by the action of biological enzymes or microorganisms. As the filler loading increased, the tensile strength and the notched Izod impact strength of PBS and the bio-composites decreased more rapidly during the compost soil burial test over 80 days This result indicated that the addition of RHF enhanced the biodegradability of PBS mainly because of the increased polymer surface created after RHF consumption by microorganisms. With increasing RHF content, the percentage weight loss of the bio-composites decreased more rapidly. This result indicates that the degradation of RHF was faster than that of PBS matrix polymer because the cellulosic materials are easily attacked by microorganisms. The results concluded that use of these bio-composites will reduce the environmental problems associated with waste pollution and the study findings support the predicted application of bio-composites as ''green-composites'' or ''eco-materials''.

D.V. Reddy et al (2006) studied the potential use of Rice Husk Ash (RHA) as a cementations material in concrete mixes. RHA is a product of burning of rice husk. The ash content is about 18-22% by weight of the rice husks. Study concluded that the cost effectiveness and enhanced durability, coupled with its energy efficient contribution to turning down the Global Thermostat make RHA a significant contributor to a holistic approach by the concrete industry to the global issue of environmental sustainability.

Didik Prasetyoko et al (2006) used White rice husk ash (RHA), an agriculture waste containing crystalline tridymite and a-cristobalite, as a silica source for zeolite Beta synthesis. Zeolite Beta was synthesized by a direct hydrothermal route, using crystalline silica of rice husk ash (RHA) as a silica source and tetraethyl ammonium hydroxide as the structure-directing agent. Rice husk ash was completely transformed to pure zeolite Beta phase after just 2 days of hydrothermal synthesis at 150°C. The transformation of RHA involves dissolution of the silica in the highly basic reaction mixture, followed by the formation of an alumino-silicate, crystallization of the metastable zeolite Na–P and finally the pure zeolite Beta phase.

Hyun-Joong Kim et al (2006), studied the effect of electron-beam (EB) irradiation on interfacial adhesion in bio flour (rice-husk flour, RHF)-filled poly(propylene) (PP) bio composite in which either only the RHF had been EB irradiated or the whole bio-composite had been EB irradiated and examined at different EB-irradiation doses. These bio-fillers had several advantages over inorganic fillers as reinforcements due to the result of their light weight, renewability, low cost, reduced abrasion during production process, eco-friendliness, and biodegradability. However, main disadvantage of using these bio-fillers as the reinforcing filler in bio-composites is the reduced physical, mechanical, and thermal properties as the bio-filler content increases. This is a result of weak interfacial adhesion between the hydrophilic bio-fillers and the hydrophobic polymer matrix. Bio-fillers mainly contain hydroxyl groups of cellulose and hemicellulose molecules which reduce the bonding and wetting of bio-composites because of poor compatibility with hydrophobic matrix (polyolefin). RHF contains the hydroxyl groups of cellulose and hemicellulose ingredients which reduce the interfacial adhesion and wetting between the RHF (hydrophilic) and PP (hydrophobic). The pretreatment of hydroxyl groups on RHF can be achieved with EB radiation. The reduction of OH groups and impurities from RHF can lead to improved interaction between the RHF surface and PP matrix. An EB radiation dose of 2 Mrad on RHF gave the highest tensile strength. Therefore, this result shows that the EB irradiation method not only increased the strength of the matrix polymer, but also improved the interaction between the bio-flour and matrix polymer at low irradiation dose. The result supports recommendation of the EB process as an environmentally friendly, new method to enhance interfacial bonding in bio-flour-polymer composites.

Hyun-Joong Kim et al (2006), studied the effects of bio-scavengers on the formaldehyde emission, bonding strength, curing behavior, and thermal decomposition properties of MF (Melamine formaldehyde) resins for engineered flooring and adhesion for wood. Four varieties of bio-scavengers, than in powder, wheat flour, rice husk flour, and charcoal, were added to MF resin at 5 wt.-%. To determine formaldehyde emission and bonding strength, they manufactured engineered floorings. MF-charcoal was most effective in reducing formaldehyde emission because of its porous nature, but its bonding strength was decreased. Than in powder and wheat flour, which contain more hydroxyl groups, showed higher bonding strength and curing degree than pure MF resin did. Although the hydroxyl groups of the bio-scavengers were effective in reducing formaldehyde emission and improve bonding strength and curing degree, rice husk flour and charcoal behaved like inorganic substances, there by disturbing the adhesion between MF resin and wood and thus reducing the bonding strength. In thermo-gravimetric analysis, MF-tannin showed the highest thermal stability in the low-temperature range from 100 to 300°C. Due to variations in the chemical and phase structures, the effect of bio-scavengers on formaldehyde emission, bonding strength and curing were different even though all the four are lingo-cellulosic byproducts. Wheat flour was the most effective bio-scavenger among the four tested in this study. Final conclusion was that rice husk flour can be used as Bio-scavenger which when added reduces the level of formaldehyde emission from the adhesives used in engineered flooring and wooden interior materials.

Hyun-Joong Kim et al (2006) used polypropylene as the matrix polymer and rice-husk flour and wood flour as the reinforcing filler to conduct Izod impact tests and he founded that the tensile properties of the composites made with the twin-screw extruding system were better than those of the composites made with the single screw extruding system, due to the improved dispersion of the filler. The tensile strength and modulus of the lignocellulosic filler-PP composite made with the twin-screw extruding system were improved in the case of the composite made without any compatibilizing agent and significantly improved in the case of the composite made with the compatibilizing agent, as compared with those made with the single-screw extruding system. The Izod impact strengths of the composites made with the two different extruding systems were almost the same, the degree of dispersion of the fillers might influence the notched impact performance, but the similar impact strength of both samples with different extruding processes might be due to the fact that impact test is not discriminating enough to reveal the difference in dispersion status of the present composites. All of the SEM micrographs of the fracture surfaces of the composites made with the twin-screw extruding system show well dispersed fillers, as compared with those made with the single screw extruding system.

Figure 4 shows the tensile strengths of lignocellulosic filler-polypropylene composites by using extruding system. Poor adhesion of filler with the matrix and agglomeration results decreased tensile strength of the composites. Twin extruding system shows the improved tensile strength of composites as compared to composites made from single extruding system. This is because of the better dispersion of the filler in the matrix system. A compatilizing agent namely Maleated polypropylene (MAPP) was used to improve the interfacial bonding between the filler and the

Figure 4.

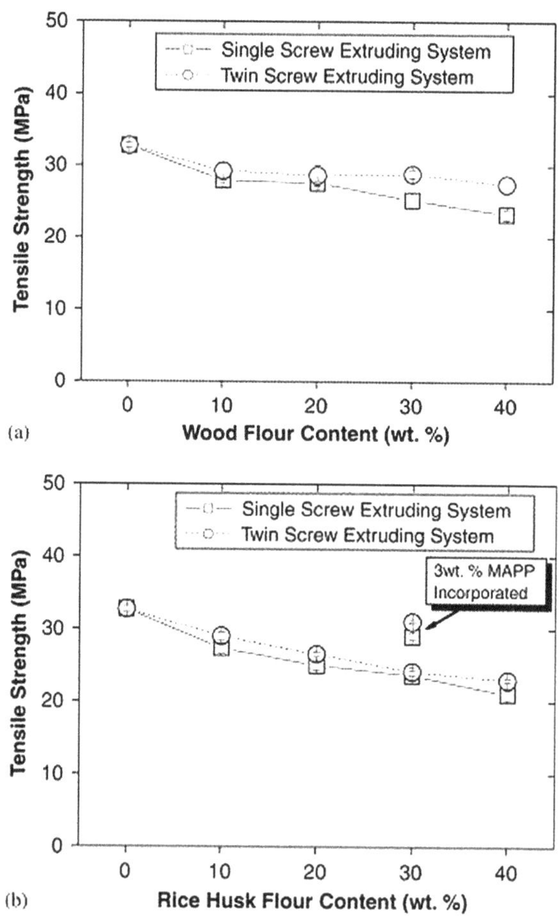

Figure 5.

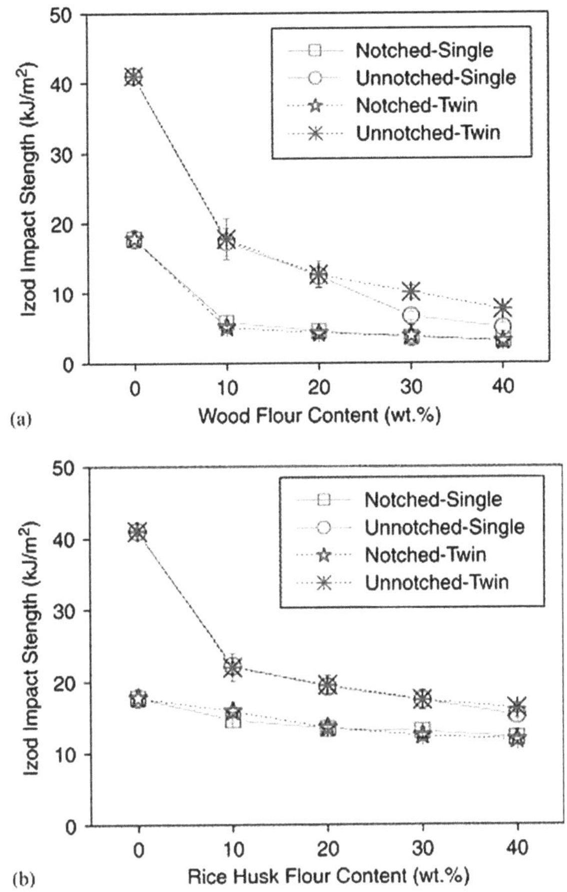

polymer matrix. From Figure 3 it has been observed that the tensile strengths of the 30 wt% lignocellulosic filler filled pp composites at a compatibilizing agent content of 3wt% represents the same propensity according to the type of extruding system as those made without the compatibilizing agent. (Kim et. al. 2006)

The Izod impact tests were performed at room temperature. Two types of specimen such as the notched and unnotched specimens were tested. Figure 5 shows the Izod impact strengths of the lignocellulosic filler-PP composites made by different extruding systems. From fig it has been observed that the Izod impact strengths of the composites made by two different extruding systems are nearly same. In case of notched specimen, the polymer matrix resists the crack propagation. On the other hand, unnotched impact strength will be influenced by the energy absorbed by the plastic flexural deformation before crack initiation. The notched specimens show lesser impact strengths as compared to the unnotched specimens (Kim et al. 2006).

Hyun-Joong kim et al (2006) investigated the effect of the addition of two different compatibilizing agents, maleic anhydride (MA)-grafted polypropylene (MAPP) and MA-grafted polyethylene (MAPE), to bio-flour-filled Polypropylene (PP) and low density polyethylene (LDPE) composites on thermal properties. He concluded that the thermal stability, degradation temperature and DTG max degradation temperature of MAPP- and MAPE-treated composites were slightly higher than those of non-treated composites. However, the thermal stability and degradation temperature of the two different base resin types of MAPE treated composites were not significantly changed.

Hyun-Joong Kim et al (2007) studied the effect of compatibilizing agents on mechanical properties and morphology of a lignocellulosic material using rice-husk flour as the reinforcing filler. The results concluded that as the filler loading increased, the composite made without any compatibilizing agent showed decrease in tensile strength and more brittleness, but the mechanical properties greatly improved by incorporating the compatibilizing agent. The poor interfacial bonding between the filler and the matrix resulted in decreased tensile strength, but modulus and strength were improved with the addition of compatibilizing agent. The compatibilizing agent had no positive effect on Izod impact strength. As the test temperature increased, the thermo plastic matrix polymer was softened and the composite showed plastic matrix deformation, leading to decreased tensile strength and modulus. Due to the strong interfacial bonding between the filler and the matrix polymer, the fracture occurred not at the interface, but at the filler particles themselves, and the composite showed more brittleness in terms of its tensile and impact properties.

Jong-sung Kim et al (2007) investigated the effects of using three additives, wheat flour (WF), rice husk (RH) and charcoal, to Melamine-formaldehyde (MF) resin, for decorative purposes and base plywood in engineered flooring in order to reduce the formaldehyde emission levels and improve the adhesion properties. In the desiccator test, the formaldehyde emission of all adhesive systems decreased with the use of additives. The charcoal additive reduced the formaldehyde emission most effectively. The same trend was shown in the perforator test results. The formaldehyde emissions of MF adhesive with charcoal (8:2) and of MF/WF/charcoal (8:1:1) were both lower than 0.15 mg/l at a press time of 160 s. The reduction effect of charcoal on formaldehyde level was attributed to a probable reduction in absorption. Formaldehyde emission levels were not reduced with increasing press time because the adhesive layers were destroyed by excessive press time which allowed adhesive to flow out on the surface of the decoratives.

M. A. Ahmadi et al (2007) made a study on the development of Mechanical properties of self-compacting and ordinary concretes with rice-husk ash (RHA) up to 180 days. In this study two different replacement percentages of cement were used by RHA, 10%, and 20%, and two different water/cementicious material ratios

(0.40 and 0.35), for both of self-compacting and normal concrete specimens. The results were compared with those of the self-compacting concrete without RHA, with compressive, flexural strength and modulus of elasticity. It was concluded that RHA provides a positive effect on the Mechanical properties at age after 60 days. The final conclusion of the result was that self-compacting concrete specimens have higher values than normal concrete specimens in all tests except that of modulus of elasticity. Also specimens with 20% replacement of cement by RHA have the best performance.

Dao Van Dong et al (2008) presented several key properties of high strength concrete using rice husk ashes (RHA). RHA obtained were used with various contents to partially replace for cement binder in high strength concrete. Key properties of concrete, including: slump, density, compressive strength, water and chloride permeability resistances, were investigated in comparison between samples without using RHA and samples using two types of RHA. The study concluded the following results: Low quality RHA can be used as filler for concrete. The acceptable content is 10% to replace for cement with an acceptance of reduction in compressive strength. High quality RHA can be used as a super pozzolanic additive for HSC.

Shazim Ali Memon et al (2008) explored the use of Rice Husk Ash (RHA) to increase the amount of fines and hence achieve self-compatibility in an economical way. The study focuses on comparison of fresh properties of SCC containing varying amounts of RHA with that containing commercially available viscosity modifying admixture. The possibility of developing low cost SCC using RHA is feasible. The utilization of RHA in SCC solves the problem of its disposal thus keeping the environment free from pollution.

Qinglin Wu et al (2008) investigated the fibre characteristics and the influences of fibre type and loading rate on HDPE crystallization behavior and composite mechanical properties Composite panels using virgin and recycled high-density polyethylene (VHDPE and RHDPE) and five types of natural fibres including four rice straw components (i.e., rice husk, rice straw leaf, rice straw stem, and whole rice straw) and wood fibre as control were made by melt compounding and compression moulding. Fibre length and aspect ratio distributions for all fibres followed a lognormal distribution after milling with two parameters defining the curve location (i.e., mean fibre length/aspect ratio) and shape (i.e., mean fibre length/aspect ratio distribution). For both VHDPE and RHDPE, rice straw fibre systems had comparable mechanical properties with those of wood composites. Increase in fibre loading led to increased moduli and decreased tensile and impact strength. Composite panels with rice husk had the smallest storage moduli, but their impact strength was comparable or better than that of other straw fibres also very little difference in mechanical properties existed among leaf, stem, and whole straw fibres. The particular recycled HDPE resin and its composites had significantly better moduli and strength properties compared to the virgin HDPE systems due to additives used during initial processing. X-ray diffraction experiments showed that introducing fibre to HDPE matrix did not change characteristic peak position, but the fibre increased crystalline thickness of HDPE system. Differential scanning calorimetry experiments showed that VHDPE had significantly larger peak heat flow during cooling run than the RHDPE, indicating higher crystallization rates for VHDPE. The use of fibre in both resin systems led to the reduced peak heat flow rate.

Wu et al. used five kinds of reinforcing material such as rice husk fiber (RHF), rice straw leaf fiber (SLF), rice straw stem fiber (SSF), rice whole straw fiber (WSF), and wood fiber (WF) for making of composites.

Although, research shows that notable differences can be observed while using 50wt% fibers for reinforcement. Figure 6a showed increased moduli of VHDPE (Virgin HDPE)/WSF composites as compared to other composites. SLF and SSF showed identical reinforcement result for composites. Figure 6b shows that the tensile strength of the VHDPE composites is lower as compared to the other four composites under same fiber loading. Also there is no notable difference was observed among all other four composites. From Figure 6c it can be observed that the RHF shows the best reinforcement result of impact strength as compared to other fibers at the fiber loading of 30 wt.

S. Ch. Turmanova et al (2008) studied the thermal stability and kinetics of non-isothermal degradation of polypropene and polypropene composites filled with 20 mass% vigorously grounded and mixed raw rice husks (RRH), black rice husks ash (BRHA), white rice husks ash (WRHA) and Aerosil Degussa (AR). This study found that the kinetics of non-isothermal degradation of the samples studied was best described by kinetic equations of n-th order (F_n mechanism). Depending on the nature of the filler used, the values of n varied from 0.66 to 1.63. The highest values

Figure 6.

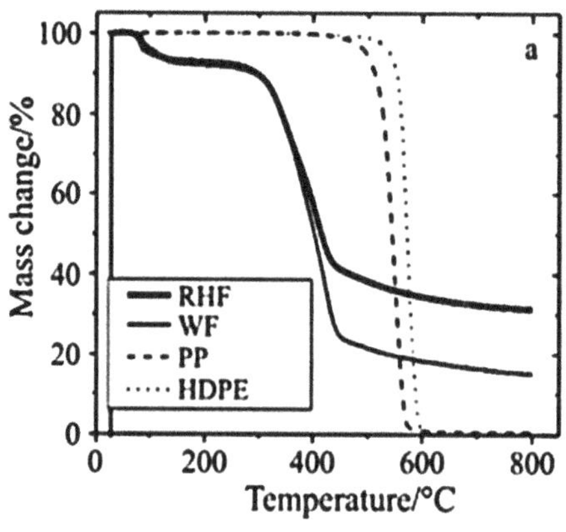

of n, E and A were observed for the composites filled with WRHA and Aerosil which have the same chemical composition (SiO_2) and porous structure. Also linear dependence was found between lnA A and E of the samples studied, known also as kinetic compensation effect. The results obtained were considered enough to conclude that the cheap raw rice husks (RRH) and the products of its thermal degradation (BRHA and WRHA), after vigorously grounding and mixing, can successfully be used as fillers for polypropene to replace the expensive synthetic additive Aerosil in the preparation of different polypropene composites.

Buzarovska A et al (2008) prepared Rice straw filled composites from poly (hydroxybutyrate-co-hydroxyvalerate) (PHBV) copolymer containing 13% (mol) hydroxyvalerate and investigated the effects of rice straw content on thermal and mechanical properties of the composites. It was found that the value of tensile modulus value almost doubled with the increase of rice straw content, while the tensile strength slightly decreased, compared to pure PHBV resin. Differential scanning calorimetry (DSC) and thermo gravimetry (TGA) results demonstrated a minor effect of the rice straw on thermal behavior of PHBV resin. This study concluded that PHBV/RS composites were expected to be developed as materials for structural application, especially for panelized components with good thermal insulation, intended for improvement of the energy efficiency in eco-buildings.

A.A. Ramezanianpour et al (2009) determined the amorphous silica content of the burnt rice husk. Attempts were made to determine the optimum temperature and duration of burning. The experiments concluded that the duration and temperature of furnace are important parameters in influencing the reactivity of RHA pozzolans. Silica in the rice husk initially exists in the amorphous form, but may become crystalline when rice husk is burnt at high temperatures. The performance of concrete with cement replacement by RHA was outstanding with respect to water and chloride ion penetrations.

Elijah T. Iyagba et al (2009) investigated the co-digestion of cow dung with rice husk for biogas production at laboratory scale this was studied at room temperature (26 - 29°C) for a period of 52 days. The results of this research suggests that rice husk does not have the potential for biogas production at the temperature range of 26 - 29°C and that the co-digestion of rice husk with cow dung at this temperature does not improve the digestibility of rice husk for biogas production.

Fidelis O. OKAFOR et al (2009) investigated the effect of RHA on some geotechnical properties of a lateritic soil classified as A-2-6 (0) or SW for sub-grade purposes. The investigation includes evaluation of properties such as compaction, consistency limits and strength of the soil with RHA content. The following conclusions may be drawn from the study: The increase in RHA content decreased the plasticity index of the soil but increased the volume stability of the soil. The addition of RHA also improved the strength property (CBR) of the soil. It was observed that 10% RHA content was optimum for the lateritic soil.

Sumin Kim et al (2009) investigated the effect of addition of rice husk on the gypsum board and assess the physio-mechanical properties, Total Volatile Organic Compound (TVOC), and in- combustibility. The study showed that because of the replacement of pore between gypsum particles by rice husk, the moisture absorption characteristic was decreased and Modulus of Rapture (MOR) of the gypsum-rice husk board was increased by up to 9.8 MPa at 30 wt%. However, MOR decreases by addition of more than 40 wt % rice husk. The modulus of elasticity (MOE) also showed similar characteristics. However, on the other hand, internal bonding strength (IB) slightly increases as the amount of rice husk is increased up to 20 wt% and decreases over 20 wt%. The incombustibility of gypsum rich husk boards decreases on increasing the rice husk content. However, up to 30 wt% of rice husk adding contents board samples was of incombustibility first class. This study concluded that Gypsum particles can be replaced up to 30 wt% by rice husk with incombustibility first class for housing materials.

Dr. Robert M. Brooks et al (2009) studied the potential of RHA-fly ash blend as a swell reduction layer between the footing of a foundation and sub-grade. In order to examine the importance of the study, a cost comparison is made for the preparation of the sub-base of a highway project with and without the admixture stabilizations. Stress strain behavior of unconfined compressive strength showed that failure stress and strains increased by 106% and 50% respectively when the flyash content was increased from 0 to 25%. When the RHA content was increased from 0 to 12%, unconfined compressive stress increased by 97% while CBR improved by 47%. Therefore, from his study it is recommended that an RHA content of 12% and a flyash content of 25% must be used for strengthening the expansive subgrade soil. A flyash content of 15% was recommended for blending into RHA for forming a swell reduction layer because of its satisfactory performance in the laboratory tests.

Dr Anbu Clemensis et al (2009) studied that RH can be utilized in the manufacturing of particleboards, ceiling boards and insulation boards. The use of biodegradable adhesives could reduce the use of synthetic adhesives based on petroleum resources and its ill effects. These materials could provide competitive composite boards for construction and at the same time, environmentally friendly also. Developing countries should make an effort to harness the potential of the abundantly available RH. As efficient utilization could serve as a revenue for the nation.

Simone M.L. Rosa et al (2009) studied the properties of rice husk flour-filled polypropylene (PP) in view of the large quantities of this agricultural product available as residue in Brazil. The rice husk flour (RHF) was characterized by SEM and particle size distribution. The properties of the composites were studied by MFI,

DMA and TGA analyses. In his study showed that most of the particles (~70%) sizes from 212 to 600 micrometers. It was observed that MFI decreased with RHF concentration because the presence of fibers restricts the flow of the polymer matrix and this restriction increases with fiber loading. For rice husk-filled polypropylene composites, it can be seen that the degradation begins below 300 °C, accompanying the profile of the rice husk alone. The temperature of the maximum degradation rate of PP (Polypropylene) in the composites seems to occur at a slightly higher temperature than that of the pristine polymer in most compositions, suggesting that the filler is inducing some kind of thermal stabilization on PP molecules. The results were as follows MFIs were found in the composites as compared to the polymer matrix. The degree of crystallinity and the crystallization temperature of the matrix were shown to increase in the composites, as well as the temperature of the maximum degradation rate of PP and the storage modulus of the Thermal and Dynamic-Mechanical Characterization of Rice-Husk Filled Polypropylene Composite.

Simone Maria Leal Rosa et al (2009) studied that thermoplastic composites filled with rice husk flour are materials that offers an alternative for using this agricultural resource, viewing the production of low dense materials with some specific properties. The study concluded that rice husk was milled and dried for using as filler in PP composites. It was verified that it is feasible to use this by-product of the rice milling process as low cost filler, in view of the properties of the obtained products. The composite stiffness was seen to increase with increasing filler loading. The tensile strengths slightly decreased, however they were improved in the presence of the coupling agent MAPP.

Segun R. Bello et al (2010) Investigated the use of rice husk as a fuel for drying in a biomass furnace dryer. The experiment concluded that Rice husk emits greater exhaust gases and produced the least energy, 2.93 kW as a result of low differential temperature, but it has longer combustion period, which could make them more suitable for long process operations. Rice husk can be used for heat process operations, which requires heat processing condition between 63-85° C for about 15 to 30 minutes such as pasteurization of milk, fruit juice and boiling of eggs.

Arshad A. Salema et al (2010) investigated the effects of feeding methods on the combustion characteristics of rice husk and their resulting ash quality. Two methods of feeding were employed, namely inclined feeding an inclined, tangential feeding. The utilization of the inclined, tangential port for feeding of rice husk into the fluidized bed combustor resulted in a higher degree of rice husk penetration and burning in the sand bed. The temperature profile during the combustion process was very stable and this resulted in the ease of temperature control. But the fly ash still retained its amorphous structure.

K. Kartini et al (2010) investigated durability performance conducted on the normal strength concrete specimens of 30 N/mm^2 containing 20% or 30% RHA by cement weight, with or without addition of superplasticizer. The study found that: Increasing the replacement of OPC with RHA in the concrete mixes resulted in lower workability. The water absorption and ISA values of RHA concrete was lower than the OPC control concrete. RHA improved the durability of concrete. OPC control concrete was more permeable than the RHA concrete. The resistance to chloride ion penetration of concrete as measured by the charge coulomb drastically enhances resistance to chloride permeability with incorporation of RHA.

Muhammad Harunur Rashid et al (2010) presented the effects of using Rice Husk Ash (RHA) as a partial cement replacement material in mortar mixes. The mechanical properties investigated were the compressive strength, and also the porosity of mortar was tested. The use of RHA significantly improves the mortar strength at the 20% replacement level. At 30% replacement level of OPC by RHA the porosity of mortar is increased at 28 and 90 days as compared to OPC mortar.

Faisal Estu Yulianto et al (2010) Studied the effectiveness of using rice husk ash (RHA) as a pozolon to enhance the lime for stabilizer material of peat soil. Besides, the effect of curing period to the behavior of stabilized peat soil is also presented. From the data and their analysis it can be concluded that The use of 10% stabilizer material (30% lime +70% rice husk ash) was able to improve physical and engineering characteristics of peat soil, i.e.: specific gravity and wet unit weight increase; water content and void ratio decrease; soil shear strength increases and its compression decreases.

Akshaya Kumar Sabat et al (2011) presented the results of a laboratory study undertaken to investigate the effect of Marble dusts on strength and durability of an expansive soil stabilized with optimum percentage of Rice Husk ash (RHA). Following conclusions were drawn from the study: The optimum percentage of RHA in stabilization of expansive soil is found out be 10%. The MDD goes on decreasing and OMC goes on increasing irrespective of the percentage of addition of Marble dust to RHA stabilized expansive soil. The UCS of the RHA stabilized expansive soil increased up to 20% addition of Marble dust. The Soaked CBR of the RHA stabilized expansive soil increased up to 20% addition of Marble dust. The swelling pressure of the expansive soil goes on decreasing irrespective of the percentage of addition of Marble dust to RHA stabilized expansive soil. The RHA stabilized expansive soil could not have survived the durability test. For best stabilization effect, the optimum proportion of Soil: Rice husk ash: Marble dust was found to be 70: 10: 20.

Can BurakSisman et al (2011) investigated the physical and mechanical properties of briquette produced from rice husk ash (RHA). To do this, six briquette classes were formed by changing the volume of RHA and cement and porous briquettes were produced for each class. Main conclusions of the study were: Dimensional deformation decreased a little with the increasing ratio of RHA in the mixture. The oven-dry bulk density of the briquette classes decreased with increasing ratios of RHA in the mixture. Water suction ratios of the briquettes classes increased with RHA. The compressive strength of the briquette classes decreased with the increasing replacement ratio of RHA in the mixture. Heat conductivity of the briquettes classes decreased with the increase in the volume of RHA in the mixture.

Haq Nawaz Bhatti et al (2011) studied free, carboxy methyl cellulose (CMC)-immobilized, polyvinyl alcohol (PVA)-alginate immobilized and chemically treated rice husk biomass was used for the bio-absorption of Ever direct Orange-3GL and Direct Blue-67 dyes. This study concluded that the HCl treatment of the rice husk biomass enhanced the bio- absorption capacity of Everdirect Orange-3GL and Direct Blue-67 dyes. Various physical and chemical modifications/treatments showed different effect on the Bio- absorption capacity of rice husk. The study showed that, rice husk can be used successfully to clean the environment and can be applied practically in industries.

S.Vivekanandan et al (2011) studied Anaerobic digestion of rice crop residues –rice chaff, rice straw and rice husk as co-substrate with cow dung . Gas chromatography was used to quantify the different component of biogas produced. The results concluded low performance of cow dung / rice husk was due to the high lignin content which contained unfavorable non lignin-carbon to nitrogen ratio (70:1). Rice husk acted as a good inoculum because it increased the number of microbes in the digestion and made the biogas process faster.

Koteswara Rao. D et al (2011) in studied the addition of rice husk ash, lime and gypsum to the expansive soil which resulted in considerable improvement in the strength characteristics of the expansive soil. This experiment found that liquid limit of the expansive soil decreases by 22% with the addition of 20% RHA+5% Lime. Free Swell Index of the expansive soil reduces by 88% with the addition of 20% RHA and 5%Lime. Unconfined compressive strength of the expansive soil increases by 548% with addition of 20% RHA, 5% lime and 3% Gypsum after 28 days curing. It was noticed that the CBR value of the expansive increased by 1350% with the addition of 20% RHA, 5% lime and 3% Gypsum after 14 days curing. The results concluded that Rice husk ash can potentially stabilize the expansive soil solely (or) mixed with lime, gypsum.

Omofuma Fabian Ehichioya et al (2011), studied the effects of particle size and particle size distribution of filler on rubber compounds. Test like moisture content, pH, ash content, hardness resistance, abrasion resistance, compression set and flex fatigue were subsequently carried out on the vulcanite. The results showed that the fillers (rice husk and wood flour) has percentage moisture of 2.5% and 6.25%, pH value of 9 and 5.5 and ash content of 12.5% and 10% respectively. pH result shows that rice husk is alkaline. As the particle size decreases, Hardness resistance for rice husk and wood flour increased; this means that the smaller the particle size, the better the hardness resistance of the material.

Luduena et al (2011) carried out multistep procedures to obtain valuable sub products, including Nano-cellulose, from rice husk. Each sub product was characterized after each step by analyzing the chemical composition and morphology. The result of this experiment showed the possibility to separate silica in the first step and then purify the resultant materials leading to Nano-cellulose production. The results concluded that the obtained sub product can be used as additive and fillers in a wide range of applications. Mohammad Ghofrani et al (2012) studied the possibility of using lignocellulosic material as reinforcing filler in Wood Plastic Composite (WPC). Recycled high density polyethylene was used as the matrix and rice-husk flour as the reinforcing filler to prepare a wood-plastic composite. Water absorption and thickness swelling of the WPC specimens was conducted after 2 and 24 hours of immersion in water to evaluate the effect of the linkage quality on the dimensional stability of the panels. It was found that increasing the HDPE percentage significantly reduces the water absorption and thickness swelling. Water absorption and thickness swelling of the specimens decreases by adding more coupling agent Meline anhydrous polythene(MAPE). Also Increasing HDPE and the coupling agent (MAPE) improved the mechanical properties of bending strength, internal bonding and MOE, all of which increased significantly. According to the results of his analysis, the optimum composition was determined as follows: 60% HDPE, 6% MAPE and density 0.8 g/cm^3.

Ephraim et al. (2012) investigated the effect of partially replacing Ordinary Portland Cement (OPC) with local additive Rice Husk Ash (RHA) which is known to be super pozzolanic in concrete at optimum replacement percentage to reduce the cost of housing. This research found that the specific gravity of RHA to be 1.55, and the density of RHA concrete was 2.043, 1.912 and 1.932kg/m3 at 10%, 20% and 25% replacement percentages respectively. RHA concrete was found to be very workable with a slump value of over 100mm. The incorporation of RHA in concrete resulted in increased water demand and enhanced strength. The compressive strength values at 28days were found to be 38.4, 36.5 and 33N/mm^2 at the same replacement percentages above. These compressive strength values compared favorably with the controlled concrete strength of 37N/mm^{2}at a mix ratio of 1:1.5:3. This concluded that the use of RHA in civil construction works will reduce environmental pollution, improve the quality of concrete and reduce its cost of production as well as solving the problem of agro-waste management by putting into use this locally found additive (RHA).

CONCLUSION

Rice husk has been used directly as a value added material for manufacturing and synthesizing new materials or as a low cost substitute material for modifying the properties of existing products. Presence of silica is an additional advantage in comparison to other byproduct materials which makes RH an important material for a wide range of manufacturing and application oriented processes. Easy availability and low price of rice husk in rice producing countries is an extra benefit towards the use of this material. Despite having high potential and suitability in so many well established uses, use of rice husk has been limited. In the competitive market, proper utilization of rice husk and its ash will be beneficial for industry. A systematic approach to this material can give birth to new industrial sectors for rice husk. Despite having so many applications but there is still a need for the improvement in manufacturing and processing techniques so as to transform this waste material to become usable and also economically viable.

REFERENCES

Ahmadi, M. A., Alidoust, O., Sadrinejad, I., & Nayeri, M. (2007). Development of Mechanical Properties of Self Compacting Concrete Contain Rice Husk Ash. *International Journal of Computer and Information Engineering*, *1*, 4.

Bello, S. R., & Adegbulugbe, T. A. (2010). Comparative Study on Utilization of Charcoal, Sawdust and Rice Husk in Biomass Furnace-Dryer. *Journal of Scientific Research and Development*, *12*, 1592.

Binod, P., Sindhu, R., Singhania, R. R., Vikram, S., Devi, L., Nagalakshmi, S., & Pandey, A. et al. (2010). Bioethanol production from rice straw: An overview. *Bioresource Technology*, *101*(13), 4767–4774. doi:10.1016/j.biortech.2009.10.079 PMID:19944601

Brooks, R. B. (2009). Soil stabilization with flyash and rice husk ash. *International Journal of Research and Review in Applied Science*, *1*, 210–217.

Buzarovska, A. (2008). Potential use of rice straw as filler in eco-composite materials, *Australian Journal of Crop Science, 1*, 37-42.

Dong, D. V., Huu, P. D., & Nguyen, N. L. (2008). Effect of rice husk ash on properties of high strength concrete. *The 3rd ACF International Conference- ACF/VCA 2008.*

Ephraim, M. E., Godwin, A. A., & Joseph, O. U. (2012). Compressive strength of concrete with rice husk ash aspartial replacement of ordinary Portland cement. *Scholarly Journal of Engineering Research*, *1*, 32–36.

Ghofrani, M., Pishan, S., Mohammadi, M. R., & Omidi, H. (2012). Study on Rice-husk/Recycled High Density Polyethylene Composites – Their Physical and Mechanical properties. *Environmental Sciences*, *9*, 99–112.

Gidde, M. R., & Jivani, A. P. (2007). Waste to Wealth - Potential of Rice Husk in India: A Literature Review. *Proceedings of the International Conference on Cleaner Technologies and Environmental Management PEC*.

Haefele, S. M., Konboon, Y., Wongboon, W., Amarante, S., Maarifat, A. A., Pfeiffer, E. M., & Knoblauch, C. (2011). Effects and fate of biochar from rice residues in rice-based systems. *Field Crops Research*, *121*(3), 430–440. doi:10.1016/j.fcr.2011.01.014

Iyagba, , & Mangibo, , & Mohammad. (2009). The study of cow dung as co-substrate with rice husk in biogas production. *Scientific Research and Essays*, *9*, 861–866.

Johnson, D. & Bhg, Y. Dato E. & Younus, N.B. (2009). Particle board from rice husk- A Brief Introduction to Renewable Materials of Construction. *JURUTERA*, 12-15.

Kartini, K., Mahmud, H. B., & Hamidah, M. S. (2010). Absorption and permeability performance of selangor rice husk ash blended grade 30 concrete. *Journal of Engineering Science and Technology*, *5*, 1–16.

Kim, H. J., & Gueneom, Y. (2001). Thermogravimetric Analysis of Rice Husk Flour for a New Raw Material of Lignocellulosic Fiber Thermoplastic Polymer composite. *Mokchae Konghak*, *29*, 59–67.

Kim, H.-J., Son, J., Park, H.-J., Lee, B.-J., & Hwang, T.-S. (2004). Rice-husk flour filled polypropylene composites; mechanical and morphological study. *Composite Structures*, *63*(3-4), 305–312. doi:10.1016/S0263-8223(03)00179-X

Kim, H.-S., Kim, H.-J., Lee, J.-W., & Choi, I.-G. (2006). Biodegradability of bio-flour filled bio degradable poly(butylene succinate) bio-composites in natural and compost soil. *Polymer Degradation & Stability*, *91*(5), 1117–1127. doi:10.1016/j.polymdegradstab.2005.07.002

Kim, H. S., Kim, S., & Kim, H.-J. (2006). Enhanced Interfacial Adhesion of Bioflour-Filled Poly(propylene) Biocomposites by Electron-Beam Irradiation. *Macromolecular Materials and Engineering*, *291*(7), 762–772. doi:10.1002/mame.200600121

Kim, H. S., Kim, S., Kim, H. J., & Yang, H. S. (2006). Thermal properties of bio-flour-fliied polyolefin composite with different compatibilizing agent type and content. *Thermochimica Acta*, *451*(1-2), 181–188. doi:10.1016/j.tca.2006.09.013

Kim, H. S., Yang, H.-S., Kim, H.-J., & Park, H.-J. (2004). Thermogravimetric analysis of rice husk flour filled thermoplastic polymer composites. *Journal of Thermal Analysis and Calorimetry*, *76*(2), 395–404. doi:10.1023/B:JTAN.0000028020.02657.9b

Kim, J. S., Eom, Y. G., Kim, S., & Kim, H.-J. (2007). Effect of natural-resource-based scavengers on the adhesion properties and formaldehyde emission of engineered flooring. *Journal of Adhesion Science and Technology*, *21*(3-4), 211–225. doi:10.1163/156856107780684602

Kim, S. (2009). Incombustibility, physico-mechanical properties and TVOC emission behavior of the gypsum–rice husk boards for wall and ceiling materials for construction. *Industrial Crops and Products*, *29*, 381–387.

Kim, S., Kim, H. J., Kim, H. S., & Lee, H. Y. (2006). Effect of Bio-Scavengers on the Curing Behavior and Bonding Properties of Melamine-Formaldehyde Resins. *Macromolecular Materials and Engineering*, *291*(9), 1027–1034. doi:10.1002/mame.200600213

Kumar, A., & Mohanta, K. (2012). Properties and Industrial Applications of Rice husk: A review. *International Journal of Emerging Technology and Advanced Engineering*, *2*, 86–90.

Lee, Y-K., Kim, S., Yang, H.S., & Kim, H.J. (2003). Mechanical Properties of Rice Husk Flour-Wood Particleboard by Urea-Formaldehyde Resins. *MokchaeKonghak*, *31*, 42–49.

Luduena, L., Fasce, D., Alvarez, V. A., & Stefani, P. M. (2011). Nanocellulose from rice husk. *BioResources*, *6*, 1440–1453.

Memon, S. A., Shaikh, M. A., & Akbar, H. (2008). Production of Low Cost Self Compacting Concrete Using Rice Husk Ash. *First International Conference on Construction In Developing Countries*.

Okafor, F. O., & Okonkwo, U. N. (2009). Effects of Rice Husk Ash on Some Geotechnical Properties of Lateritic Soil. *Leonardo Electronic Journal of Practices and Technologies*, *15*, 67–74.

Omofuma, F. E., Adeniye, S. A., & Adeleke, A. E. (2011). A Case Study of Rice Husk and Wood Flour. *World Applied Sciences Journal*, *14*, 1347–1352.

Park, , Han, Kim, Yang, & Kim. (2005). Evaluation of the impact on manufacturing temperature and time in the production process of bio-composites. *MokchaeKonghak, 33*, 29–37.

Prachayawarakorn, J., & Yaembunying, N. (2005). Effect of recycling on properties of rice husk-filled-polypropylene*; songklanakarinJ. Sci.Technol, 27*, 343–352.

Prasetyoko, D., Ramli, Z., Endud, S., Hamdan, H., & Sulikowski, B. (2006). Conversion of rice husk ash to zeolite beta. *Waste Management (New York, N.Y.), 26*(10), 1173–1179. doi:10.1016/j.wasman.2005.09.009 PMID:16274981

Ramezanianpour, A., Mahdi Khani, M., & Ahmadibeni, G. (2009). The Effect of Rice Husk Ash on Mechanical Properties and Durability of Sustainable Concretes. *International Journal of Civil Engineering*, *7*, 83–91.

Rao, K., Pranav, D., & Anusha, P. R. T. (2011). Stabilization of expansive soil with rice husk ash, lime and gypsum –an experimental study *International Journal of Engineering Science and Technology*, *3*, 8076–8085.

Rashid, M.H., & Molla, M.K., & Ahmed, T. (2010). Mortar Incorporating Rice Husk Ash: Strength and Porosity. *European Journal of Scientific Research, 40*, 471–477.

Reddy, D. V. (2006) Durability Characteristics of Rice Husk Ash-Modified Reinforced Concrete.*Fourth LACCEI International Latin American and Caribbean Conference for Engineering and Technology*.

Waswa-Sabuni, B., Syagga, P.M., Dulo, S.O., & Kamau, G.N. (2002). Rice Husk Ash Cement – An alternative pozzolanacement for Kenyan building Industry. *Journal of Civil Engineering, 8*, 13 – 26.

Rosa, S. M. L., Santos, E. F., Ferreira, C. A., Marli, S., & Nachtigall, B. (2009). Studies on properties of rice husk-filled-PP composite. *Materials Research*, *12*(3), 333–338. doi:10.1590/S1516-14392009000300014

Rosa, S.M.L., Santos, E.F., Ferreira, C.A., Marli, S. & Nachtigall, B. (2009). Thermal and Dynamic mechanical characterization of rice husk filled Polypropylene Composites. *Macromolecular Reaearch, 17*, 8-13.

Rozainee, M., Ngo, S.P., & Arshad, A., Salema, & Tan, K.G. (2010). Effect of feeding methods on the rice husk ashquality in a fluidised bed combustor. *Emirates Journal for Engineering Research*, *15*, 1–12.

Sabat, A.K. & Nanda, R.P. (2011). Effect of marble dust on strength and durability of Rice husk ashstabilised expansive soil. *International Journal of Civil and Structural Engineering*, *1*, 939-948.

Safa, Y., & Bhatti, H. N. (2011). Adsorptive removal of direct dyes by low cost rice husk: Effect of treatments and modifications. *African Journal of Biotechnology*, *10*, 3128–3142.

Sisman, C. B., & Gezer, E. (2011). Effects of rice husk ash on characteristics of the briquette produced for masonry units. *Scientific Research and Essays*, *6*, 984–992.

Turmanova, S. Ch., & Genieva, S. D. (2008). Non isothermal degradation kinetics of filled with rise husk ash polypropene composites. *Express Polymer Letters*, *2*(2), 133–146. doi:10.3144/expresspolymlett.2008.18

Vivekanandan, S., & Kamaraj, G. (2011). Effect of co-digestion of cow dung with rice chaff, rice straw and rice husk in biogas production using anaerobic digestion. *International Research Journal of Biotechnology*, *2*, 114–118.

Yang, H. S., Kim, H. J., Park, H. J., Lee, B. J., & Hwang, T. S. (2007). Effect of compatibilizing agents on rice-husk flour reinforced polypropylene composite. *Composite Structures*, *77*(1), 45–55. doi:10.1016/j.compstruct.2005.06.005

Yang, H. S., Wolcon, M. P., Kim, H.-S., & Kim, H.-J. (2005). Thermal properties of lignocellulosic filler-thermoplastic polymer bio-composites. *Journal of Thermal Analysis and Calorimetry*, *82*(1), 157–160. doi:10.1007/s10973-005-0857-5

Yang, H. S., Wolcott, M. P., Kim, H. S., Kim, S., & Kim, H. J. (2006). Properties of lignocellulosic material filled polypropylene bio-composite made with different manufacturing process. *Polymer Testing*, *25*(5), 668–676. doi:10.1016/j.polymertesting.2006.03.013

Yang, H-S. (2004). *Role of rice husk flour as reinforcing filler for thermoplastic polymer bio composite*. Doctoral Thesis.

Yao, F., Wu, Q., Lei, Y., & Xu, Y. (2008). Rice straw fiber-reinforced high-density polyethylene composite: Effect of fiber type and loading. *Industrial Crops and Products*, *28*(1), 63–42. doi:10.1016/j.indcrop.2008.01.007

Yulianto, F. E., & Mochtar, N. E. (2010). Mixing of Rice Husk Ash (RHA) and lime for peat soil stabilization. *Proceedings of the First Makassar International Conference on Civil Engineering* (MICCE2010).

Chapter 9
Techno-Economic and Life Cycle Assessment for the Production of Green Composites

Siddharth Jain
College of Engineering Roorkee, India

Xiaolei Zhang
Queen's University Belfast, UK

ABSTRACT

Botanically, green composites belong to an economically important seed plant family that includes maize, wheat, rice, and sorghum known as Saccharum offi cinarum. There are so many natural fibers available in the environment such as rice husk, hemp fibers, flax fibers, bamboo fibers, coconut fiber, coconut coir, grawia optiva and many others also. Life Cycle Assessment (LCA) is a process to estimate the environmental feature and potential impacts related to a product, by organizing a directory of pertinent inputs and outputs of a product system, assessing the potential environmental impacts related with the said inputs and outputs, explaining the results of the inventory analysis and impact evaluation phases in connection to the objectives of the study. Particularly Bagasse, an agricultural residue not only becomes a problem from the environmental point of view, but also affects the profitability of the sugarcane industries. This chapter discusses the properties, processing methods and various other aspects including economic and environmental aspects related to green composites.

DOI: 10.4018/978-1-5225-0424-5.ch009

INTRODUCTION

Important efforts to protect the environment are focused on finding alternatives to replace synthetic materials, with a growing array of natural materials. The number of research works aiming to develop polymers or composites with natural materials is constantly increasing; one way to accomplish this is by combining the properties of different materials taking advantage of the biopolymer's characteristics.

In present scenario green composites have become most versatile and useful materials as compared to the synthetic fiber reinforced composites. The main reason of popularity of green composites resides in their biodegradability, light weight and high strength. The another important thing regarding the green composite is that these materials utilizes the agriculture wastes such as wheat straw, rice husk, banana fibers, coconut fibers etc and used as reinforcements in various matrix materials like thermoplastic polymer, thermoset polymer matrix and also biopolymer matrix materials. Utilization of these wastes results in an ultimate disposal and manufacturing of a new class of materials such as green composites. The another most important point is related to the environmental issue. This environmental issues are also resolved by the green composites. As these composites use the various agricultural wastes as reinforcements in the polymer matrix system as described earlier in this chapter.

Biomass is mainly composed of organic matter derived from plant sources and the very exclusive process such as "photosynthesis" enables trees and plants to store the solar energy into the chemical bonds of their respective structural components. During the photosynthesis process, the carbon dioxide (CO2) from the blanket of air present in the atmosphere vigorously reacts with the universal solvent, water from the earth to produce arbohydrates (mainly sugars in the form of glucose) and this constitutes the building block of biomass.

The essential raw materials of photosynthesis, water and CO2 on entering the cells of dorsal side of leaf produces simple sugar and oxygen. Since the earth's biomass exists in a thin layer called biosphere, where the life is supported and stores enormous energy constantly which is replenished by flowing energy from the sun as a result of photosynthesis. In this chapter we tried to assess the life cycle of green composites. Also the various factors associated or which are useful for the life cycle assessment of the green composites have also reported in the current chapter.

ECONOMIC FACTORS OF WOODY BIOMASS

Biomass generally has two main categories: "virgin biomass" which mainly comprises forestry and energy crops and "waste biomass" leading from the forest thinning, wood residues, recycling, sewage, municipal wastes, food and animal wastes as well as the domestic waste. Despite the advent of modern fossil energy technologies, the biomass still regarded as the vital source of energy for human beings and also for the advancement of raw materials used especially in the present era of the developing world. According to a recent estimation, it has been noted that the biomass production is about eight times higher than the total annual world consumption of energy from all other sources available on earth. According to literature reports in 2003, the world's population uses only a 7% of the estimated annual production of biomass on the basis of new reading of the production rate (Koren and Bisesi 2003; Berndes et al. 2003).

The energy generated from biomass combustion is used as the basic heat source for all the processes and the heat energy is used to vaporize the working fluid in the medium available. The vapour is stretched downward in the turbine to produce mechanical energy which is further converted into electricity through hydroelectricity and geothermal energy as an alternative source of energy. During the process, an electric boiler is utilized for the preliminary investigation of the whole system and the energy liberated by the combustion of biomass lies in the range of 8 MJ/Kg for wet greenwood to 55 MJ/kg for oven dried plant material; while a 55 MJ/kg is generated from methane combustion and 23–30 MJ/kg for coal burning (Twidell 1998).

Basically, the biomass-based energy production is considered to be a carbon neutral process, i.e. the amount of carbon emissions released after combustion are wholly taken up by the plants during their catabolic activity of growth. This results in no net gaining of carbon dioxide by the atmosphere which proves the law of conservation of energy. If the forest and agricultural residues or wastes are allowed to decompose naturally on their own, the same amount of carbon emissions as biomass-based energy will be released into the atmosphere. The use of biomass as a source of fuel has much wider implications in terms of social, economic, biophysical, biological and environmental aspects. However, the excessive deforestation, i.e. cutting of the trees for fuel needs leads to a reduction in the biodiversity of plant species and also destructs habitat for wildlife, land degradation, soil erosion, etc. The loss of soil can be covered by the use of crop residues and overgrazing increases soil erosion and thus reduces the agricultural production and consumption. Also, the use of biomass fuels gives rise to high levels of indoor air pollution caused from various sources affects human health in a very indigenous way.

In recent years, due to the rapid development and existence of the "peak oil" theory into reality, the renewable carbon, i.e. the base of fuels for energy production has been playing a vital role in today's world economy. Further, in order to depend completely on the carbon-based economy and also to provide energy fully to the current growing population, the research and development efforts are continued to transform the existing fuel-wood technology into a high-tech liquid biofuel technology.

Also, a continuous supply of funds has been provided for the research activities to meet the requirements of the international protocols and guidelines of various agencies such as Kyoto Protocol on the Climatic Changes, Reducing Emissions from Degradation and forest Degradation (REDD) and Cleaner Development Mechanisms at smaller village scales level (Gibbs et al. 2007; Woodhouse 2006a, b). The burning of biomass in the atmosphere, especially the fuel-wood, has served as a major source of energy production according to most of the recorded history. D.O. Hall indicates that biomass produces only a 14% of all energy consumed on worldwide range (Hall 1991). In all the developing countries, fuel-wood produces up to 95% of energy that is consumed yearly. The most dominant use of biomass energy is for cooking and heating and also for some other rural industrial activities including beer brewing, brick firing and pottery making. Other uses of biomass include medicine, food, building materials, household utensils and toys. While biomass fuel is essential for survival in many activities, its use is burdened with lots of problems. Its use is inefficient as it generates domestic indoor air pollution, resulting in various health problems leading to deadly diseases. It is normally women who are said to be affected the most, since they spend most of their time in cooking inside the dwelling. The gathering of fuel-wood is also labour demanding and excessive use of wood results in soil erosion as mentioned above.

There are some major environmental problems arising in the world due to biomass consumption. The scarcity in fuel-wood has nowadays resulted in the people of third world countries to rely on the enormous crop residues and animal dung as an alternative sources of fuel, where households are forced to purchase wood from vendors for domestic use. In such a situation, finding the necessary cash to purchase wood or an alternative energy sources, creates an additional burden on the people residing in rural areas. During the decline in woody biomass, a huge array of the use of this versatile resource is affected to its maximum. This means that as the woody biomass supply diminishes rapidly, the availability of all the artefacts that comes from trees are also affected due to the uprising circumstances. Since, the woody biomass serves as an important source of energy that is currently the most significant source of sustainable as well as renewable mode of energy production in today's world.

The woody biomass, due to its importance and continued dependence of limited, primarily fixed land occupancy are further burdening the available woodland resources in order to meet the energy needs of the ever growing population. Also in recent years, the occurrences of the continuous changes in woodland occupancy are significantly altering the overall biomass production and subsequent energy generation. Due to such unreliable statistics, the modelling of a structure to meet the domestic energy demands at a local level is becoming a challenge (Banks et al. 1996). In 2010, the extensive and global use of woody biomass for energy was about

3.8 Gm3/year (30 EJ/year), which consisted mainly of 1.9 Gm3/year (16 EJ/year) for household fuel-wood and 1.9 Gm3/year (14 EJ/year) for large-scale industrial use in general. During the same period, the world's primary energy consumption was estimated at 541 EJ/year and world's renewable primary energy consumption was observed to be 71 EJ/year, according to International Energy Association (IEA) (2013, http://www.iea.org). Hence in 2010, the woody biomass formed roughly 9% of the world primary energy consumption and 65% of world renewable primary energy consumption. Despite the widespread uses of woody biomass for energy, the current consumptions are still substantially below the existing resource potentials available exclusively (Openshaw 2011).

The woody biomass energy potentials do not depend only on the available woody biomass resources but also on the competition between the factors such as alternative uses of those resources and alternative sources of energy in a very consistent manner (Radetzki 1997; Sedjo 1997). These effects can be depicted and separated by using the concept of supply and demand curves which has been defining its importance. The energy wood supply curve defines the amount of woody biomass which is made available for large-scale energy production at various hypothetical energy wood prices, i.e. it summarizes all the relevant information and data regarding its application from the biomass sector needed to model large-scale energy wood uses. On the other hand, the energy wood demand curve defines the desired amount of woody biomass required for large-scale energy production at various hypothetical energy wood costs.

Environmental Factors of Woody Biomass

The woody biomass is a prevailing attractive feedstock that can be sustainably obtained from nature through the process of photosynthesis for bio-ethanol production (Arato et al. 2005; Zhu et al. 2010). The hybrid poplars in well-managed plantations, native lodgepole pine represents a major wood species from forest thinning of the unmanageable forests that are available in large volumes. This requires value-added utilizations to diminish expensive thinning cost for sustainable healthy forest and ecosystem management exclusively in the environment. Thus the intensive utilization of lodge pole pine for bio-ethanol provides an important sector of the feedstock supply which in other words contributes to future economy based on biofuels.

The woody biomass possesses high fibre with strong physical characteristics in addition to significant amount of lignocellulose material than any other feedstock such as agricultural residues, grasses and agricultural waste which makes it more obstinate to enzymatic destruction leading to serious threat (Sassner et al. 2008; Shi et al. 2009). This gives an idea that the woody biomass research should emphasize majorly the upstream processing, i.e. the pretreatment and also the size reduction phenomenon to overcome the inherent recalcitrance which further enhances the subsequent enzymatic saccharification of polysaccharides. The chemical pretreatments are commonly capable of improving, generating the enzymatic digestibility of biomass by means of diminishing the non-cellulosic constituents (Chen et al. 2009; Rawat et al. 2013) increasing the size of pore (Grethlein 1985) and breaking down fibre crystallization in a very consistent order (Kamireddy et al. 2013).

The CO2 emissions released during the combustion of woody biomass from short rotation of plantations were recently captured by some plants during their growth process. Therefore, it is the very standard convention to assume that burning biomass generates zero GHG emissions. However, emissions from fertilizers use (N2O) and management activities represent a net contribution to the stock of GHG in the atmosphere on a wide range. While considering the emissions from long-distance transport, it is not possible to count all the emissions from fertilizers or from other local management activities, because of the lack of reliable data and also the exact information is not estimated yet. In this way, the biomass is exempted from any carbon-related taxes. This implies that a power plant that generates BECCS electricity receives a financial support which is equal to the value of the tax for capturing and storing CO2 and pays tax only on emissions from the international transport of woody biomass. BECCS power generation firms are eagerly willing to demand biomass subject to the optimality condition obligatory. This states that, for a given price of electricity, the higher the tax is, the higher will be the price of biomass that they are willing to pay. This suggests that the regional social planner may be willing to pay a price higher than the global marginal cost of biomass production, if the global demand of biomass is exceeding the global maximum endowment. Even if the carbon tax increases the marginal production, the cost of biomass remains the same when there are limitations for production. However, the value of biomass increases with the carbon tax and thus BECCS firms are willing to pay a higher price in the international market as well. A firm in the forestry sector captures all the rent as overall hinders are done to the BECCS firms. This is a peculiar outcome of the non-cooperative interaction in the environment. According to different settings, with strategic coalition formation, a group of importing countries would have the incentive to form and motivate a cartel to extract a part of rents from the forestry sectors of exporting regions (Rose et al. 2012).

CONCLUSION

In conclusion, we reviewed the methods available for the estimation, analysis, production and consumption of biomass and related products for the fulfilment of various forms of energy needs in the current world. The work presented here broadens the understanding of economic analysis of the operational and transportation costs in addition to technological innovations required for the production and consumption of biomass. Further work in this field is to explore and enhance the individual web-based options for serving the information to various practitioners working in various fields like woodland dynamics, socio-economic and energy security domains. A thorough understanding of these factors not only entrench poverty, unemployment but also have terrible implications for a nation's economy from rural backgrounds. Also, the continued dependence of rural lifestyle on biomass resources to meet the sustenance and livelihood in poor economic conditions are exerting unsustainable pressure on the limited resources that are available. For example, the diminishing of fuel-wood supplies is making the rural people to spend more time to collect wood from the forests and in this way, they spend less time on food preparation and other activities such as farming, childcare, housekeeping, sanitation, socializing and education. The other issue of concern includes the high cost of wood purchasing from vendors and the personal security in and around the places where wood is collected.

REFERENCES

Arato, C., Pye, E. K., & Gjennestad, G. (2005). The lignol approach to biorefining of woody biomass to produce ethanol and chemicals. *Applied Biochemistry and Biotechnology*, *123*(1-3), 871–882. doi:10.1385/ABAB:123:1-3:0871 PMID:15930566

Banks, D. I., Griffin, N. J., Shackleton, C. M., Shackleton, S. E., & Mavrandonis, J. M. (1996). Wood supply and demand around two rural settlements in semi-arid Savanna, South Africa. *Biomass and Bioenergy*, *11*(4), 319–331. doi:10.1016/0961-9534(96)00031-1

Berndes, G., Hoogwijk, M., & Van den Broek, R. (2003). The contribution of biomass in the future global energy supply: A review of 17 studies. *Biomass and Bioenergy*, *25*(1), 1–28. doi:10.1016/S0961-9534(02)00185-X

Chen, M., Zhao, J., & Xia, L. (2009). Comparison of four different chemical pretreatments of corn stover for enhancing enzymatic digestibility. *Biomass and Bioenergy*, *33*(10), 1381–1385. doi:10.1016/j.biombioe.2009.05.025

Gibbs, H. K., Brown, S., Niles, J. O., & Foley, J. A. (2007). Monitoring and estimating tropical forest carbon stocks: Making REDD a reality. *Environmental Research Letters*, *2*(4), 1–13. doi:10.1088/1748-9326/2/4/045023

Grethlein, H. E. (1985). The effect of pore size distribution on the rate of enzymatic hydrolysis of cellulosic substrates. *Nature Biotechnology*, *3*(2), 155–160. doi:10.1038/nbt0285-155

Hall, D. O. (1991) *Biomass for energy in the developing countries (current role, potential, problems and prospects)*. Pergamon Press. Retrieved from http://www.iea.org

Joshi, S. V., Drzal, L. T., Mohanty, A. K., & Arora, S. (2004). Are natural fiber composites environmentally superior to glass fiber reinforced composites? *Composites. Part A, Applied Science and Manufacturing*, *35*(3), 371–376. doi:10.1016/j.compositesa.2003.09.016

Kamireddy, S. R., Li, J., Abbina, S., Berti, M., Tucker, M., & Ji, Y. (2013). Converting forage sorghum and sunn hemp into biofuels through dilute acid pretreatment. *Industrial Crops and Products*, *49*, 598–609.

Koren, H., & Bisesi, M. S. (2003). *Handbook of environmental health: pollutant interactions in air, water, and soil*. Boca Raton, FL: Lewis Publishers.

Openshaw, K. (2011). Supply of woody biomass, especially in the tropics: Is demand outstripping sustainable supply? *International Review (Steubenville, Ohio)*, *13*, 487–499.

Radetzki, M. (1997). The economics of biomass in industrialized countries: An overview. *Energy Policy*, *25*(6), 545–554. doi:10.1016/S0301-4215(97)00043-8

Rawat, R., Kumbhar, B., & Tewari, L. (2013). Optimization of alkali pretreatment for bio-conversion of poplar (Populus deltoides) biomass into fermentable sugars using response surface methodology. *Industrial Crops and Products*, *44*, 220–226. doi:10.1016/j.indcrop.2012.10.029

Rose, S. K., Ahammad, H., Eickhout, B., Fisher, B., Kurosawas, A., Rao, S., & van Vuuren, D. P. et al. (2012). Land-based mitigation in climate stabilization. *Energy Econ*, *34*(1), 365–380. doi:10.1016/j.eneco.2011.06.004

Sassner, P., Galbe, M., & Zacchi, G. (2008). Techno-economic evaluation of bioethanol production from three different lignocellulosic materials. *Biomass and Bioenergy*, *32*(5), 422–430. doi:10.1016/j.biombioe.2007.10.014

Sedjo, R. (1997). The economics of forest-based biomass supply. *Energy Policy*, *25*(6), 559–566. doi:10.1016/S0301-4215(97)00045-1

Shi, J., Sharma-Shivappa, R. R., Chinn, M., & Howell, N. (2009). Effect of microbial pretreatment on enzymatic hydrolysis and fermentation of cotton stalks for ethanol production. *Biomass and Bioenergy*, *33*(1), 88–96. doi:10.1016/j.biombioe.2008.04.016

Twidell, J. (1998). Biomass energy. *Renew Energy World*, *1*, 38–39.

Woodhouse, I. H. (2006a). *Introduction to microwave remote sensing*. New York: CRC Press.

Woodhouse, I. H. (2006b). Predicting backscatter-biomass and height-biomass trends using a macroecology model. *IEEE Transactions on Geoscience and Remote Sensing*, *44*(4), 871–877. doi:10.1109/TGRS.2006.872356

Zhu, W., Zhu, J., Gleisner, R., & Pan, X. (2010). On energy consumption for size-reduction and yields from subsequent enzymatic saccharification of pretreated lodgepole pine. *Bioresource Technology*, *101*(8), 2782–2792. doi:10.1016/j.biortech.2009.10.076 PMID:20006490

Chapter 10
Banana Fiber Reinforcement and Application in Composites: A Review

Abhinav Shandilya
Federal Mogul Goetze India Ltd., India

Ayush Gupta
College of Engineering Roorkee, India

Deepak Verma
Graphic Era Hill University, India

ABSTRACT

The growing awareness about sustainable development, environmental ecology and new legislations has led researchers to focus attention on bio fibres reinforced composites. In this field research has been done on many fibres but fibres such as banana, coir, bagasse, jute have gained importance in the recent decades. The main advantage of the natural fibre based composites materials being their low cost, easy availability, low density, acceptable specific properties, ease of separation, enhanced energy recovery, C02 neutrality, biodegradability and recyclability in nature. The attention is being given to the development of natural fibre composites is to explore value-added application avenues for their use and also for a sustainable and economical use of easily available natural material in hand. Agricultural waste is a very good example of such naturally available material and it can also be used to prepare composite materials for commercial use this has a very significant advantage over other natural fibres as its abundance and because of almost no cost.

DOI: 10.4018/978-1-5225-0424-5.ch010

INTRODUCTION

Composites: A Definition

A composite material is defined as a material made by combining two or more materials to give a unique combination of properties, one of which is made up of stiff, long fibres and the other, a binder or 'matrix' which holds the fibres in place.

According to Jartiz 1965, "Composites are multifunctional material systems that provide characteristics not obtainable from any discrete material. They are cohesive structures made by physically combining two or more compatible materials, different in composition and characteristics and sometimes in form."

This definition is can be contradicted as it allows to any mixture of materials to be classified as a composite without specifying the laws which should distinguish it from other very banal, meaningless mixtures.

Beghezan (1966) defined as "The composites are compound materials which differ from alloys by the fact that the individual components retain their characteristics but are so incorporated together as such to take advantage only of their attributes and not of their short comings, in order to obtain improved materials."

Kelly (1967) stated that "the composites should not be regarded simple as a combination of two materials. In the broader significance; the combination has its own distinctive properties. In terms of strength, resistance to heat or some other desirable quality, it is better than either of the components alone or radically different from either of them."

Composites Constituents

Most composites consist of a bulk material referred as the "matrix" and a "reinforcement" of some kind. The reinforcement is added primarily to increase the strength and stiffness of the matrix.

Importance of Matrix in a Composite

A matrix material of composites is required to perform the following functions:

1. To bind together the fires of the reinforcement by the virtue of its cohesive and adhesive characteristics.
2. To protect the fibres from environment and handling.
3. To disperse the fibres to maintain the desired fibre orientation and spacing.

Table 1. Properties of some vegetable fibres

Fibre	Cellulose Content (%)	Lignin Content (%)	Dia (μm)	UTS (MN/m2)	Elongation Max. (%)	Elastic Modulus
Banana	64	5	50-250	700-780	3.7	27-32
Sisal	70	12	50-200	530-630	5.1	17-22
Pineapple	85	12	20-80	360-749	2.8	24-35
Coir	37	42	100-450	106-175	47	3-6
Talipot	68	28	80-800	143-263	5.1	10-13
Polymer	40-50	42	70-1300	180-250	2.8	4-6

4. To transfer stresses on the fibres by adhesion or friction across the fibre-matrix interface when the composite is under load thus to avoid propagation of cracks and subsequent failure of composites.
5. To be chemically and thermally compatible with the reinforcing fibres.
6. To be compatible with the manufacturing methods which are available to fabricate the desired composite components (Mathews et al. 1994).

Reinforcement

The main objective of the reinforcement is to enhance the mechanical properties of the resin system. All of the distinct fibres that are used in composites have distinct properties and they also affect the properties of the composite in different ways.

Table 1 shows the properties of some vegetable fibres used in India for composites (Rai et al.).

Fibrous Composites

A fibre is defined by its length which is much greater than its cross-sectional dimensions. Fibbers are very effective in improving the fracture resistance of the matrix as reinforcement having a long dimension opposes the growth of cracks normal to the reinforcement that otherwise may lead to failure, particularly with a brittle matrix. Fibres, mostly because of their small cross- sectional area, are not directly usable in engineering applications. They are, therefore, embedded in matrix system to form fibrous composites. The matrix serves as a binder to bind fibres together, transfer loads to the fibres and also to protect them against environmental attack and damage due to handling. Fibrous composite can be subdivided into three categories namely

1. Continuous fibre (large aspect ratio),
2. Discontinuous (short) fibre (low aspect ratio), and
3. Hybrid.

Continuous Fibres

Continuous fibre composites can be either single layered or multi-layered. The single layered continuous fibre composites can be either unidirectional or woven. The multi-layered composites are generally referred to as laminates. The material response of a continuous fibre composite is generally orthotropic.

Discontinuous Fibres

Material systems composed of discontinuous fibres reinforcements are considered single layer composites. The discontinuities may produce material responses that are anisotropic in nature.

Hybrid Fibres

These are the combination of more than one fibre.

NATURAL FIBRE COMPOSITES

The interest in natural fibre reinforced composite materials is growing rapidly in industrial applications as well as fundamental research. The natural fibres are renewable, cheap, completely or partially recyclable, and biodegradable in nature. Plants such as such as flax, cotton, hemp, jute, sisal, kenaf, bamboo, banana, etc. are serving as the source of lingo-cellulose fibres are more applied as the reinforcement in the composite. Their availability, renewability, low density and low price have made them economical alternative to glass and carbon fibres. The natural fibre composites are not only environment friendly but they also provide satisfactory mechanical strength. The composites are increasingly being used in transportation and construction sector.

Banana fibre, a lingo-cellulosic fibre, obtained from the pseudo stem of banana plant (Musa sepientum), is a bast fibre with relatively good mechanical properties. The ‘‘pseudo-stem’’ is a clustered, cylindrical aggregation of leaf, stalk and bases. Banana fibre at present is a waste product of banana cultivation and is not properly utilized. The extraction process of banana fibre is not a common practice so the price of banana fibre is high as compared to other natural fibres (Mohanty et al. 2001).

Table 2. Botanical composition of banana pseudo stems fibres

Sl. No.	Constituents	Percentage
1.	Cellulose	31.27 ± 3.61
2.	Hemicellulose	14.98 ±2.03
3.	Lignin	15.07±0.66
4.	Extractives	4.46±0.11
5.	Moisture	9.74±1.42
6.	Ashes	8.65±0.10

Useful application of the fibre can raise the demand which will lead to systematic way of extraction of fibres and lead to the fall in prices of the fibre.

Bast fibres like banana are complex in structure. They are generally lingo cellulosic in nature consisting of helically wound micro fibrils in an amorphous matrix of lignin and hemicelluloses. The cellulose content of the fibre serves as

Figure 1. Schematic representation for the preparation of commingled composite
Adapted from (Paul et al. 2008)

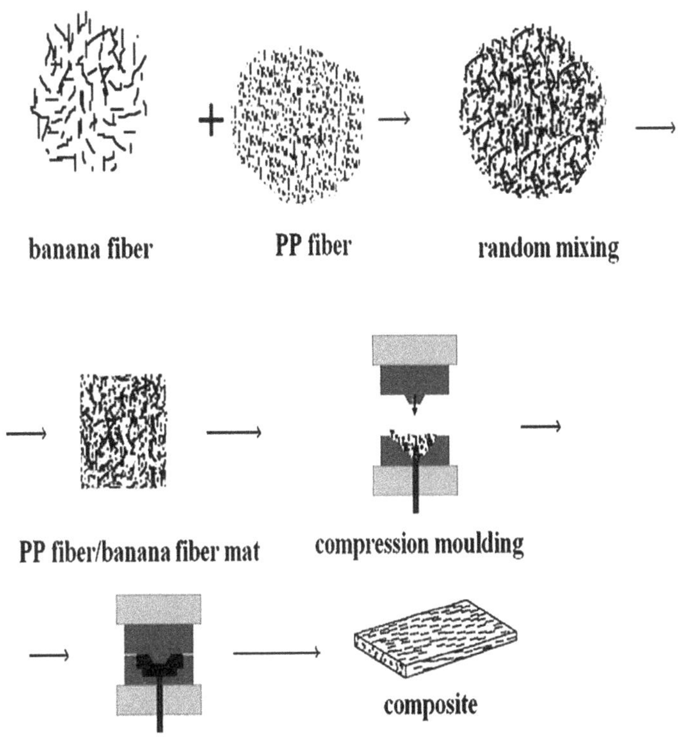

the deciding factor for mechanical properties coupled with the micro fibril angle. High cellulosic content and low micro fibril angle are desirable to obtain the desired mechanical properties. Lignin plays an important role in natural decay resistance as they are associated with the hemicelluloses. The composition of banana fibres as determined by Bilba et.al (2007) is as given in Table 2. The X-ray analysis of banana fibres reveals that cellulose crystallites are arranged in helices with helix angles of 11-12° (Kulkarni et al. 1983). Marias et.al (2005) reported a spiral angle of 11° for banana fibres. The schematic representation of the preparation of the composite is given in Figure 1.

Figure 2 shows the picture of banana fibres. In the past researchers have done substantial work on the use of banana fibre as are inforcing agent in composite materials. This has led to formation of some very novel materials with good physical and chemical properties. As Rajesh Ghosh et al. (2011) studied the influence of different volume fraction of the banana fibres in the composite. In his experiment banana fibres were modified with the help of 5% NaOH solution. The results showed that with the increase in the fibre fraction, the tensile strength have increased after an initial dip. At 35% of fibre volume fraction, an increase of 38.6% in tensile strength (95Mpa) was noted from the initial value of 69Mpa of pure vinyl ester resin. The initial specific tensile strength of 3.2 Gpa increased by 65% to 5.2 Gpa at 35% volume fraction of fibre.

Sherely Annie Paul et al (2008) used periodical method to estimate the thermal conductivity, thermal diffusivity and specific heat of polypropylene (PP)/banana fibre commingled composites at room temperature. The thermo-physical properties

Figure 2. Banana fibre

of the composites were investigated with respect to banana fibre loading and for different chemical treatments given to the banana fibre. It was found that thermal conductivity and thermal diffusivity of the composites decrease while increasing fibre loading. But the specific heat of the fibre composites didn't show significant change. The uses of chemically treated banana fibres lead to an increase in the thermo-physical properties of the composites irrespective of the nature of the chemical treatments. The benzoylated treated fibre composites showed the highest values of thermal conductivity and thermal diffusivity. It was also observed that NaOH concentration has an influence on the thermo-physical properties of the composites. A 10% NaOH treated banana fibre composites showed better thermo-physical properties than 2% NaOH treated banana fibre composites.

Aluminium silicates with different concentration (5, 7.5 and 10%) were added as filler (to fill the spaces between banana fibres) to the treated banana fibre with 20% polyurethane and then pressed under 50 kN. The results showed that addition of aluminium silicate decreased in strength properties of the composites while it increased water absorption. This could be attributed to that aluminium silicate interfere with the fibre bonds, forming weak fibre- filler bonds instead of strong fibre – fibre bonds.(Ullmann et al. 1991).

Robin Zuluaga et al (2009) used four different alkaline treatments for isolation of cellulose micro fibrils from vascular bundles of banana rachis and a comparative analysis was carried out. The isolated cellulose micro fibrils were characterized with the use of high performance anion exchange chromatography for neutral sugar composition. The attenuated total reflection Fourier transforms infrared spectroscopy, transmission electron microscopy, X-ray and electron diffraction and solid-state 13C NMR. The cellulose micro fibrils were treated with peroxide alkaline, peroxide alkaline–hydrochloric acid or 5 wt. % potassium hydroxide. The fibres had an average diameter of 3–5 nm. Although the interpretation of their structure was difficult because of the low crystalline, X-ray diffraction, 13C NMR and ATR-FTIR results suggested that cellulose micro fibrils from banana rachis could be either interpreted as cellulose IV1 or cellulose Ib. The specimens treated with more concentrated KOH solution (18 wt. %) were still micro fibrillated but their structure was converted to cellulose II. The SEM micrograph (Figure 3) illustrates the architecture of the cross-section area of vascular bundles isolated from banana rachis. As observed in other natural fibres (Engels et al. 1998-Zhong et. al. 2001), they are constituted by elementary fibres (Figure 3a). Each elementary fibre consists of helical spirals, namely the macro fibrils (mf). As shown in Figure 3b, a thick cell wall, generally containing on-cellulosic polysaccharides, is observed. This cell wall had a width ranging from 1 to 2 lm (Ganan et al. 2008). Also in Figure. 3b, the middle lamella (ml), largely consisting of pertinacious substances, can be seen. That can be recognized as a thin layer between two adjacent elementary fibres (Dickinson 2000).

Figure 3. (a) SEM micrograph of the vascular bundles in banana rachis and (b) micrograph at higher magnification. cw, cell wall; ml, middle lamella.
Adapted from Zuluaga et al. 2009

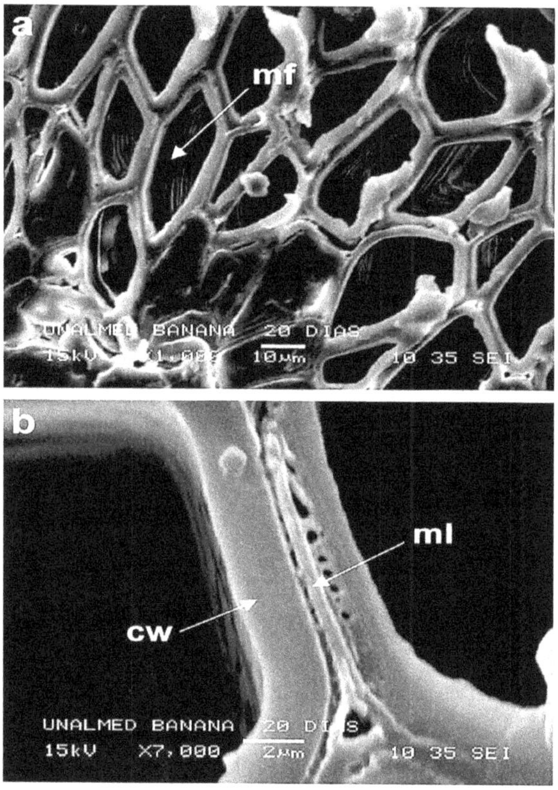

Figure 4 shows thee diffraction patterns of PA, PA–HCl and KOH-5 samples. They both exhibit several broad peaks. By comparison with the spectrum of native cellulose I the peak at the lowest angle can be described an overlapping of 1-10 and IĪ0 reflections. With increasing diffraction angle, 102 and 200 and finally 004 reflections are observed. The spectra in particular, the overlapping of 1-10 and IĪ0, suggest that the structure of the micro-fibrils corresponded to cellulose I, quite resembling cellulose IV1, a less ordered form of cellulose I The micro-fibrils thus consist of cellulose II crystallites. This spectrum displays a peak at about 12, indexed 1-10, and two more intense reflections should have similar intensities. The fact that, in this spectrum, the intensity of IĪ0 is slightly higher than that of 200 reflections may be explained by a preferential orientation of the micro-fibrils during the drying of the film.

Figure 4. X-ray diffraction spectra recorded from films of cellulose micro fibrils prepared by different treatments: peroxide alkaline (PA), peroxide alkaline–hydrochloric acid (PA–HCl), potassium hydroxide 5 wt% (KOH-5), and potassium hydroxide 18 wt% (KOH-18).
Adapted from (Zuluaga et al. 2009)

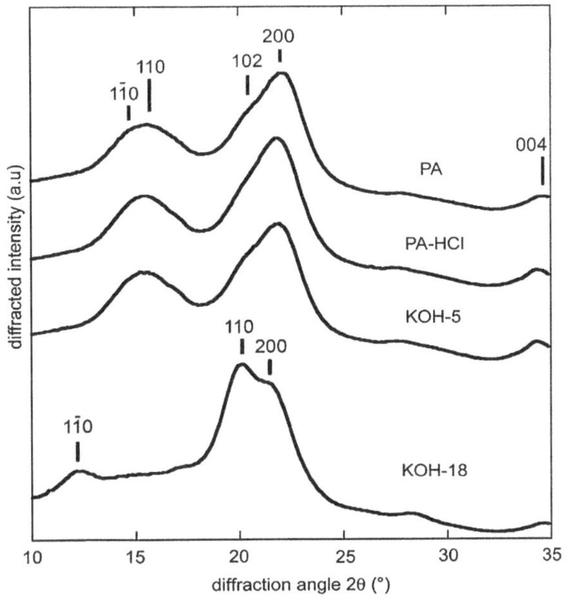

A comparison of the morphology of cellulose micro fibrils isolated by different chemical treatments is presented in Figure 5, where TEM micrographs recorded from negatively stained specimens are shown. The micrographs reveal that the degree of aggregation and shape of the fibrils are different. Selected area of electron diffraction patterns recorded at low temperature on bundles of unstained cellulose micro fibrils treated with 5 wt% and 18 wt% KOH, respectively, are shown in Figure 6.

Richard Mpon et al 2012 optimized graft copolymerisation of fibres from banana trunk. Banana trunk sheets were opened out and dried for several weeks in air. Thus pulp was obtained by the nitric acid process with a yield of 37.7% while fibres were obtained according to the modified standard Japanese method for cellulose in wood for pulp (JIS 8007) with a yield of 65% with respect to oven dried plant material. The average diameter, Young's modulus, tensile strength and strain at break-off of Single fibre obtained by the JIS method was 11.0 μm and 7.05 GPa, 81.7 MPa and 5.2% respectively. Modification of the fibres was carried out by grafting ethyl acrylate in the presence of ammonium nitrate cerium (IV). Optimisation of the copolymerisation reaction conditions was studied by measuring the rate of conversion,

Figure 5. TEM micrographs of negatively stained preparations of cellulose micro fibrils isolated after different treatments: (a) peroxide alkaline (PA); (b) peroxide alkaline–hydrochloric acid (PA–HCl); (c) potassium hydroxide 5 wt% (KOH-5) and (d) potassium hydroxide 18 wt% (KOH-18).
(Adapted from Zuluaga et al. 2009).

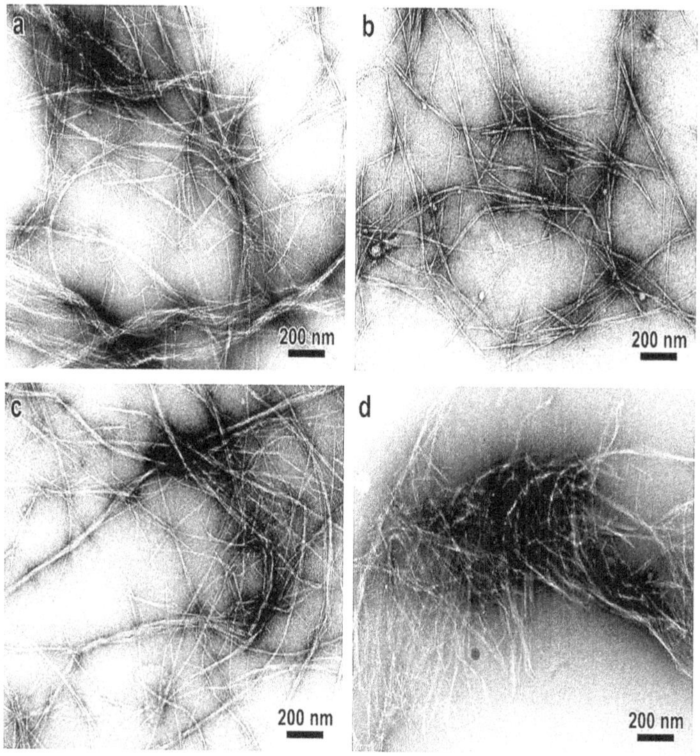

the rate of grafting and the grafting efficiency. The results showed that at low values of ceric ion concentration (0.04 M), at ambient temperature, after three hours and at a concentration of 0.2 M ethyl acrylate.

D. Brindha et al (2012) investigated the effect of fibres from banana varieties (kadhali, pooven, mondhan and rasthali) on physio-chemical properties. The fibres form leaf sheaths of each variety were extracted using the decorticating machine. The fibres obtained were 60-110cm long. Fibres extracted from Banana varieties scoured with 2% NaOH solution at 1000°C for 45min and analysed for their physio-chemical and mechanical properties. The moisture regain, cellulose content, modulus and tenacity of the fibres of all varieties were increased after scouring process. The fibre from the variety rasthali sought to have high moisture content (13.21%), cel-

Figure 6. TEM images of unstained bundles of cellulose micro fibrils prepared using alkaline treatments KOH-5 (a) and KOH-18 (b). Inset: corresponding cellulose I and cellulose II fibre electron diffraction patterns, respectively, recorded at low temperature from the selected circular areas indicated in the image.
(Adapted from Zuluaga et al. 2009)

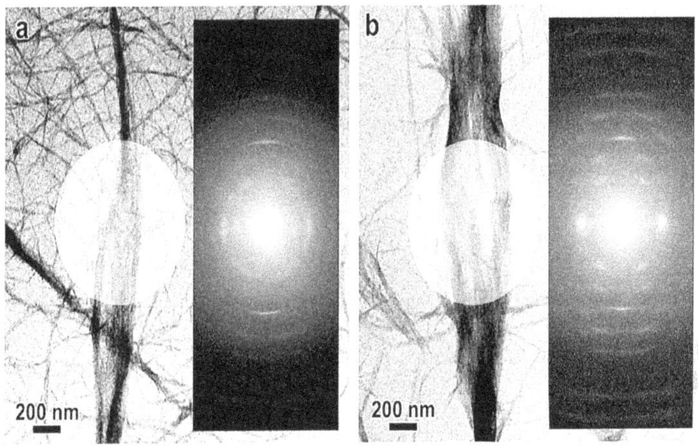

lulose content (83.02%), modulus (3293.16gf/den) and tenacity (48.66gf/den) when treated with 2% NaOH solution. Thus, the scouring can increase the fiber quality for their usage in industries like paper and textile.

Youssef Habibi et al (2009) determined the chemical composition of Lignocellulosic fibres extracted from Egyptian industrial crops, viz. cotton stalk, rice straw, bagasse, and banana plant waste. Composite materials were processed from these natural lignocellulosic fibres using low density polyethylene and acid stearic as compatipilizer, or maleate low density polyethylene. It was found that cotton stalk and banana plant waste fibres exhibited the highest lignin content, in contrast with the cellulose being the lowest. A high amount of minerals was reported for banana plant waste and rice straw fibres. After grounding, fibers around 300-500 μm long and 40 μm in diameter were used to process composite materials using low density polyethylene (LDPE) or maleated LDPE (MLDPE) as matrix. The degree of crystallinity of LDPE-based composites was found to slightly increase when adding lingo cellulosic fibres, mainly in the case of lignin- and hemicellulose-rich cotton stalk fibres. A significant increase of the strength was reported upon fibres addition when using MLDPE as matrix showing the beneficial effect of this copolymer as compatibilizer.

J.L. Guimarães et al (2010) presented results on the characterization of corn starch by X-ray powder diffraction (shown in Figure 7) and thermal analysis, as well as processing and characterization of starch–banana/sugarcane bagasse fibre

Figure 7. X-ray diffraction patterns of (a) starch 3001, (b) banana fiber and their composites (c) A and (d) B
Adapted from Guimarães et. al. 2010

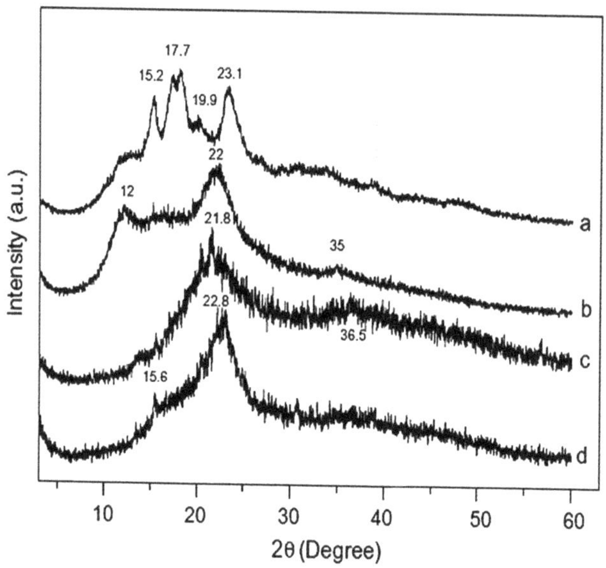

composites. Thermal analysis (shown in Figure 8) of Amidex- 3001 showed good thermal stability. Fracto-graphic analysis of starch composites containing 70 wt.% Amidex- 3001 and 30% commercial glycerol showed matrix cracks between smooth and rough surfaces and dimples in rough regions, resulting in the composite sample to be ductile and not properly homogenized. Starch/banana and starch/bagasse fibre composites were prepared by compression moulding with two different processing methods. XRD studies showed structural changes in both the fibre composites. Without much variation in their crystalline index (20–21%).Compared with banana fibre composites, fracto-graphs of bagasse fibre composites showed a large number of fibre pull-outs and fibres lying perpendicular to the fracture surface, and these seemed to explain the tendencies observed in their tensile properties. Good bonding between the bagasse fibres and the matrix was shown by the starch coating on these fibres, compared with the free surface of banana fibres.

Figure 7 shows the X-ray powder diffraction pattern of the starch sample Amidex-3001. The main diffractions peaks, cantered at 15.2, 17.2, 19.9, 23.3 and 26.5of 2h (inter planar distances ‘‘d” of5.82, 5.15, 4.46, 3.82 and 3.36 Å, respectively), are characteristic of an A-type starch, and the crystalline structure is typical of cereals Figure 7(b–d) shows the X-ray diffraction patterns of banana fibres and their composite with 10 and 60 wt. % fibre content. Banana fibres are partially crystalline, as seen by the three diffraction peaks at 10, 25 and 35 in 2h (d = 8.84, 3.56

Figure 8. Thermal analysis (TGA/DTA) curves obtained for starch (Amidex-3001). Adapted from
(Adapted from Guimarães et. al. 2010)

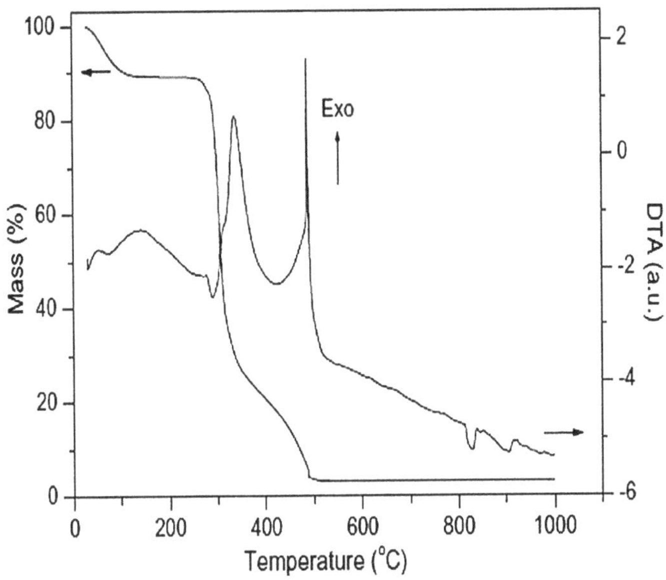

and 2.56 Å), corresponding to the cellulose I structure. On the other hand, the composites show much lower crystallinity (20–21%) compared to that of banana fibres (39%), as expected with the plasticisation of starch. The composite with 10% fibre content shows a diffraction peak at 2h = 22 which comes from the crystalline region of the banana fibre. This diffraction peak has a wide shoulder from 15 to 22, values which incidentally were in the same region of the starch diffraction peak. In contrast, in the composite with 60% fibre, the diffraction peak at 2h = 22 belongs to the fibre, with no contribution coming from the starch. In the case of composite "A" (Figure 7c), the diffraction peak disappears between 2h values of 5 and 20, which can be associated with the condensation of starch, its cross linking with glycerol and the consequent destruction of the starch's crystalline region. The third diffraction peak, located between 20 and 30 of 2h, can be attributed to the contribution of the native structure of cellulose (cellulose I). On the other hand, in the case of composite "B" (Figure 7d), a narrower diffraction peak centered at 22.85 of 2h and a shoulder (small peak) at 35 of 2h were observed, indicating the presence of loose fibre. However, the peak at 12.5 of 2h was missing, suggesting that changes had occurred in the fibre structure of both composites. Furthermore, there was not much variation between the crystallinity indices (20–21%) of both composites.

Figure 8 shows the TGA/DTA curves obtained for Amidex-3001, showing a mass loss of 10.8% up to 200 0C due to the presence of hydrated/adsorbed water, which is associated with a broad endothermic peak centered at 1400C. The mass loss of 86.1% between 315 and 4950C due to the burning of organic matter is also associated with two endothermic peaks at 339 and 4870C. The remai1lning 3.1% is due to the presence of inorganic material in the starch, probably silicates. TGA/DTA studies of both types of fibers used here have been carried out and reported elsewhere.

Figure 9a shows rough fracture surface with a large amount of cracks and fibres lying parallel to the fracture surface (straight arrows). Figure 9b is the same as Figure 9a, showing a pulled out fibre (dotted arrow) and fibre fracture, pull-outs and some voids. Figure 9c shows clean fibres (without coating of matrix - straight arrow) that are lying parallel to the fracture surface, some pull-outs (dotted arrows), voids (dash and dot arrow) and ruptured fibres perpendicular to the crack (dashed arrow), while at higher magnification there is a ruptured fibre lying perpendicular to the fracture surface (Figure 9d). Also, fibre rupture occurred probably before the matrix failure was completed, as indicated at different magnifications in Figure 9c and d. Figure 9e shows more fibres lying parallel to the fracture surface (straight arrows), residues of a split parallel fibre (dotted arrow) and pull-out fibre/fibre breakage (dash and dot arrow). All these impair the strengthening effect in the composite. Generally, fibres break at the fracture surface itself without any broken fibres protruding from that surface or any resin sticking to the fibres, when good bonding existed between the fibre and the matrix. On the other hand, weaker bonding is exemplified by the longer pull-out fibres, as in Figure. 9c.

Bakiyalakshmi K et al. (2012) investigated the bio-sorption of Copper (II) ions onto chitosan (CS)/sisal fibre (SF)/banana fibre (BF) hybrid composites. Chitosan/sisal/banana fibre hybrid composite were prepared by solution mixing method. The prepared composite was confirmed by FTIR, X-ray, DSC and SEM measurements. The efficiency of the adsorption was evaluated by varying pH of the solution, contact time and adsorbent dose. At pH=5 the maximum uptake of Cu (II) ions by CS/SF/BF composite was found tobe169 mg/g. The equilibrium data was analysed by Freundlich, Langmuir, Tempkin, Dubinin- Radushkevich (D-R) isotherm models. The Freundlich adsorption isotherm proves to best fit, suggesting multilayer layer adsorption on heterogeneous surface. The kinetic model indicated that adsorption of copper (II) ion from solution by CS/SF/BF hybrid composite corresponded to pseudo second order reaction. Also, Intra particle diffusion study suggested that the diffusion of copper ions is very fast in the beginning and then stabilizes slowly. The results showed that the prepared composite provides high efficiency in removing copper (II) ions from aqueous solution.

Figure 9. SEM tensile fractographs of fractographs of 45% Starch (Amidex-3001)–30% crude glycerine 5% banana fibre composite. (a) rough fracture surface with lot of cracks and fibres lying parallel to the crack surface; (b) same a, pulled out fibre and fibre fracture, pull-outs and some voids; (c) fibres lying parallel to the crack surface, some pull-outs, ruptured fibres perpendicular to the crack and clean fibre surface without coating by the matrix; (d) ruptured perpendicular fibre at higher magnification; (e) more fibres lying parallel to the crack region, residues of a split parallel fibre and pull-out fibre/fibre breakage.

Reproduced with permission from Elseveir ltd. (Adapted from Guimarães et. al. 2010)

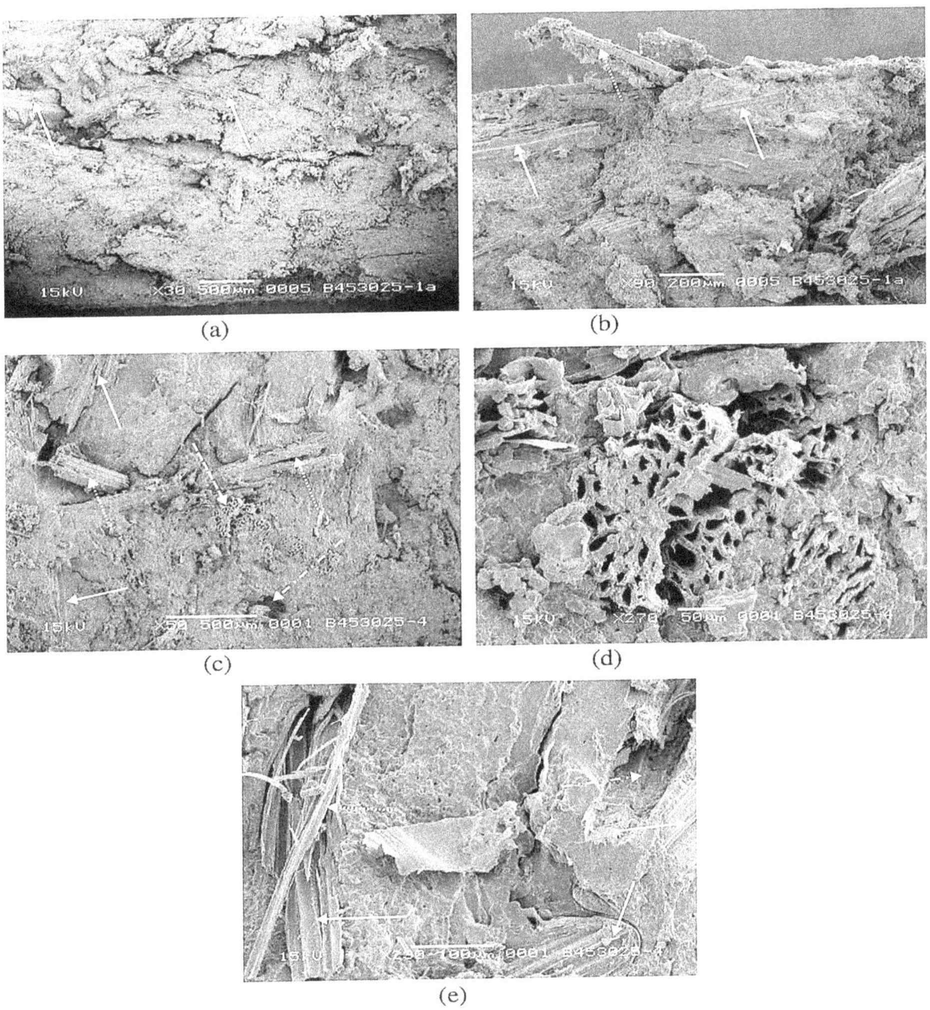

BANANA FIBER REINFORCED WITH THERMOPLASTIC COMPOSITES

Tensile, flexure, impact, and fracture surface study of woven pseudo stem banana fibre reinforced with epoxy composite by hand layup method was investigated by Sapuan et al (2006). Tensile test showed that the ultimate strength of banana reinforced epoxy is increased from 25 to 47MPa when compared with unreinforced epoxy. Also the Young's modulus increased from 1300 to 1850 MPa. Also the flexure strength increased to 75 from 45MPa. The fracture surface study through scanning electron microscope showed that the banana fibre composite exhibits ductile type of failure with minimum plastic deformation.

Zainuddin et al. (2008) found that the decomposition temperature of banana pseudo stem fibre/un-plasticized polyvinyl chloride decreases from 279°C to 256°C on the addition of banana filler from 10-40%. It was also predicted that the thermal stability of acrylic modified composites was found to be more stable than unmodified banana pseudostem fibre/un-plasticized polyvinyl chloride composite.

Paul et al (2008) observed that the effect of fibre loading on banana/polypropylene (PP) composite. The thermal diffusivity and thermal conductivity were found to be decreasing on increasing fibre loading (i.e., from 0.24W/m K for neat polypropylene matrix to 0.217 and 0.157W/m K for 0.10 and 0.50 of volume fraction). It was also predicted that all the chemical treatments (alkali, silane, permanganate, and benzoyl chloride) increased thermal diffusivity and conductivity in comparison with untreated fibre.

It can be seen in the Figure 10a that in unmodified composites the tensile rupture was accompanied by the de-bonding of the fibres which leaves holes that indicated a weak adhesion between the fibre and the matrix. But for NaOH treated and benzoylated composites (Figure. 10 b and c), a significant improvement in the fibre/matrix adhesion was noticed as shown by the absence of holes and de-bonding of the fibers. The fractured surface of the treated fibre composites shows fibre break a greater than pull-out, which in turn indicates better interfacial strength.

BANANA FIBRES REINFORCED WITH CEMENT COMPOSITES

H. Savastano Jr. et al (2005) examined the microstructure of composite materials containing fibrous wastes. Both secondary back-scattered electron imaging and energy dispersive X-ray spectroscopy was used for compositional analysis. The results showed that Sisal and Eucalyptus grandis pulps showed satisfactory bonding to the cement matrix, with fibre pull-out predominating as indicated by high values

Figure 10. Tensile fracture surface of the PP/banana composite: (a) untreated; (b) NaOH treated; (c) benzoylated; and (d) KMnO4 treated.
Reproduced with permission from Elseveir ltd.

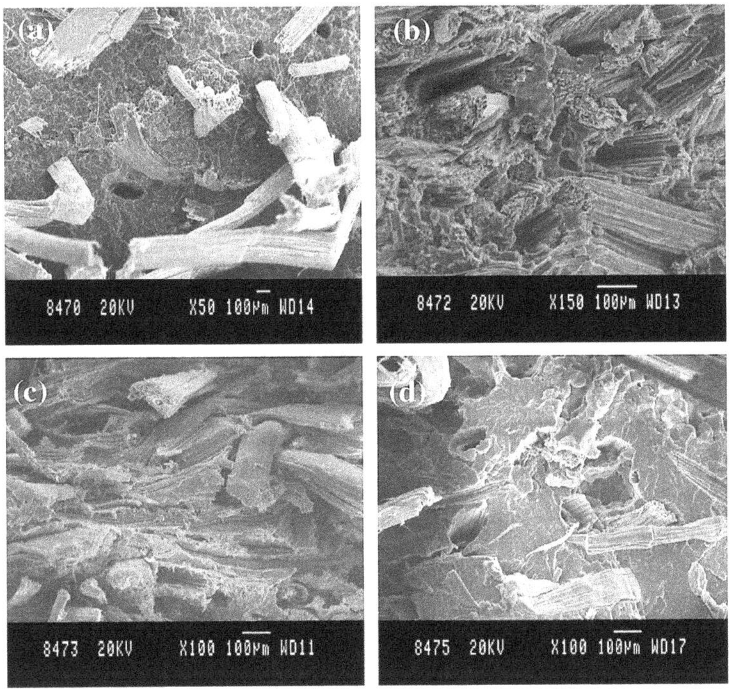

of energy absorption. In contrast banana pulp reinforced composites exhibited fibre fracture as the main failure mechanism. In all analysed composites, partial fibre de-bonding and matrix micro-cracking were dominant at the interfaces.

H Savastano Jr et al (2000) evaluated the performance of thin fibre-cement elements produced from alternative raw materials using the Hats-check process, to be used in low-cost housing. Sisal and banana fibres were prepared using mechanical and kraft pulping procedures while residual Eucalyptus grandis pulp was obtained from a commercial pulp mill. Ordinary Portland cement was used in the process. Composites were prepared using a slurry vacuum de-watering process, pressing and air-curing. Fibre contents of 8–12% by mass, moduli of rupture (MOR) up to 23 MPa and fracture toughness (FT) values in the range of 0.6–1.7 kJ/m2 were obtained at 28 days. After 12 months of exposure under temperate and tropical conditions, the MOR of the BFS-based composites had decreased to values in the range of 6.6–10.1 MPa. FT values remained stable or even increased with the weathering exposure.

H. Savastano Jr. et al (2000) compared the properties of different fibre reinforced cement based composites. The composites were prepared using kraft pulps from sisal and banana waste and from Eucalyptus grandis pulp mill residues. Conventional chemical pulping conditions for the non-wood strands and a slurry vacuum de-watering method was used for composite preparation followed by air-curing. Mechanical testing showed that optimum performance of the various waste fibre reinforced composites was obtained at a fibre content of around 12% by mass, with flexural streOngth values of about 20 MPa and fracture toughness values in the range of 1.0–1.5 kJ m−2. Experimental results showed that, of the waste fibres studied, E. grandis is the preferred reinforcement for low-cost fibre-cement.

BANANA FIBRE REINFORCED WITH THERMOSET POLYMER COMPOSITES

Singh V. K et al (2012) studied the banana fibre and silica powder reinforced composite material. Reinforced with epoxy resin (CY 230) and hardener (HY 951). Silica particle (2 wt. %) and banana fibre with different weight percentage was used. The results showed that the density of banana filled composite is 1.097 gm/cm^3 for 10wt% banana filled composites, 1.147 gm/cm^3 and for 10%banana-2%silica wt. % reinforced composite which was found to be lower than the pure epoxy resin which is 1.176gm/cm^3.It was noticed that the ultimate tensile strength for the banana-silica composite is 27.618 MPa but it is still less than that of the epoxy resin, which has a ultimate tensile strength of 43.790 MPa. About 12% increase in modulus of elasticity was found due to addition of 10wt% of banana fibres. The ultimate compressive strength of 2wt% of silica and 10 wt. % of banana fibre is about 63.2 MPa. An increase of impact strength from 22.07 Nm to 23.683 Nm has been noticed due to addition of 10 wt. % of banana and 2 wt. % of silica to the epoxy composite. It was found that hardness of neat epoxy resin (CY-230+ 8 wt. % of HY-951) is 57.8. The hardness of the fabricated composite made of epoxy resin and 10 wt. % of banana fibre and 2 wt. % of silica is 60.4. Hardness values were measured on the Rockwell M-Scale

M. Thiruchitrambalam et al (2009) investigated the effect of alkali and SLS (Sodium Lauryl Sulphate) treatment on Banana/Kenaf Hybrid composites and woven hybrid composites. The fibers are treated with 10% of sodium hydroxide (NaOH) and 10% Sodium Lauryl Sulphate (SLS) for 30 minutes. Woven banana and kenaf fibre reinforced unsaturated polyester (USP) composites were fabricated by moulding technique. The fibre content in the composite was constant at 40%. Experimental results showed that the SLS treatment provide better improvement in the tensile strength about 13% and 10% for non-woven and woven hybrid composites when compared with alkali treatment. The percentage increase in flexure strength of alkali and SLS treated are 12% and 10% for non-woven and woven composites. It is due to increase in the interface adhesion between fibre and matrix, which is due to removal of foreign particle and decrease in the lignin level in fibres by SLS treatment when compared with alkali. The increase in impact strength because of SLS treatment was 23% for random mix hybrid composite and 16% for woven hybrid composites.

Laly A. Pothan et al (2012) studied the variations in tensile and impact properties of banana fibre reinforced polyester composites caused by the addition of glass fibre. The results showed that the volume fraction of 0.11 glasses mixed with banana fibre gives 54.5% increase in the tensile strength and 196% increase in the impact strength of the composites. Linear increase in tensile strength was noted as a result of the increase of glass. The tensile strength shows the highest value when a glass volume fraction of 0.17 was used and an interleaving arrangement of glass and banana fibre was followed. When lower volume fraction of glass was used, an intimate mixture of banana fibre and glass shows the highest tensile strength. The impact strength shows the highest value when a glass volume fraction of 0.11 was used.

Rafah A. Nassif et al (2010) investigated the effect of the chemical treatment of banana fibres on the physical properties of the composites. Banana fibres were treated with 10% sodium hydroxide and some physical tests were carried out like: dielectric strength, dielectric constant, and thermal conductivity. The results showed that the chemical treatment improved the dielectric strengthened thermal conductivity by about 29.37% and 139% respectively as compared with untreated fibre composites. Finally, the dielectric constant value of the treated fibre composite was found to be lower than the untreated fibre composite and virgin unsaturated polyester.

Lina Herrera-Estrada et al (2009) investigated the process for production of banana fibre reinforced composite materials with a thermoset, suitable for automotive and transportation industry application. Banana fibres were used in the form of short non-woven fibres (4-9 mm). The fibres were immersed in 6% NaOH solution for 2 hr at room temperature. The results show that the flexural strength of banana fibre/eco-polyester composites was 40.16 MPa, which was 14.78% higher than the strength of banana fibre/epoxy. The higher flexural strength and modulus observed in the banana fibre/eco-polyester composites was related to improved fibre/matrix interaction. Compressive properties were also found to be dependent upon the fibre/matrix interaction, which improves with alkaline pre-treatment for an epoxy matrix and degraded with such treatment in the eco-polyester matrix. Thus the highest compressive strength of 122.11 MPa of the banana fibre/epoxy composite is attained after fibre pre-treatment and is 38.35% higher than the observed strength without the treatment. On the contrary the highest compressive strength in banana fibre/eco-polyester composite is122.88 MPa and is achieved without fibre pre-treatment, the use of alkaline substance yields 31.07% lower properties.

Mishra et al. (2000) used maleic anhydride as compatiblizer for novolac/fiber (banana, hemp, sisal) composites. Their mechanical properties and swelling behaviour (steam and water) was studied. Treatment with maleic anhydride was used in the process. It was suggested that the use of maleic anhydride reduces the water absorption and steam absorption with increase in mechanical properties like Young's modulus, hardness, and impact strength. The study showed that a better quality of plant fibre-reinforced composites can be achieved by maleic anhydride treatment.

Joseph et al (2002) compared the mechanical properties of phenol formaldehyde composites reinforced with banana fibres and glass fibres. The predicted optimum lengths of fibre for maximum value of tensile stress were30 and 40 mm, respectively. The tensile strength and Young's modulus value of banana/phenol formaldehyde (PF) composites increases to 400 and 320%, respectively, when fibre loading is 48% as compared with neat resin. By increasing the fibre loading to 45% the flexural modulus was also increased to about 25% for banana/PF composites.

Uday Kiran et al. (2007) found that the tensile strength of banana fibre reinforced with polyester composite to be 59MPa for a fibre length of 30mm and 51%of fibre weight.

Haneefa et al. (2008) concluded that the hybridization of banana fibre and glass fibre increased the tensile strength and Young's modulus of the composite. The increase in volume fraction of glass fibre increased the properties because of greater compatibility of glass fibre with polystyrene matrix.

Pothan et al (1997, 2003, 2006)).in their work on dynamic analysis of banana fibre reinforced with polyester composites found that a better fibre/matrix bonding occurs at 40% fibre loading, and de-bonding occurs at 10 and 20% respectively also observed through SEM. When the fibres were added into the polyester matrix the loss of modulus peak had a double humped nature. At lower frequencies the peak of the loss modulus curve was around 120°C.With the increase in frequency, the peak of the loss modulus curve, which corresponds to the glass transition temperature, is also found to shift to a higher temperature. The role of fibre/matrix interaction on the dynamic mechanical properties with chemical modification on the surface was also analysed. It was found that there are two peak values for loss modulus curve and damping curve around 80°C and130°C, respectively. The peaks around 130°C are associated with the glass transition temperature 'Tg' of the matrix and that around 80°C is due to the transition of interlayer. Further the study of static and dynamic mechanical properties of banana and glass fibre reinforced polyester composites was also carried out. The composites were prepared in woven form. It was found that high tensile strength can be obtained using two layers of fabric. Four-layers of fabric show two peaks and one shoulder. Increasing the number of layers made the second relaxation peak visible. Damping values are found to be lower with the incorporation of more layers.

BANANA FIBRE REINFORCED WITH BIODEGRADABLE COMPOSITES

Rakesh Kumar et al (2008) prepared banana fibre reinforced soy protein composites. Soy protein composites were prepared by incorporating different volume fractions of alkali-treated and untreated fibres into soy protein isolate (SPI) with different amounts of glycerol (25%–50%) as plasticizer. Composites thus prepared were characterized in terms of mechanical properties, SEM and water resistance. The results showed that at 0.3 volume fraction, tensile strength and modulus of alkali treated fibre reinforced soy protein composites increased to 82% and 96.3%, respectively, compared to soy protein film without fibres. Water resistance of the composites increased significantly with the addition of glutaraldehyde which acted as cross-linking agent. Biodegradability of the composites has also been tested in the contaminated environment and the composites were found to be 100% biodegradable.

Sanjay K. Chattopadhyay et al (2011) studied Various composites of polypropylene (PP) produced using natural fibres such as pineapple leaf fibre, banana fibre, and bamboo fibre for their degree and rate of aerobic biodegradation. Composites used contained 10, 15, and 50% volume fractions of pineapple leaf fibre, banana fibre, and bamboo fibre, respectively. All the composites exhibited partial biodegradation in the range of 5–15% depending on the fibre content. Degradation had not taken place in the covalent ester linkages between the natural fibre and the MA-g-PP compatiblizer but in the areas of the fibres which have remained only physically embedded in the resin matrix. Thus natural fibres reinforced PP composites are not excellent biodegradable material, but they can address the management of waste plastics by reducing the amount of polymer content used that in turn will reduce the generation of non-biodegradable polymeric wastes.

CONCLUSION

The use of banana fibre as reinforcing agent in composite materials was reviewed from the view of replacing mineral inorganic fillers for reduction in the use of petroleum based non-renewable resources. The review takes in account structural, morphological and mechanical properties as well as different surface treatments for banana fibre based composites. Due to good mechanical properties banana fibre based composites have a great potential to be used in sectors like automobile, construction etc. India being one of the largest banana producing countries of the world would find it attractive for the economy if banana fibre based composites are to be used for commercial purposes. These composites can be further economical if more efficient process for composite preparation and fibre separation are designed. Thus the systematic use and further research in the field of banana fibre will increase its further uses and also of its composites.

ACKNOWLEDGMENT

The authors are thankful and grateful to the Elseveir Ltd for granting permission to reproduce figures and tables to be included in this chapter from their journals.

REFERENCES

Bakiyalakshmi, K., & Sudha, P. N. (2012). Adsorption of Copper (II) ion onto chitosan/sisal/banana fiber hybrid Composite. *International Journal Environmental Sciences*, 3.

Brindha, D., Vinodhini, S., Alarmelumangai, K., & Malathy, N. S. (2012). Physico-Chemical Properties of Fibers rom Banana Varieties After Scouring. *Indian Journal of Fundamental and Applied Life Sciences*, *2*, 217–221.

Chattopadhyay, S. K., Singh, S., Pramanik, N., Niyogi, U., Khandal, R. K., Uppaluri, R., & Ghoshal, A. K. (2011). Biodegradability studies on natural fibers reinforced polypropylene composites. *Journal of Applied Polymer Science*, *121*(4), 2226–2232. doi:10.1002/app.33828

Dickinson, W. C. (2000). *In integrative plant anatomy* (pp. 15–21). San Diego, CA: Harcourt Academic press.

Engels, F. M., & Jungs, H. G. (1998). Alfalfa stem tissues: Cell-wall development and signification. *Annals of Botany*, *82*(5), 561–568. doi:10.1006/anbo.1998.0705

Gañán, P., Zuluaga, R., Cruz, J., Vélez, J., Retegí, A., & Mondragon, I. (2008). Elucidation of the fibrous structure of Musaceae maturate rachis. *Cellulose (London, England)*, *15*(1), 131–139. doi:10.1007/s10570-007-9150-z

Ghosh, R., Reena, G., Ramakrishna, A., Raju, B.L. (2011). Effect of fibre volume fraction on the tensile strength of Banana fibre reinforced vinyl ester resin composites. *International Journal of Advanced Engineering Sciences and Technologies*, *4*, 89 – 91.

Gritsch, C. S., Kleist, G., & Murphy, R. J. (2004, August19). Developmental changes in cell wall structure of phloem fibres of the bamboo Dendrocalamusasper. *Annals of Botany*, *94*(4), 497–505. doi:10.1093/aob/mch169 PMID:15319227

Guimarães, J. L., Wypych, F., Saul, C. K., Ramos, L. P., & Satyanarayana, K. G. (2010). Studies of the processing and characterization of corn starch and its composites with banana and sugarcane fibers from Brazil. *Carbohydrate Polymers*, *80*(1), 130–138. doi:10.1016/j.carbpol.2009.11.002

Habibi, Y., Zawawy, W. K., Ibrahim, M. M., & Dufresne, A. (2009). Processing andcharacterization of reinforced polyethylene composites made with lignocellulosic fibers from Egyptian agro-industrial residues. *Composites Science and Technology*, *68*(7-8), 1877–1885. doi:10.1016/j.compscitech.2008.01.008

Haneefa, A., Bindu, P., Arvind, I., & Thomas, S. (2008). Studies of Tensile and Flexural Properties of Short Banana/Glass Hybrid Fiber Reinforced Polystyrene Composites. *Journal of Composite Materials*, *42*(15), 1471–1489. doi:10.1177/0021998308092194

Herrera-Estrada, L., Pillay, S. & Vaidya, U. (n.d.). *Banana Fibre Composites For Automotive and Transportation Applications*. Department of Material Science and Engineering, University of Alabama at Birmingham.

Jinnah, M., Mariatti, M., & Abubacker, A. (2008). Effect of Chemical Surface Modification on the Properties of Woven Banana Reinforced Un-saturated Polyester Composites. *Journal of Reinforced Plastics and Composites*, *41*, 2371–2380.

John, M. J., Francis, B., Varghese, K. T., & Thomas, S. (2008). Effect of Chemical Modification on Properties of Hybrid Fiber Bio-composites. *Composites. Part A, Applied Science and Manufacturing*, *39*(2), 352–363. doi:10.1016/j.compositesa.2007.10.002

Joseph, S., Sreekala, M. S., Oommen, Z., & Thomas, S. (2002). A Comparison of the Mechanical Properties of Phenol Formaldehyde Composites Reinforced with Banana Fibers and Glass Fibers. *Journal of Composite Science and Technology.*, *62*(14), 1857–1868. doi:10.1016/S0266-3538(02)00098-2

Kelly, A. (1967). The Nature of Composite Materials. *Scientific American Magazine, 217*(B), 161.

Ketty, B. (2007). Study of banana and coconut fibers: Botanical composition, thermal degradation and textural observations. *Bioresource Technology*, *98*(1), 58–68. doi:10.1016/j.biortech.2005.11.030 PMID:16442281

Kulkarni, A. G., Satyanarayana, K. G., Rohtagi, P. K., & Vijayam, K. (1983). Mechanical Properties of Banana Fibers. *Journal of Materials Science*, *18*(8), 2290–2296. doi:10.1007/BF00541832

Kumar, R., Mishra, S., & Varma, I. K. (2008). Banana fiber-reinforced biodegradable soy protein composites. *Frontiers of Chemistry in China*, (September), 243–250.

Majhi, S. K., Nayak, S. K., Mohanty, S., & Unnikrishnan, L. (2010). Mechanical and Fracture Behavior of Banana Fiber Reinforced Polylactic Acid Biocomposites. *International Journal of Plastics Technology*, *14*(S1), 57–75. doi:10.1007/s12588-010-0010-6

Maleque, M. A., Belal, E. Y., & Sapuan, S. A. (2006). Mechanical Properties of Study of Pseudo-stem Banana Fiber Reinforced Epoxy Composite. *The Arabian Journal for Science and Engineering*, *32*, 359–363.

Maries, I., Neelakantan, N. R., Oommen, Z., Joseph, K., & Thomas, S. (2005). A study of the mechanical properties of randomly oriented short banana and sisal hybrid fiber reinforced polyester composites. *Journal of Applied Polymer Science*, *96*(5), 1699–1709. doi:10.1002/app.21636

Mathews, F. L., & Rawlings, R. D. (1994). Engineering Composites. Chapman and Hall.

Mishra, S., Naik, J. B., & Patil, Y. P. (2000). The Compatibilising Effect of Maleic Anhydride on Swelling and Mechanical Properties of Plant Fiber Reinforced Novolac Composites. *Journal of Composite Science & Technology*, *60*(9), 1729–1735. doi:10.1016/S0266-3538(00)00056-7

Mohanty, A. K., Mishra, M., & Drzal, L. T. (2001). Surface modifications of natural fibers and performance of the resulting biocomposites: An overview. *Composite Interfaces*, *8*(5), 313–343. doi:10.1163/156855401753255422

Mubashirunnisa, A., Vijayalakshmi, K., Gomathi, T., & Sudha, P. N. (2012). Development of banana/glass short hybrid fiber reinforced nano-chitosan polymer compositics. *Scholars Research Library Der Pharmacia Lettre*, *4*, 1162–1168.

Mukhopadhyay, S., Fanguerio, R., Arpac, Y., & Sentrik, U. (2008). Banana Fibers _ Variability and Fracture Behavior. *Journal of Engineered Fibers and Fabrics*, *3*, 39–45.

Nassif. (2010). Effect of Chemical Treatment on The Some Electrical And Thermal Properties For Unsaturated Polyester Composites Using Banana Fibers. *Eng. & Tech. Journal, 28*, 10.

Paul, S. A., Boudene, A., Joseph, K., & Thomas, S. (2008). Effect of Fiber Loading and Chemical Treatments on Thermo Physical Properties of Banana Fiber/Polypropylene Commingled Composite Materials. *Composites. Part A, Applied Science and Manufacturing*, *39*(9), 1582–1588. doi:10.1016/j.compositesa.2008.06.004

Paul, S. A., Boudenne, A., Ibos, L., Candau, Y., Joseph, K., & Thomas, S. (2008). Effect of fiber loading and chemical treatments on thermo physical properties of banana fiber/polypropylene commingled composite materials. *Composites. Part A, Applied Science and Manufacturing*, *39*(9), 1582–1588. doi:10.1016/j.compositesa.2008.06.004

Pothan, L. A., George, C. N., Jacob, M., & Thomas, S. (2007). Effect of Chemical Modification on the Mechanical and Electrical Properties of Banana Fiber Polyester Composites. *Composite Material*, *41*(19), 2371–2386. doi:10.1177/0021998307075456

Pothan, L. A., Oommen, Z., & Thomas, S. (2003). Dynamic Mechanical Analysis of Banana Fiber Reinforced with Polyester Composites. *Journal of Composite Science and Technology*, *63*(2), 283–293. doi:10.1016/S0266-3538(02)00254-3

Pothan, L. A., Thomas, S., & George, J. (2012). Tensile and Impact Properties of Banana Fibre/Glass Fibre Hybrid Polyester Composites.*ICCM Conference*.

Pothan, L. A., Thomas, S., & Groeninckx, G. (2006). The Role of Fibre/Matrix Interactions on the Dynamic Mechanical Properties of Chemically Modified Banana Fiber/Polyester Composites. *Composite Part A*, *36*(9), 1260–1269. doi:10.1016/j.compositesa.2005.09.001

Pothan, L. A., Thomas, S., & Habler, R. (2005). The Static and Dynamic Mechanical Properties of Banana and Glass Fiber Woven Fabric Reinforced Polyester Composites. *Journal of Composite Materials*, *39*(11), 1007–1025. doi:10.1177/0021998305048737

Pothan, L. A., Thomas, S., & Neelakantan, N. R. (1997). Short Banana Fiber Reinforced Polyester Composites: Mechanical Properties, Failure and Aging Characteristics. *Journal of Reinforced Plastics and Composites*, *16*, 1012–1025.

Rai, A., & Jha, C. N. (n.d.). Natural fibre composite and its potential as building material. *BMTPC Seminar.*

Sapuan, S. M., Leeni, A., Harimi, M., & Beng, Y. K. (2006). Mechanical Properties of Woven Banana Fiber Reinforced Epoxy Composites. *Journal of Materials and Design., 27*(8), 689–693. doi:10.1016/j.matdes.2004.12.016

Satyanarayana, K. G., Pillai, C. K. S., Sukumaran, K., Rohatgi, P. K., & Vijayan, K. (1982). Structure property studies of fibres from various parts of the coconut tree. *Journal of Materials Science, 17*(8), 2453–2462.

Savastano, H. Jr, Warden, P. G., & Coutts, R. S. P. (2000). Brazilian waste fibres as reinforcement for cement-based composites. *Cement and Concrete Composites, 22*(5), 379–384. doi:10.1016/S0958-9465(00)00034-2

Savastano, H. Jr, Warden, P. G., & Coutts, R. S. P. (2003). Potential of alternative fibre cements as building materials for developing areas. *Cement and Concrete Composites, 25*(6), 585–592. doi:10.1016/S0958-9465(02)00071-9

Savastano, H. Jr, Warden, P. G., & Coutts, R. S. P. (2005). Microstructure and mechanical properties of waste fibre–cement composites. *Cement and Concrete Composites, 27*(5), 583–592. doi:10.1016/j.cemconcomp.2004.09.009

Singh, V. K, & Gope, P. C., Sakshi, C., & Bisht, D. S. (2012). Mechanical Behaviour of Banana Fiber Based Hybrid Bio Composites. *J. Mater. Environ. Sci., 3*, 185–194.

Thiruchitrambalam, M., Alavudeen, A., Athijayamani, A., Venkateshwaran, N., & Elaya Perumal, A. (2009). Improving Mechanical Properties of Banana/Kenaf Polyester Hybrid Composites using Sodium Laulryl Sulfate Treatment. *Materials Physics and Mechanics, 8*, 165–173.

Uday Kiran, C., Ramachandra Reddy, G., Dabade, B. M., & Rajesham, S. (2007). Tensile Properties of Sun Hemp, Banana and Sisal Fiber Reinforced with Polyester Composites. *Journal of Reinforced Plastics and Composites, 26*(10), 1043–1050. doi:10.1177/0731684407079423

Zainuddin, E. S., Sapuan, S. M., Abdan, K., & Mohamad, M. T. M. (2008). Thermal Degradation of Banana Pseudo-stem Filled Unplasticized Polyvinyl Chloride Composites. *Journal of Materials and Design, 30*(3), 557–562. doi:10.1016/j.matdes.2008.05.060

Zhong, R., Burk, D. H., & Ye, Z. (2001). Fibers. A model for studying cell differentiation, cell elongation, and cell wall biosynthesis. *Plant Physiology*, *126*(2), 477–479. doi:10.1104/pp.126.2.477 PMID:11402177

Zuluaga, R., Putaux, J. L., Cruz, J., Vélez, J., Mondragon, I., & Gañán, P. (2009). Cellulose micro fibrils from banana rachis: Effect of alkaline treatments onstructural and morphological features. J. *Carbohydrate Polymers*, *76*(1), 51–59. doi:10.1016/j.carbpol.2008.09.024

Chapter 11
Bamboo Fiber–Reinforced Composites

Irem Sanal
Atlim University, Turkey

ABSTRACT

Nowadays, there has been an increased interest in the applications of bio-composites based on natural fibers, with the increasing emphasis on materials and processes which are environmental friendly and sustainable. Environmental friendly, fully biodegradable reinforced polymers or 'green' composite materials will play a major role in making the products of the future to protect our environment. The use of biodegradable and environment-friendly plant-based natural fibers has been a promising choice for polymers to make them 'greener'. In addition to being obtained from renewable sources, natural fibers suitable for composites are biodegradable and have enhanced properties. Bamboo is an excellent example for the development of sustainable natural fibers, since it can grow very fast per day, and the fibers of bamboo have excellent mechanical performance. Additionally, research in the development of bamboo-reinforced composites should be increased in the future, considering their enhanced properties, economical benefits and environmental friendly nature.

INTRODUCTION

Nowadays, the concept of "environmental friendly materials" has become very important due to the need to protect our environment. The definition of "environmental friendly materials" should include "safe" materials for human and other

DOI: 10.4018/978-1-5225-0424-5.ch011

life forms as well. Longer-term use of materials should not result in the emission of toxicities, which are highly damaging for both environment and for all living organisms, especially us, human. Both short and long-term utilization of materials should be needed to be characterized carefully; safer and more environmental friendly materials should be selected for applications. Contamination of wasted materials after expiration of their useful lives is utmost importance. However, the long-term characteristics of many kinds of materials are not well understood, especially their effects on environment.

Considering these, the most suitable material selection should be composite material. The main reason for choosing composite materials is that, using a sole material will not be sufficient, in order to satisfy a wide range of requirements such as, being environmentally friendly, human friendly and biodegradable. Throughout the various combinations of different composite materials, the most appropriate fiber-matrix pair to be considered an "environmental friendly material" is the combination of natural fibers and biodegradable polymer matrix. Biodegradable polymers have been developed in recent years and many kinds have appeared.

Among the natural fibers, special attention should be paid to bamboo fibers due to their environmental sustainability, mechanical properties, and recyclability have been utilized as reinforced polymer matrix composite in construction industries. (Zakikhani et al 2014)

Polymer composite materials have been used throughout very different application areas such as aircraft, automotive and submarine due to their enhanced mechanical and thermal properties when compared to other materials. However, the production and processing procedures of these materials will cause harm on the environment. Whereas, natural fibers are environmentally friendly and restorable materials. Natural fibers usually have low density and reasonable properties, in terms of mechanical performance, apart from being more economical. Therefore, usage of synthetic fibers is considered to be replaced with natural fibers, to conserve the economy and environment.

Natural fibers are divided into three categories basically, in terms of their origin:

1. Plant fibers (sisal, hemp, flax, bamboo, etc.),
2. Animals parts involving protein (silk, hair, wool, etc.)
3. Minerals

Out of other natural fibers, bamboo fiber is a good candidate for natural fibers in composite materials. Bamboo is economically advantageous with high growth rate, besides it helps reduction of carbon dioxide emission in atmosphere, making it one of the most important natural fiber. Bamboo has several advantages in terms of light weight, high strength, stiffness, biodegradability, and even its roots and leaves keep the soil together and protect it against the sun respectively (Janssen 2000). These enhanced material properties of bamboo, gives the opportunity of this materials to be used in building construction, furthermore it can also be used as reinforcement as composite materials with the proper extraction of fibers.

Polymer Composites

Definition of polymers can be made as a lengthy molecular formation made up of many systems where the basic systems are made of carbon, hydrogen, and oxygen. Polymers are products of petroleum, natural gas and fossil fuels. Currently when the polymer materials are produced, they are directed to major consuming industries such as, clothing, fabric and paints, or even to commercial products in construction, packaging, automobile, agriculture, furniture, electrical and general engineering.

Polymer materials can be further classified into three types where their molecular arrangement differs chemically, thermosetting, thermoplastic, elastomer.

1. **Thermosetting:** The molecular chains of thermoset polymers are characterized by three-dimensional closely-meshed crosslinking. This means that they can hardly be re-shaped and melted after hardening. However, they offer good chemical resistance and a high level of thermal stability, as well as being hard and brittle.
2. **Thermoplastic:** Thermoplastics are polymers in which the molecular chains are not crosslinked. They consequently demonstrate plastic elastic behaviour and are meltable and weldable. This formability can be repeated as often as required as long as the material is not thermally damaged by overheating.
3. **Elastomers:** Elastomers are characterized by wide-meshed crosslinking of the "knotted" molecular chains. This type of crosslinking means that the materials are elastically malleable. (Netravali and Chabba 2003)

Considering the industrial applications, a polymer is often used together with other reinforcement material in order to form a composite.

A composite is usually made up of two or more materials out of which one is the binding material, also called matrix and the other is the fiber. The advantage of composite materials over conventional materials are largely from their higher specific strength, stiffness and fatigue characteristics, which enables structural design to be more versatile.

Therefore, a composite is formed when two or more constituent materials are combined together. Generally, the matrix is made up of polymers, metals, alloys, inorganic cements or glass and the reinforcement consists of other materials either of natural, synthetic, metallic, organic or inorganic composition. Composites are normally reinforced materials consisting of two or more phases. Normal composites have two phases, that is, the matrix which is the continuous phase and the particulates or fibers which form the dispersed phase.

Considering polymer composites, generally, carbon and glass fibers are the most frequently used fibers to reinforce polymer matrices. They have great attractiveness due to their enhanced mechanical and thermal performance. Therefore, carbon or glass fiber reinforced polymers have been used for several years in many applications, which can vary from aerospace components to household goods. Composites gives the opportunity of lightweight and environmental-friendly design. Even though, the end-of-life disposal of these composites is not known well yet and is causing environmental pollution, beside the sophisticated production methods and high costs for the production of these synthetic fibers. Undoubtedly, these facts continue limiting their use for the preparation of fiber-reinforced composite materials.

Nowadays, an awareness of the importance of using biomaterials and sustainable natural resources are a necessity and they will play a crucial role in the near future (e.g., environmental protection and recycling targets). Accordingly, there is a rising attention in the renewable, green materials, to which category natural fibers clearly belong.

These new composites reinforced with natural fibers are rising as a realistic alternative to glass-reinforced composites in some applications (Wambua et al. 2003). Even though, there are some hindrance that have to be conquered, such as moisture sensitivity and thermal degradation of the of natural fibers and the necessity of low temperature processing (Jindal et al. 1986). Interface zone and bonding strength between the reinforcing fiber and the matrix has also a significant effect on the performance of the composite since transfer of stress and load distribution efficiency at the interface are determined by the adhesion between the components

Natural Fiber-Reinforced Polymer Composites

Nowadays, natural fiber reinforced composites are finding much attraction as a replacement for other types of reinforced polymer composites. Advantages of using natural fibers as reinforcement in polymers are their non-abrasive nature and low energy consumption. Natural fibers withdraw carbon dioxide from the atmosphere, hence provide an advantageous contribution to the carbon emission in the atmosphere. Disposal of natural fiber composites are very straightforward, making it at upmost importance, since they can be easily composted or recycled at the end of their life cycle. In addition to economic benefits of natural fiber compared to synthetic fibers, these fibers offer high security if used for automotive applications as well (Flemming et al. 1995). Additionally, the specific mechanical properties of natural fibers are founded out to be comparable to those of traditional reinforcements

(Defoirdt et al. 2010) (Biswas et al. 2011). As a result, the constitutive properties of natural fibers can satisfy the requests of the global market especially for those industries concerned in light-weight material applications. Therefore, they can be potential replacement for non-renewable synthetic fibers. However, high moisture absorption, poor wettability and insufficient adhesion between untreated fiber and polymer matrix lead to de-bonding at fiber-matrix interface.

Bamboo Materials

Since the very beginning of technological progress, bamboo materials has been part of human development, such as being a protective shelter, used in household stuff and other artifacts. In the East, it is known as the plant of a thousand uses due to its excellent physical, chemical and mechanical characteristics (Datschefski 2001). Bamboo belongs to the plant family and it is a fast growing plant by having little time to be re-grow without any necessity of replanting. Bamboo allows the infiltration of rain into the soil, they help control erosion, sedimentation and recovery of carbon dioxide from the atmosphere (Pereiara 2007). Therefore, bamboo can be a possible choice of sustainable material selection for application in industrial design.

Bamboo Structure

The main important part in the manufacturing of bamboo fibers is to know the microstructure of a bamboo culm, which consists of vascular bundles and tissues.

Bamboo culms are hollow, and every culm from inner side is divided by several diaphragms which are seen as rings on the outside. The part between two rings is called "Internode" where branches grow (Janssen 2000). Length of the gap between each node can be different, depending on the type of the species. The microstructure of a bamboo culm consists of many vascular bundles which are embedded in parenchyma tissue and distributed through the wall thickness. Generally, tissues keep the vascular bundles in the longitudinal direction. The number of vascular bundles' concentration is high in the regions close to the outside of the bamboo culm wall, and this amount is basically reduced on the inner parts. They involve vessels, sclerenchyma cells, fiber strand and sieve tubes with companion cells (Liese 1971). The performance of a bamboo culm is defined by its vascular bundles, where strength is mainly developed. Therefore, it is needed to use a suitable processing method to separate the tissues from fiber strands and vascular bundles without causing any damage on the extracted fibers.

The structure of a bamboo culm transverse section is characterized by numerous vascular bundles embedded in the parenchymatous ground tissue (Grosser and Liese 1971). The culm tissue consists of two cell types: parenchyma cells and vascular

Figure 1. Structure of bamboo culm

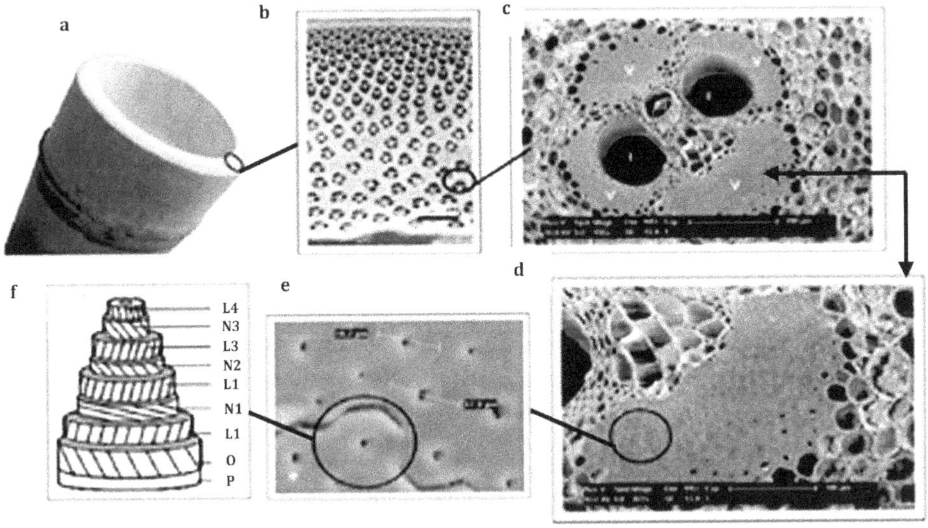

bundles. The parenchyma cells are mostly thin-walled and connected to each other by numerous simple pits. Pits are located predominantly on the longitudinal walls. The horizontal walls are scarcely pitted. The structure of a bamboo culm is shown in Figure 1. The size of the vascular bundle is large in the inner and middle layer but smaller and denser in the outer layer as shown in Figure 2.

Among the very well recognized natural fibers, bamboo has low density and high mechanical strength, that is, it has high specific stiffness and strength, making it one of the most favorable combinations. In previous researches, bamboo fibers found out to be comparable with glass fibers in terms of specific properties (Trujillo et al. 2010). According to several authors, (Liese 1998) (Parasmewaran and Liese 1976) (Parasmewaran and Liese 1980) bamboo fiber bundles are distributed densely in the outer region of the culm wall and sparsely in the inner region, and

Figure 2. Bamboo vascular bundle

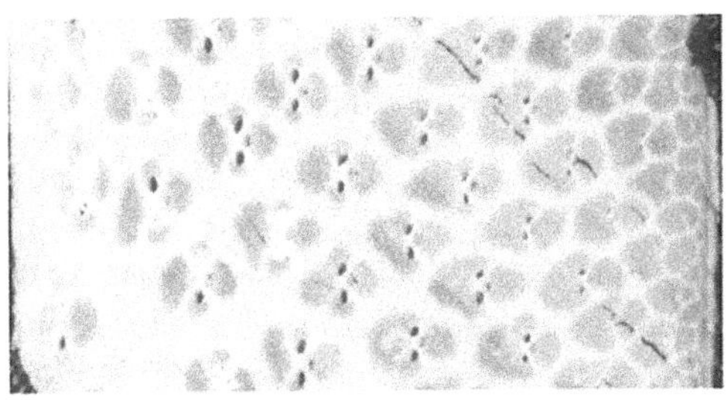

also concentrated in the upper part of the culm compared with the base (Figure 1(b)). Elementary fibers in such a bundle consist of thick and thin layers with different fiber orientation (Figure 1(f)). In the thick layers, the fibrils are oriented at a small angle to the fiber axis, whereas the thin ones show mostly a more transverse orientation. This structure does not exist in the cell walls of fibers of normal wood and leads to an extremely high tensile strength of the culm (Osorio et al 2010).

As is it can be seen from Figure 1 that the lignifying cell construction of the bamboo and its technical conditions are very similar to the original texture of wood. Whereas wood has got a hard center and becomes weaker towards the outer parts, the bamboo is in its outer parts hard and in its inner parts weak, what causes a much more stable construction. From the inner parts of the tube towards the outer parts you can realize a continuous accumulation. The more stable fiber structures are most dense where you find the highest stress.

There are many differences between bamboo and wood. In bamboo, there are no rays or knots, which give bamboo a far more evenly distributed stresses throughout its length. Bamboo is a hollow tube, sometimes with thin walls, and consequently it is more difficult to join bamboo than pieces of wood. Bamboo does not contain the same chemical extractives as wood, and can therefore be glued very well [Janssen 1995]. Bamboo's diameter, thickness, and internodal length have a macroscopically graded structure while the fiber distribution exhibits a microscopically graded architecture, which lead to favorable properties of bamboo [Amada et al. 1998].

Bamboo Fibers

Bamboo is a material, having both grass and also wood characters. Bamboos are very strong in their longitudinal direction, since strong fiber bundles penetrates their body from the bottom to the top although their pulp is shorter than ~2mm. Therefore, the average length of bamboo fibers is about 2 mm, and average diameter between 10 and 20 μm. However, hardness of the bamboo fibers can be different, since it depends on the number of fiber bundles and the manner of their scattering.

Bamboo is consisting of cellulose, hemicellulose and lignin. In bamboo, cellulose and hemicellulose are present in the form of holo-cellulose which amounts to more than 50% of the total chemical constituents. Other most plentiful ingredient of bamboo is lignin. Lignin works as a binder and acts as the matrix for the cellulose fibers. Lignin is mostly taking part in load carrying as part of a composite material (Jain et al. 1992).

In Table 1, chemical constituents, density and microfiber angle of bamboo and some of the other plant species are listed. The bamboo has 60% cellulose and a considerably high percentage of lignin (about 32%) (Jain et al. 1992).

Table 1. Composition of bamboo and some other natural fibers

Types of Fiber	Microfibril Angle (Deg)	Cellulose (%)	Lignin (%)
Coir	30-49	43	45
Banana	11	65	5
Sisal	20-25	70	12
Jute	8.1	63	11.7
Bamboo	2-10	60.8	32.2

(Jain et al 1992)

Bamboo materials also contain some other organic composition in addition to cellulose and lignin. It contains about 2-6% starch, 2% deoxidized saccharide, 2-4% fat, and 0.8-6% protein. The carbohydrate content of bamboo plays an important role in its durability and service life. Durability of bamboo against mold, fungal and borers attack is strongly associated with its chemical composition. Bamboo is known to be susceptible to fungal and insect attack. The natural durability of bamboo varies between 1 and 36 months depending on the species and climatic condition (Liese, 1980). The presence of large amounts of starch makes bamboo highly susceptible to attack by staining fungi and powder-post beetles [Mathew and Nair 1990]. It is noteworthy that even in 12 year old culms starch was present in the whole culm, especially in the longitudinal cells of the ground parenchyma (Liese and Weiner 1997).

Bamboo culms were extracted from a typical bamboo plantation in Colombia, specifically from the Coffee Region. Technical fibers, also called 'fiber bundles,' which will be referred to as fibers in this article, were extracted from the bamboo culms using a novel purely mechanical process (neither chemicals nor high temperature were used during the extraction). The maximum length of the extracted fibers is the internode length. For a 48-month-old culm, the internode lengths range between 20 and 35 cm (London et al. 2002). Figure 3 shows a group of mechanically extracted fibers whose diameter ranges between 90 and 250 mm; the main diameter concentration is around 150 mm (Osorio et al. 2010).

Bamboo fibers are attracting more and more attention from researchers, and often called 'natural glass fiber' (Okubo et al. 2004). Bamboo fibers are focused as one of substitution for natural plant fibers having many advantages such as low cost, low density, ecologically friendly, sustainability and biodegradability.

Compared to other natural fibers, bamboo fibers are very brittle due to their thicker diameter as well as chemical composition containing a high content of lignin and hemicelluloses. These compositions play an important role in fiber bundle strength and individual fiber strength. Additionally, they also affect fiber rigidity, fiber swelling and moisture absorption (Bogoeva-Gaceva et al. 2007).

Figure 3. Group of mechanically extracted fibers diameter ranges between 90 and 250 mm
(Osorio et al. 2010)

Morphology of Bamboo Fibers

Relation between fiber nano- and micro-structure and the mechanical properties of the bamboo fibers are very important, prior to their applications in the composites. Therefore, the performance of fibers of different species, age, geographical location and location in the culm can be explained, allowing a better selection of the material or even in future extraction of better suited bamboo plants.

Polymer composites and bamboo fibers are different and incompatible with each other in terms of their polarity structures, the using fiber treatment or agents making them more compatible (such as Sodium Hydroxide - NaOH or polypropylene with maleic anhydride –PPMA) will help to reduce interfacial tensions and improve the adhesion between the matrix and fiber. By using fiber treatment or compatibilizing agents are expected to cause the material having an appropriate behaviour of the many applications based on mechanical forces required for different application related in which the mechanical behaviour of samples need to be known through the tensile tests, bending, impact and fatigue.

To evaluate and compare the material performance of bamboo, age, moisture content and diameter of the tube are of up most importance. Comparing the different results of investigation of the strength properties of bamboo, so you can see that there is a big fluctuation of the results, although they all tested the same species of bamboo, the guadua angustifolia.

Figure 4 shows a detail of the primary wall of elementary bamboo fibers of the species Guadua Angustifolia. The 90-degree orientation of the nano-fibers will provide some off-axis mechanical performance.

Figure 4. Detail of primary wall

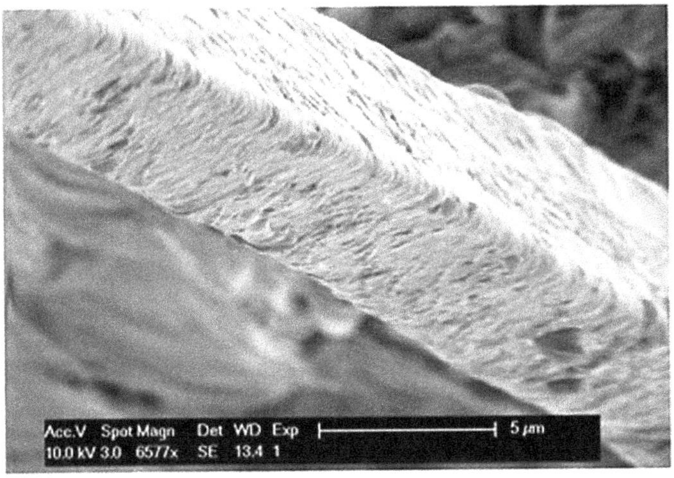

Figure 5 shows the secondary wall, after the primary wall has been cut out. The alignment and orientation of the micro-fibers are close to 0 degrees. As the secondary wall is the most important layer in a bamboo fiber, this is why bamboo fibers have high longitudinal stiffness and strength. (van Vuure et al. 2011)

According to the previous study of Kazuya (Kazuya 2004) the bamboo column consists of many vascular bundles and xylem; a vascular bundle includes four sheaths of fibers, two vessels and some sieve tubes and xylem surrounds each vascular

Figure 5. Secondary wall

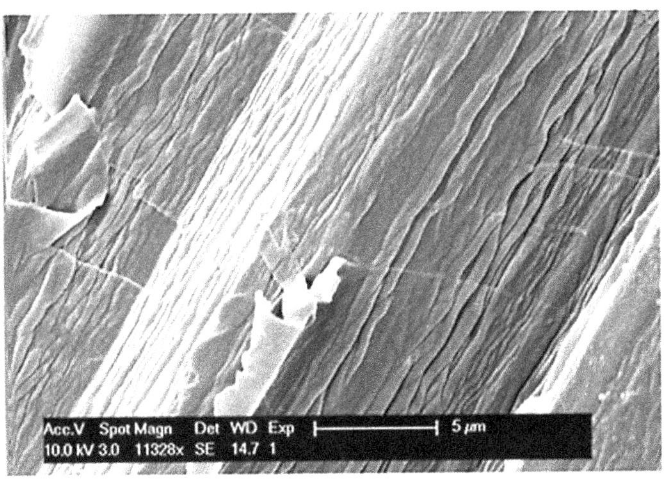

Figure 6. (a) and (b) cross-section of bamboo

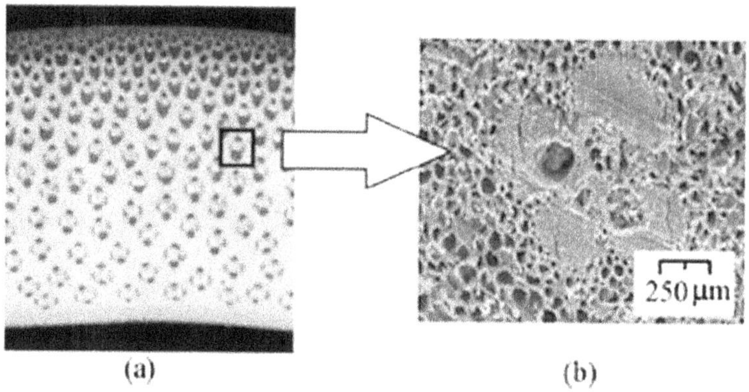

bundle. The sheath consists of many single fibers whose diameter is 10–20 mm each in average. Figure 6 (a) and (b) show a cross-section of bamboo.

Considering the previous studies on bamboo-fiber reinforced polymers, it was stated that tensile modulus increased with increasing bamboo content, whereas tensile strength decreased with increasing bamboo content.

Mechanical Properties of Bamboo Fibers

The mechanical analysis is the study of a material's behavior when subjected to loads. The mechanical properties mainly provided by the cellulose content, which is influenced by many factors such as fibers volume fraction, fiber length, fiber aspect ratio, fiber-matrix adhesion or fiber orientation. Much research has been carried out on bamboo fiber reinforced composites reported that mechanical properties of bamboo vary due to the different testing methods used and the samples tested.

The work that is needed for the punch of a bamboo tube is nearly the same whether the punch hits the knot or the internodium. But the breaking conditions itself are totally different. If the punch hits the knot the tube will burst in axial stripes; that means a break as a result of the effort of the strength vertical to the fibres. If the punch hits the internodium you will find the actual break; that means break as a result of the effort of the tension strength in direction of the fibre. Examinations of the load-carrying capacity of the bamboo by Dr. Simon Eicher, Otto- Graf-Institut are given in Table 2.

The chemistry of bamboo is important in determining its utilization potential. Several studies have investigated the chemical composition of bamboo. But systematic and thorough research on a commercially important bamboo species is needed

Table 2. Load-carrying capacity of bamboo

Load-Carrying Capacity of Bamboo	
Compressive strength	5,6kN/cm^2
Compressive strength	1840 kN/cm^2
Average bending strength	7,4 kN/cm^2
Average bending strength at perfect drying	10 kN/cm^2
Average elasic modulus of bending	1790kN/cm^2
Average elasical modulus of tension	1900 kN/cm^2

to determine utilization potential for the products such as medium density fiberboard (MDF). Most of previous studies provide either only general information of several bamboo species or focuses on only one aspect of one species.

Furthermore, the amount of each chemical composition of bamboo varies with age, height, and layer, the chemical compositions of bamboo are correlated with its physical and mechanical properties. Such variation can lead to obvious physical and mechanical properties changes during the growth and maturation of bamboo

Aging of a bamboo culm influences physical, chemical, and mechanical properties, and consequently its processing and utilization. The physical and mechanical properties of bamboo vary with the age of the bamboo and the height of the culm (Chauhan 2000). In general, properties of bamboo drop from the top portion to the bottom. The increase in weight is cumulative and directly related with age. Strength properties are reported to decrease in older culms (Zhou 1981). Limaye (1948; 1952) found that older culms of Dendrocalamus strictus became 40-50 percent stronger and stiffer than young ones. Maximum values were found in 3-6 year old culms. Sekhar et al. (1962) found highest values in 3-4 year old culms of Bambusa nutans. There is also variation in strength properties along the culm height as well. Compressive strength tends to increase with height (Liese 1987; Sattar et al. 1994). The strength increases from the central to the outer part. There is more than 100 percent variation in strength from the inner to the outer layers (Narayanamurti and Bist 1947).

Moisture Absorption of Bamboo Fibers

Bamboo fibers absorb water, due to their cellulosic fiber structure. This absorption behavior is observed in bamboo fibers, especially when they are soaked in water or exposed to humid conditions. Thus, this absorption ability of fibers, may affect the mechanical performance of bamboo. Additionally, according to the results of previous studies, moisture absorption caused poor interfacial bonding between bamboo fibers and hydrophobic matrix polymers. Many previous studies examined the effects of moisture absorption on the mechanical properties of bamboo in detail. The results indicated that both the compressive and flexural strength of bamboo remained steady under dry conditions. When bamboo was soaked in water or exposed to humidity, the tensile modulus and tensile strength decreased accordingly. In specimens coated with one layer of epoxy, these reductions in tensile modulus and tensile strength were also observed.

Therefore, generally moisture absorption of bamboo fibers may lead to a decrease in the elastic modulus and a slight increase in the tensile strength, and in order to overcome this issue, bamboo fibers extracted by alkali treatment and steam explosion can be used, instead of untreated fibers. The flexural strength of bamboo fibers is improved by increasing the bamboo content and also the surface treatment.

Extraction of Bamboo Fibers

It was mentioned previously that bamboo materials are very strong in their longitudinal direction, due to the fibers longitudinally aligned in their body. As a result of this high strength property, bamboo fibers are often called 'natural glass fibers'. It is necessary to develop a process to fabricate bamboo composites as well as to extract qualitatively controlled fibers from bamboo trees, in order to benefit from this high strength of performance of bamboo fibers. However, it is difficult to extract bamboo fibers having its superior mechanical properties, since these fibers are often brittle compared with other natural fibers, due to its lignin content. Therefore, a well arranged process should be chosen to extract the bamboo fibers for reinforcement of composite materials. (Zakikhani et al. 2014)

Several methods have been used for extraction process of bamboo fibers, considering their different industrial applications. Considering various extraction procedures for bamboo fibers, some extraction methods has benefits over others, such as "steam explosion" mechanical method and "chemical methods". These two methods are able to remove lignin from bamboo fibers, affecting the microstructure of bamboo and bamboo fibers extracted using these procedures are usually short in length. The length of extracted fibers could be controlled in the retting process. and in comparison with "steam explosion" and "chemical process" the retting and rolling mill methods produced long fibers. In the crushing and grinding methods the extracted fibers could be used in the form of particles for the crystallization of the matrix.

In addition to that, alkali treatment in the category of "chemical process" could remove lignin from fibers and improve the strength of interfacial zone of composite. The strength mainly depends on the interfacial condition, and poor adhesion at the interface becomes a defect and leads to poor strength. In the chemical retting methods more lignin from bamboo fibers were removed. In general, these extraction methods have been performed based on the applications of bamboo in various fields of study. (Zakikhani et al. 2014)

Bamboo Fiber Reinforced Composites and Design Applications

Recently, the application of bamboo has been strengthened to accomplish bamboo as an eco-friendly and renewable fiber. The fast growing and renewability of bamboo lead to an evolution in theoretical and applied research on bamboo based products, particularly in housing, furniture, packaging, transport, etc. These bamboo fiber-reinforced composites have replaced traditional materials in terms of indoor as well as outdoor applications. Developed products from bamboo composites defeated the deficiency such as dimensional stability, longevity, weather resistant, high impact resistant, low maintenance, non-toxic, and low flame spread in conventional composites.

The development of economical approach, eco-friendly, less utilization of renewable building materials, reduction in environmental pollution and conservation of energy has been well affected by present days bamboo based composite industry.

For example, bamboo based composites used for water surfing are commonly known as surfboards. These surfboards are light in weight, with particular design and water proof surfaces. These decks are multilayered bamboo boards with epoxy matrix with specialty over glass boards so that it cannot get bends thus maintaining its shape for long uses.

Several other bamboo composite products are also available in the markets. The created composites based on bamboo have been used in many applications such as transport, furniture, packing and other fields. It is reported that bamboo composites can be used as beam, frame, and joints in structural upgradation (Sen and Reddy, 2011). Research is still going on for producing durable furniture, bicycle, tricycles and car bodies by using bamboo composites.

A broad change from economic perspective is needed in composites; the only option is use of plant natural fibers. Bamboo can serve as an excellent source to replace the existing crises of cost and availability of raw material.

Comparison of Bamboo Fiber Reinforced Composites with Conventional Composites

The production of large amount of synthetic fiber reinforced composites, such as glass/carbon fiber reinforced polymer composites, conventional composites and petroleum based plastics have brought serious danger to our environment. The recycling of glass fiber reinforced composites is not easy and safe for environment, even after recycling these composites, very fewer fractions of them is incinerated. Furthermore, synthetic fibers have higher densities (1–2.8 gm/cm−3) than natural fibers (0.5–1.5 gm/cm3). Therefore, various products made by synthetic fibers used in different industries affect directly the weight of those components.

These problems have to be overcome in order to develop sustainable and eco-friendly materials. Mainly development of materials, derived from the plant source which have a capability of rapid growth, bamboo fibers will be one of the leading examples as these fibers are derived from the sources which is known for its rapid growth and sustainable nature. Many research is going on in order to replace conventional and synthetic materials with eco-friendly materials such as bamboo fibers, being one of the sustainable and economical materials to be exploited in field of biocomposites.

Biocomposites can be produced from bamboo fibers for indoor and outdoor applications to replace materials/products generally fabricated from glass fibers based composites and conventional composites. As mentioned previously, bamboo fiber based polymer composites possess high strength to weight ratio, dimensional stability, durability, and amenability to be engineered to any complex shape or size at low cost of production as compared to conventional composites. Therefore, there is a large opportunity for bamboo fibers to replace or reduce the applications of glass fiber content in composites and conventional composites, and to be utilized commercially on industrial scale.

SOLUTIONS AND RECOMMENDATIONS

Bio-composites can supplement and eventually replace petroleum-based, toxic composite materials in most of the different application, offering especially environmental and humanitarian advantages. However, several critical issues regarding bio-fibers are the importance of surface treatment, for making it more reactive and making the matrix-fiber interface zone stronger.

- For the case of bamboo fiber-reinforced polymer composites also have some limitations of susceptibility to microbial and environmental challenges that preclude their use for product standardization and repeatability.
- Additionally, effective surface treatment of bamboo fibers reduces the protective layer for breakdown due to oxidation and consequent increased strength of the reinforced composite.
- The cylindrical shape of the bamboo fiber can be a limitation for its direct use in several engineering systems. A more flexible alternative is extracting the bamboo fibers from the culm and using them as reinforcement of polymeric matrices.

- Another limitation of bamboo-fiber reinforced polymers are their high cost due to bamboo fiber extraction and processing. Further research and increased demand/supply should overcome this limitation.

Development of appropriate processing techniques are also very important, to satisfy the performance needs.

FUTURE RESEARCH DIRECTIONS

Bamboo fibers have been widely used in composite applications for socio-economic empowerment of peoples. Manufacturing of bamboo fiber based composites, using different matrix types has been promoted cost effective and eco-friendly bio-composites. Undoubtedly, these development of bio-composites directly affected the market values of bamboo. Therefore, investigation of fundamental, mechanical, and physical properties of bamboo fibers are necessary, in order to design such composites. A considerable effort should be made by researchers in various different uses of bamboo fiber as reinforcement in polymer composites.

Bamboo fiber is obtained from a source which is known for its renewability in terms of fast growth and enhanced mechanical properties. The application of bamboo fibers for fabrication of bio-composites by using advanced technology transforms future of coming generation in terms of environment and human well-being.

The strengthening effects on the bamboo fibers containing various matrices such as polystyrene, polyester and epoxy resins should be studied comprehensively in the future, since the economic value, light weight, high specific strength and non-hazardous nature of bamboo fibers are up most attractive properties of these material. These properties of bamboo fibers should make researchers to work in the direction of bamboo-reinforced composite technology. Therefore, bamboo fiber based composites should be examined and tested in order to understand their potential use in automotive, household and some other industries, if they can be used as replacement for the non-renewable, costly synthetic fibers in composite materials.

Successfully designed and engineered products from the bamboo fibers can help in making new revolution to sustain our natural resources and environment. It will be an alternative way to develop the bio-composites which can be particularly used for daily needs of common people whether it is household furniture, house, fencing, decking, flooring, and light weight car components or sports equipment. Their economic benefits, environmental gains, easy availability and aesthetic designs will be the main motivation to transform traditional applications and materials to sustainable futures.

CONCLUSION

Industrial development towards eco-friendly industrial products (such as bio composites) and their process of manufacturing will lead to the sustainable tomorrow for future generations. High performance, low cost, environmental-friendly and biodegradable materials and renewable plant materials can form new paths for sustainable and eco-efficient advanced technology products. These products will undoubtedly eliminate synthetic/petroleum based products presently dominated in market which are diminishing natural petroleum feedstock.

Environmental friendly bio-composites from plant-derived fibers and crop-derived plastics are promising materials of the new generation and will definitely be of upmost importance to the materials world, as a solution for protecting our environment. Additionally, natural fibers and bio-composites will lead to the sustainable, eco-friendly and well-designed industrial products which can be replacement for petroleum based products in future. Therefore, use of materials from renewable resources is attaining great importance and leading industries of different application fields are seeking to replace their traditional toxic materials with composites derived from natural fibers and biopolymers.

Among the well-known natural fibers, bamboo has one of the most favorable combinations of low density and high mechanical strength, high specific stiffness and strength, in addition to having rapid growth rates of production. These aspects very well fit the new generation needs of materials world toward the use of green, renewable and sustainable resources, which warrant the further development of bamboo fiber-reinforced polymers.

However, many research has already been carried out on bamboo fibers and bamboo fiber-reinforced composites, it still needed to do more research and innovation in this area to overcome potential challenges ahead. Therefore, further methodical research on bamboo fiber as a reinforced composite could lead to a promising future for bamboo composites as replacements for synthetic fibers.

BACKGROUND

Bamboo fibers have significant advantages over other natural fibers such as fast growing, high strength, and fixing the carbon dioxide, furthermore it is light weight, biodegradable, and economically advantageous. Therefore, there is a great interest in using bamboo fiber as a reinforced composite material in different applications, especially in polymer composites. Bamboo fibers and bamboo fiber reinforced composites have ability to be used in more applications.

REFERENCES

Biswas, S., Ahsan, Q., Verpoest, I., & Hasan, M. (2011). Effect of Span Length on the Tensile Properties of Natural Fibers. *Advanced Materials Research, 264-265*, 445–450. doi:10.4028/www.scientific.net/AMR.264-265.445

Bogoeva-Gaceva, G., Avella, M., Malinconico, M., Buzarovska, A., Grozdanov, A., Gentile, G., & Errico, M. E. (2007). Natural fiber eco-composites. *Polymer Composites*, *28*(1), 98–107. doi:10.1002/pc.20270

Datschefski, E. (2001). *The Total Beauty of Sustainable Products*. New York: Rotovision.

Defoirdt, N., Biswas, S., De Vriese, L., Tran, L. Q. N., Van Acker, J., Ahsan, Q., & Verpoest, I. et al. (2010). Assessment of the tensile properties of coir, bamboo and jute fiber. *Composites. Part A, Applied Science and Manufacturing*, *41*(5), 588–595. doi:10.1016/j.compositesa.2010.01.005

Flemming, M., Ziegmann, G., & Roth, S. (1995). Einfuhrung, Faserverbundbauweisen. In Fasern und Matrices. Springer-Verlag.

Jain, S., Kumar, R., & Jindal, U. C. (1992). Mechanical behaviour of bamboo and bamboo composite. *Journal of Materials Science*, *27*(4), 598–604.

Janssen, J. J. (2000). *Designing and building with bamboo*. International Network for Bamboo and Rattan China.

Jindal, U. C. (1986). Development and testing of bamboo-fibers reinforced plastic composites. *Journal of Composite Materials*, *20*(1), 19–29. doi:10.1177/002199838602000102

Liese, D. G. W. (1971). On the Anatomy of Asian Bamboos, with Special Reference to their Vascular Bundles. *Wood Science and Technology*, *5*(4), 290–312. doi:10.1007/BF00365061

Liese, W. (2010). *The anatomy of bamboo culms*. INBAR Technical Report No. 18. International Network for Bamboo and Rattan.

London, X., Camayo, G., Rian, N., & Lopez, Y. (2002). Characterization of the anatomy of Guadua angustifolia (Poaceae: Bambusoideae) culms. *Bamboo Sci Cult: J Am Bamboo Soc*, *16*, 18–31.

Netravali, A. N., & Chabba, S. (2003). Composites get greener. *Materials Today*, *6*(4), 22–29. doi:10.1016/S1369-7021(03)00427-9

Okubo, K., Fujii, T., & Yamamoto, Y. (2004). Development of bamboo-based polymer composites and their mechanical properties. *Composite: Part A*, *35*(3), 377–383. doi:10.1016/j.compositesa.2003.09.017

Osorio, E. T. L., Van Vuure, A. W., & Verpoest, I. (2010). Morphological aspect and mechanical properties of single bamboo fibers and flextural characterization of bamboo/epoxy composites. *Journal of Reinforced Plastics and Composites*, *30*(5), 396–408. doi:10.1177/0731684410397683

Parasmewaran, N., & Liese, W. (1976). On the fine structure of bamboo fibers. *Wood Science and Technology*, *10*, 231–246.

Parasmewaran, N., & Liese, W. (1980). Ultrastructural aspects of bamboo cells. *Cellul Chem Technol*, *14*, 587–609.

Pereira, M. A. R., & Beraldo, A. L. (2007). Bambu de Corpo e Alma. Bauru: Canal.

Sen, T., Reddy, H.N.J., (2011). Application of Sisal, Bamboo, Coir and Jute Natural Composites in Structural Upgradation. *International Journal of Innovation, Management and Technology*, *2*(3).

Trujillo, E. E., Osorio, L., Fuentes, C., Van Vuure, A., & Verpoest, I. (2010) Bamboo fiber thermoplastic for transport applications. In *Proceedings of SAMPE Europe 31 International Technical Conference and Forum*.

van Vuure, A. W., Osorio, L., Trujillo, E., Fuentes, C., & Verpoest, I. (2011). Long bamboo fibre composites. In *Proceedings of the ICCM 18*.

Wambua, P., Ivens, J., & Verpoest, I. (2003). Natural fibers: Can they replace glass in fiber reinforced plastics? *Composites Science and Technology*, *63*(9), 259–1264. doi:10.1016/S0266-3538(03)00096-4

Zakikhani, P., Zahari, R., Sultan, M.T.H, & Majid, D.L. (2014). Bamboo Fiber Extraction and Its Reinforced Polymer Composite Material. *International Journal of Chemical, Molecular, Nuclear, Materials and Metallurgical Engineering*, *8*(4).

Chapter 12
Coir Fiber-Reinforced Composites

Irem Sanal
Atilim University, Turkey

ABSTRACT

Nowadays, fiber-reinforced polymer composites have played a significant role in many different fields of applications, regarding their high specific strength and high modulus. The fiber which serves as reinforcement mechanism in polymer composites may be either synthetic or natural. Natural fibers are not only strong and lightweight but also very economical and environmental friendly. Natural fibers as reinforcement are stated to be a major step taken in promoting environmental protection and sustainability. There are many types of natural cellulose fibers but the thickest and most resistant of all commercial natural fibers, coir/coconut fiber is a coarse, short fiber extracted from the outer shell of coconuts. Coir/coconut fibers have the highest concentrations of lignin, making it most suitable for applications where slow biodegradability is required. This chapter has been written with an aim to explore the potential of the coconut/coir fiber reinforced polymer composites in terms of their performance, surface treatments/modifications and areas of application.

INTRODUCTION

The advantage of composite materials over conventional materials are mainly due to their higher specific strength, stiffness and fatigue characteristics, which enables structural design to be more versatile. By definition, composite materials consist of

DOI: 10.4018/978-1-5225-0424-5.ch012

two or more constituents with physically separable phases (Lilholt and Lawther, 2000) (Mueller and Krobjilowski 2003), having notably different and enhanced properties.

Processing of composites using natural fibers as reinforcement has increased dramatically in recent years. (Li, 2008) Fiber-reinforced composites consist of fiber as reinforcement and a polymer as matrix. Studies about the utilization of natural fibers as reinforcement in polymeric composites are increasing due to the improvements that fibers can provide to the product, regarding the increasing environmental consciousness and awareness of the need for sustainable development.

Natural fiber composite materials have gained importance and popularity due to their lightweight, high strength, stiffness, corrosion resistance, and lower impact on the environment (Satyanarayan et al. 1981). Because of their quality, durability and other advantages, they are used to make a wide variety of floor mats, yarn, rope, etc. (Mohanty et al 2000). In some cases, natural fibers can be obtained as a by-product of the coconut fruit, for example. The traditional products consume only a small percentage of the total world production of coconut husks that is generated by the food industry and coastal touristic regions. Thus, research and development efforts have been carried out to find new uses for coir, including its utilization as reinforcement in polymer composites in automotive parts, household and electrical applications. However, a high level of moisture absorption and insufficient adhesion between untreated fibers and the polymer matrix may lead to bio-composites presenting high water absorption and poor mechanical properties that reduce their use in electrical devices.

Several fiber surface treatment methods have been studied to improve the adhesion between coconut fibers and the surrounding matrix, as well as, to reduce water absorption and increase mechanical properties. Nowadays, effective methods based on chemical treatment (Samadi et al 1986) (Murali et al. 2007) such as dewaxing and grafting are used to increase the surface area available for contact with the matrix, but these methods are expensive (like silane agents) and they may cause serious damage to the environment (acids or alkalis agents). Alkalization is the main technique used on natural fiber to remove hemicelluloses (Joseph et al 2002) of fiber surfaces and it has been employed as a less harmful treatment to the environment and cheaper than other methods proposed. However, its use is still controversial, some are favorable (Bledzki and Gassan, 1999) others suggest controlled application (Luo and Netravali,1999) mainly due to chemical wastes generated. Physical treatments (cold plasma treatment, corona treatment) (Ràcz and Hargitai, 2000) have been proposed as eco-friendly processes for superficial modification of the fibers, but these are usually complex and very expensive methodologies. Research on an effective low cost treatment of natural fibers is necessary since the cost of the raw material, i.e. the natural fiber, is very attractive to the market.

Despite the attractiveness of natural fiber reinforced polymer matrix composites and the problem related to fiber–matrix adhesion, manufacturing processes have not been explored in order to enhance industrial productivity of such composites. Injection molding is the main method used by the plastic industry due to its high efficiency and low cost. The combination of injection techniques and natural fiber, especially coconut fiber, to manufacture industrial products has not been sufficiently investigated. As far as these issues are concerned, a balance between performance and technological applications of composite material might be achieved through proper methodology.

BACKGROUND

Increasing concern about global warming and depleting petroleum reserve have made scientists focus more on the use of natural fibers such as bagasse, coir, sisal, jute etc. as reinforcement materials in composites. This has resulted in emerging potential for natural fiber composites to become future replacements as these composites provide general advantages which include low cost, easy availability, low density, acceptable specific properties, ease of separation, enhanced energy recovery, carbon-dioxide neutrality, biodegradability and recyclability (De Rosa et al., 2010).

In this chapter, a detailed review has been made to make use of coir/coconut, a natural fiber as reinforcing material especially for polymer composites.

NATURAL FIBER-REINFORCED COMPOSITES

The application of natural fiber-reinforced composites has been extended to almost all fields. Natural fibers are hydrophilic in nature as they are derived from lignocellulose, which contain strongly polarized hydroxyl groups. These fibers, therefore, are inherently incompatible with hydrophobic thermoplastics, such as polyolefin.

Natural fiber reinforced polymer composites have successfully proven their high qualities in various fields of many applications. Over the past two decades, natural fibers have received considerable attention as a substitute for synthetic fiber reinforcements in polymer composites. As replacements for conventional synthetic fibers like aramid and glass fibers, natural fibers are increasingly used for reinforcement in thermoplastics due to their low density, good thermal insulation and mechanical properties, reduced tool wear, unlimited availability, low price, and problem-free disposal (Ayrilmis and Ashori, 2015). Additional advantageous properties of natural fibers can be found in Table 1.

Table 1. Properties of natural fibers

Specific Gravity	Low
Cost	Low
Renewability	Yes
Recyclability	Yes
Energy consumption	Low
Distribution	Wide
CO2 neutral	Yes
Abrasion to machines	No
Health risk when inhaled	No
Disposal	Biodegradable

(Bledzki and Gassan, 1999)

By stating natural fiber composite materials, it was meant that a composite material is reinforced with fibers, particles or platelets of natural otherwise renewable resources, in difference to for example carbon backbone or agamid have to be synthesized.

Natural fibers can be divided into two basic categories:

1. Animal fibers and
2. Plant cellulose fibers, as can be seen in Figure 1.

Figure 1.

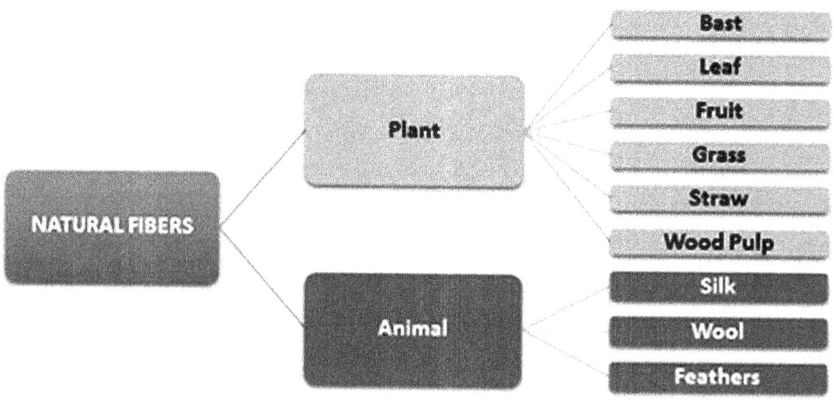

Plants that produce natural fibers are categorized into primary and secondary depending on the utilization. Primary plants are grown for their fibers while secondary plants are plants where the fibers are extracted from the waste product. There are 6 major types of fibers namely; bast fibers, leaf fibers, fruit fibers, grass fibers, straw fibers and other types (wood and roots etc.) (Figure 1). There are thousands of natural fibers available and therefore there are many research interests in utilization of natural fibers to improve the properties of composites. (Faruk et al. 2012)

Natural fiber composites mostly consist fibers of jute, cotton, hemp and non-conventional fibers such as coir and many empty fruit bunches. Natural fiber thermoplastic composites are attractive as they are cheaper, stiffer, paintable and also can be given the look of wood in addition to all this they have more life- cycle. Natural fiber composites are attractive to industry because of their low density and ecological advantages over conventional composites. Natural fibers also offer economic and environmental advantages over traditional inorganic reinforcements and fillers. As a result of these advantages, natural fiber reinforced thermoplastic composites are gaining popularity in automotive and non-structural construction applications. These composites are gaining importance due to their non-carcinogenic and bio-degradable nature. Natural fiber composites are very cost effective material especially in building and construction purpose packaging, automobile and railway coach interiors and storage devices. These can be potential candidates for replacement of high cost glass fiber for low load bearing applications.

The major limitations of using these fibers as reinforcements in such matrices include poor interfacial adhesion between polar-hydrophilic fiber and nonpolar hydrophobic matrix, and difficulties in mixing due to poor wetting of the fiber with the matrix. This in turn would lead to composites with weak interface. There are many parameters which affect the performance of a natural fiber-reinforced composite. The degree and type of adhesion cannot be estimated quantitatively even though its importance is well recognized. Aspect ratio has a considerable effect on composite properties, hence it is important to conserve fiber length as much as possible during composite processing operations. Fiber aspect ratio must be in the range of 100–200 for optimum effectiveness. Fiber orientation has a significant effect on composite properties. During processing, the fibers tend to orient along the flow direction causing mechanical properties to vary in different directions.

The development of strength in a composite also depends on the existence of a strong interface. The fiber/ matrix interface in fiber-reinforced composites transfers externally applied loads to fibers themselves. Load applied directly to the matrix at the surface of the composites is transferred to the fibers nearest the surface and continues from fiber to fiber via matrix and interface. If the interface is weak, effective load distribution is not achieved and the mechanical properties of the composites are impaired. On the other hand, a strong interface can assure that the composite is able to bear load even after several fibers are broken because the load can be transferred to the intact portions of broken as well as unbroken fibers. A poor interface is also a drawback in situations other than external mechanical loading, e.g. because of differential thermal expansions of fiber and matrix, premature failure can occur at a weak interface when the composite is subjected to thermal stress. Thus, adhesion between fiber and matrix is a major factor in determining the response of the interface and its integrity under stress.

A better understanding of the chemical composition and surface adhesive bonding of natural fiber is necessary for developing natural fiber-reinforced composites. The components of natural fibers include cellulose, hemicelluloses, lignin, pectin, waxes and water soluble substances.

COIR/COCONUT FIBERS

The coconut husk is available in large quantities as residue from coconut production in many areas. Coir is a lingo-cellulosic natural fiber. It is a seed-hair fiber obtained from the outer shell, or husk, of the coconut. It is resistant to abrasion and can be dyed. Total world coir fiber production is 250,000 tones. The coir fiber industry is particularly important in some areas of the developing world. Over 50% of the coir fiber produced annually throughout the world is consumed in the countries of origin, mainly India (Harisha et al. 2009). Because of its hard-wearing quality, durability and other advantages, it is used for making a wide variety of floor furnishing materials, yarn, rope etc (Satyanarayana et al. 1982). However, these traditional coir products consume only a small percentage of the potential total world production of coconut husk. Hence, research and development efforts have been underway to find new use areas for coir, including utilization of coir as reinforcement in polymer composites (Owolabi et al., 1985) (Varma et al. 1985) (Varma et al. 1986) (Prasad et al. 1983) (Greethamma et al 1998) (Paul and Thomas 1997) (Abdul Khail and Rozman 2000).

Coconut fiber is extracted from the outer shell of a coconut. The common name, scientific name and plant family of coconut fiber are coir, cocos nucifera and arecaceae (Palm), respectively. There are two types of coconut fibers, brown fiber extracted from matured coconuts and white fibers extracted from immature coconuts. Brown fibers are thick, strong and have high abrasion resistance, while white fibers are smoother and finer, but also weaker. Coconut fibers are commercially available in three forms, namely bristle (long fibers), mattress (relatively short) and decorticated (mixed fibers). Coconut tree, coconut and coconut fibers can be seen in Figure 2. (https://en.wikipedia.org/wiki/Coconut).

Figure 2.

These different types of fibers have different uses depending upon the requirement. In engineering, brown fibers are mostly used. There are many general advantages of coconut fibers e.g. they are moth-proof, resistant to fungi and rot, provide excellent insulation against temperature and sound, not easily combustible, flame-retardant, unaffected by moisture and dampness, tough and durable, resilient, springs back to shape even after constant use, totally static free and easy to clean.

According to official website of International Year for Natural Fibers 2009, approximately, 500,000 tones of coconut fibers are produced annually worldwide, mainly in India and Sri Lanka. Its total value is estimated at $100 million. India and Sri Lanka are also the main exporters, followed by Thailand, Vietnam, the Philippines and Indonesia. Around half of the coconut fibers produced is exported in the form of raw fiber.

Coir Fibre Extraction

The effectiveness of the wet processes investigated by the project team, such as bleaching and dyeing of coir, are strongly dependent on the procedures used to extract fibres from the husks and the pretreatment given the fibres. Both state-of-the art and commonly used technologies for fibre extraction are described.

The traditional production of fibres from the husks is a laborious and time-consuming process. This is highly polluting of surface waters and results in the accumulation of large dumps of pith. After manual separation of the nut from the husk, the husks are processed by various retting techniques, and generally in ponds of brackish waters (for three to six months) or in salt backwaters or lagoons. This requires 10-12 months of anaerobic (bacterial) fermentation. By retting the fibres they are softened and can be decorticated and extracted by beating, which is usually done by hand. After hackling, washing and drying (in the shade) the fibres are loosened manually and cleaned.

Traditional practices of this kind yield the highest quality of (white) fibre for spinning and weaving. Retted fibres from green husks are the most suitable fibres for dyeing and bleaching.

For the production of coarser brown yarns, shorter periods of retting may be applied. These find an increasing outlet in geotextile applications,

Alternatively, mechanical processing using either defibering or decorticating equipment can be used to process the husks after only five days of immersion in water tanks. Crushing the husk in a breaker opens the fibres. By using revolving "drums" the coarse long fibres are separated from the short woody parts and the pith. The stronger fibres are washed, cleaned, dried, hackled and combed. The quality of the fibre is greatly affected by these procedures.

Properties of Coir/Coconut Fibers

The mechanical properties of a natural fiber-reinforced composite depend on many parameters, such as fiber strength, modulus, fiber length and orientation, in addition to the fiber-matrix interfacial bond strength. A strong fiber-matrix interface bond is critical for high mechanical properties of composites. A good interfacial bond is required for effective stress transfer from the matrix to the fiber whereby maximum utilization of the fiber strength in the composite is achieved (Karnani et al 1997). Modification to the fiber also improves resistance to moisture induced degradation of the interface and the composite properties (Joseph et al. 2000). In addition, factors like processing conditions/techniques have significant influence on the mechanical properties of fiber reinforced composites (George et al. 2001).

The general advantages of coconut fibers include moth-proof; resistant to fungi and rot, provide excellent insulation against temperature and sound, flame-retardant, unaffected by moisture and dampness, tough and durable, resilient, spring back to shape even after constant use. Coconut fiber is the toughest fiber, having 21.5 MPa tensile strength, amongst natural fibers (Munawar et al. 2007). They are also capable of taking strain 4–6 times more than that of other fibers (Munawar et al. 2007 and Satyanarayana et al. 1990).

Abiola (Abiola, 2008) evaluated the mechanical properties (load-extension and stress–strain curves, Young's modulus, yield stress, stress and strain at break) of inner and outer coconut fibers experimentally, and the results were verified by finite element method using a commercial software ABAQUS. The author found that the inner coconut fiber had a higher mechanical strength as compared to that of outer fiber, but the outer coconut fiber had a higher elongation property which enables it to absorb or withstand higher stretching energy.

Ramakrishna and Sundararajan (Ramakrishna and Sundararajan, 2005) investigated the variation in chemical composition and tensile strength of four natural fibers, i.e. coconut, sisal, jute and hibiscus cannabinus fibers, when subjected to alternate wetting and drying and continuous immersion for 60 days in water, saturated lime and sodium hydroxide. Chemical composition of all fibers changed because of immersion in the considered solutions. Continuous immersion was found to be critical due to the loss of their tensile strength. However, coconut fibers were reported best for retaining a good percentage of its original tensile strength in all tested conditions.

It was stated in a previous study by Rao and Rao that the mechanical properties of coir fibers reinforced composites are expected to depend on the content or volume fraction of the fibers in the composite (Rao and Rao, 2007). Even a small change in the physical nature of fibers for a given volume content of fibers may result in distinguished changes in the overall mechanical properties of composites.

Treatment of Coir/ Coconut Fibers

Major problem of using natural fibers with polymers is the poor interfacial bonding between the fiber and matrix. Compared to other natural fibers, coir (Geethamma et al., 2005, Espert et al., 2004, Brahmakumar et al., 2005 and Lovino et al., 2008) is quite new in reinforcing polymer matrices and they possess high weather resistance due to higher lignin content and absorb less water compared to other fibers due to lower cellulose content. Coir/coconut high level of moisture absorption, poor wettability and insufficient adhesion between untreated fiber and the polymer matrix lead to debonding with age. In order to improve the above qualities, adequate surface modification is required. (Rahman and Khan 2007)

The green coconut fibers are chemically treated with two different types of chemicals namely H2O2 and NaOH at varies concentration levels. The purpose of chemical treatment is to remove the moisture content of green coconut fiber and to increase the tensile strength of green coconut fiber.

Mostly, chemical treatment is used to improve the adhesion between the polymer matrix and surface of the fiber and the strength of the fiber. Water absorption capacity of the fibers will also be reduced and helps to increase the mechanical properties.

Treatment with alkaline or mercerization is one of the most used chemical treatment for natural fibers. This treatment will remove some amount of lignin, oil and wax from the external layer of the fiber cell; it decomposes cellulose into small segments and exposes the short length crystallites. (John and Anandjiwala, 2008) Adding of NAOH (Aqueous sodium hydroxide) to coconut fiber assist the ionization to alkoxide from hydroxyl group.

$$\text{Fiber} + \text{NaOH-OH} = \text{Fiber} - \text{Na-O} + H_2O$$

Figure 3.

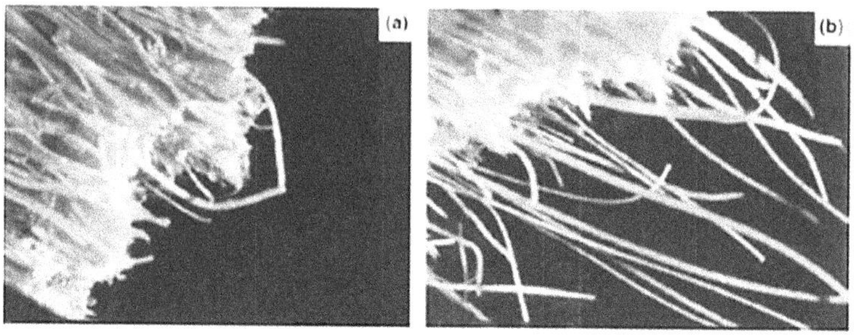

Thus process of alkaline directly effects the cellulosic fiber and the degree of polymerization and the pulling out the lignin and hem cellulosic compounds. This treatment has two effects on fibers.

1. Surface roughness is increased to result good mechanical properties.
2. It effects fiber strength and stiffness of it.

Brahamakumar et al. (Brahamakumar et al., 2005) studied the effect of natural waxy surface layer of coir fiber-reinforced polyethylene composites. To assess the effectiveness of waxy layer on interfacial bonding, coconut fibers with wax-free surface and surface-modified wax-free fibers obtained by grafting of an isocyanate derivative of cardanol (CTDIC) were also used. The natural waxy surface layer of coconut fiber was found to provide a strong interfacial bonding between the fiber and the polyethylene matrix. Removal of the waxy layer resulted in a weak interfacial bonding which increased the critical fiber length by 100% and decreased the composite tensile strength and modulus by 40 and 60%, respectively. The waxy layer due to its polymeric nature showed a stronger effect on fiber/matrix bonding than by grafted layer of a C15 long alkyl chain molecule onto the wax-free fiber. Figure 3 (a) and (b) show the optical fractographs of the as received and wax-free fiber composites containing 20-mm long coir fibers.

Large fiber pullout can be seen for the wax-free fiber-reinforced composites due to poor interfacial bonding, and this does not occur for the composites made from as-received coir fibers. Calado and Barreto (Calado and Barreto, 2000) reported the effect of surface treatment on the surface topography of coir fibers. The chemical treatment comprised of two steps and the first was the removal of lignin from the surface of the fibers by treatment with 2% sodium sulphite solution. The second step was to treat the lignin free fibers with acetic anhy dride. This treatment was

Figure 4.

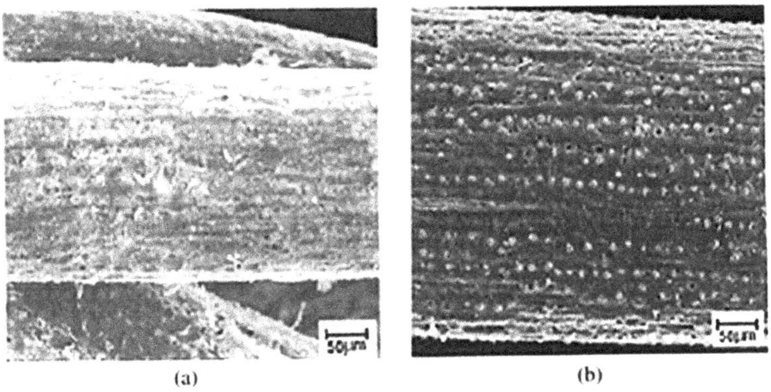

done to reduce the number of hydroxyl groups on the surface of the fibers. Reduction in the number of hydroxyl groups reduces the polarity and improves the compatibility with common thermosetting matrices. Figure 4 (a) and (b) show the clear difference found between the surface of the untreated and treated fibers.

The untreated fibers have an outer surface layer and this layer is completely removed by the chemical treatment with sodium sulphate and acetic anhydride. On removal of this outer layer, a rougher but more ordered structure is revealed. The ordered white dots found on the surface of the treated fibers can be seen in Figure 4 (b) and were identified as silica-rich material. Mohanty et al. (Mohanty et al., 2001) studied the influence of surface modification on the performances of coir–polyester amide (PEA bio composites at a fiber content of 50 wt.%). Coir fibers in the form of unidirectional nonwoven mat were used in this study. The tensile strengths for coir/PEA composites reported were in the range of 29–38 MPa. The surface modification methods involved were alkali treatment, cyanoethylation, acetylation, and grafting. Mercerization of coir fibers has been found to change the surface topography of the fibers and their crystallographic structure. The physical and chemical behavior of coir as a result of the chemical modification was characterized by FTIR spectra (Samal et al., 1995. The influence of delignification and alkali treatment on the fine structure of coir fibers was investigated by Sreenivasan et al. (Sreenivasan et al., 1996). Surface modifications of coir fibers by graft copolymerization with methyl methacrylate and acrylonitrile have also been reported (Rout et al., 1999). Rout et al. (Rout et al., 2001) have reported the scanning electron microscopy (SEM) studies of the alkali-treated coir fibers. They have reported on the progressive changes in the surface of coir fibers by SEM due to the removal of tyloses from the surface as a result of alkali treatment. Silva et al. (Silva et al., 2000) investigated the mechanical and thermal characterization of coir fiber. The effect

Figure 5.

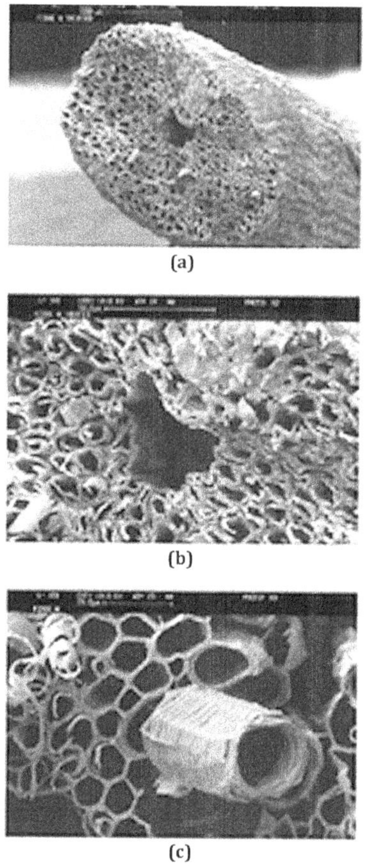

of alkali treatment was evaluated and morphological changes were studied by SEM. Thermogravimetric analysis showed an increase in water content and a slight decrease in thermal stability due to alkali treatment. Figure 5 shows the oval cross section of the coir fibers. In Figure 5 (a), a fiber treated for 48 h in NaOH is shown for which the central lumen was observed at higher magnification in Figure 5 (b). The cellulose helical spirals emerging from the fractured cross section are shown in Figure 5.

Coir/Coconut Fiber Reinforced Composites

Among the natural fibers, the coir fiber has remarkable interest in the composite industry owing to its hard-wearing quality and high hardness (not fragile like glass fiber), good acoustic resistance, moth-proof, not toxic, resistant to microbial and

Table 2. Chemical Compositon of Coir

Items	Percentages
Water Soluble	5.25%
Pectin and Related Compounds	3.00%
Hemi-cellulose	0.25%
Lignin	45.84%
Cellulose	43.44%
Ash	2.22%

fungi degradation, and not easily combustible (Mohanty et al., 2005). The coir fibers are also more resistant to moisture than other natural fibers and withstand heat and salt water (Rijswijk et al., 2001).

Coir is a natural fiber extracted from the husk of coconut fruit and the husk consists of coir fiber and a corky tissue called pith. It consists of water, fibers and small amounts of soluble solids. The Table 2 shows the percentage contribution of each of these fibers in coir.

Because of the high lignin content coir is more durable when compared to other natural fibers. With increasing emphasis on fuel efficiency, natural fibers such as coir based composites enjoying wider applications in automobiles and railway coaches & buses for public transport system. There exists an excellent opportunity in fabricating coir based composites towards a wide array of applications in building and construction such boards and blocks as reconstituted wood, flooring tiles etc. Value added novel applications of natural fibers and coir based composites would not go in a long way in improving the quality of life of people engaged in coir cultivation, but would also ensure international market for cheaper substitution. Natural fibers have the advantages of low density, low cost and biodegradability. However, the main disadvantages of natural fibers and matrix and the relative high moisture sorption. Therefore, chemical treatments are considered in modifying the fiber surface properties.

Although coconut palms grow throughout the world's tropical regions, the vast majority of the commercially produced coir fibers come from Indonesia, Philippines, India, Brazil, Sri Lanka, Thailand, Vietnam, and Malaysia (Arancon, 2007). Coconut palms are grown in an area of about 12.05 million ha in the world and the total production has been 61.1 million nuts per annum in recent years (Arancon, 2007). The high increase in the consumption of the coconuts and the industrialization of the processing of the coconut water has increased the generation of green coconut trash, which corresponds to around 85% of the weight of the fruit. The coir fiber is used for making a wide variety of floor furnishing materials, yarns, ropes, mats, mattresses, brushes, sacking, caulking boats, rugs, geo-textiles, and insulation panels.

Thermal Characterization of Coir/Coconut Fiber Reinforced Composites

Reddy et al. studied the thermal conductivity of characteristics of cow dung powder filled glass/polyester hybrid composite. It was found that as the volume fraction of cow dung powder increased, the thermal conductivity of the composite was decreased. Ramanaiah et al. studied on the mechanical and thermal properties of polyester composites reinforced with waste grass broom fiber. They varied the volume fraction of fibers in the composites from 0.163 to 0.358. It was observed that the thermal conductivity of the composite decreased gradually as the volume fraction of fiber increased. And a quite opposite result of increase of thermal conductivity was found when the temperature increased. Specific heat capacity had also measured and found a similar tendency of that thermal conductivity. Volume fraction and temperature had no influence on thermal diffusivity. Fernandez et al. made cork polyethylene composite and studied its characteristics. Its thermal and acoustic properties, impact strength, hardness, dimensional stability were tested. The obtained results showed that this natural-based composite exhibits improved characteristics such as low water absorption, impact resistance, fire resistance, and insulation properties than conventional materials. Idicula et al. carried out the experiment on the thermo physical characteristics of polyester composites reinforced with natural fibre. Thermal conductivity, thermal diffusivity, and specific heat of banana/sisal /polyester hybrid composer was evaluvated. They concluded that with the incorporation of the banana/sisal fiber the effective thermal conductivity of the composite reduces. However, there are no detailed thermal characterization studies on coir/coconut fiber reinforced polymer composites. Therefore, future research should also be focused on the characterization of thermal properties of coir/coconut fiber reinforced polymer composites as well.

Mechanical Characterization of Coir/ Coconut Fiber Reinforced Composites

About the Mechanical Properties of Natural Fiber Based Polymer Composite The growing factors like environmental challenges, biodegradability, non-toxicity etc., leads the researchers to focus their studies on exploring the features of natural materials like natural fibers. A lot of research is going on to make use of natural fibers as a reinforcing material in the polymer matrix composites. There are a lot of challenges faced by the researchers to make the natural fiber suitable for their needs due to its hydrophilic nature, thermal and chemical instability. But now a days natural fibers can replace synthetic fibers to some extent by making them compatible with polymer matrixes by some surface modification techniques. Rout et al. studied the significance of surface treatment on the coir reinforced polyester composites. The coir fiber was subjected to alkali treatment, vinyl grafting, and bleaching before adding them with general purpose polyester resin. The mechanical characteristics like tensile strength, bending and impact strength were increased because of surface treatment. Bleached fiber composite (at 650C) showed better flexural strength. NaOH treated fiber/polyester composite exhibited better tensile strength. Because of the chemical treatments of fibers, the water absorption tendency of composite was reduced Biswas et al. carried out a study on the significance of fiber length on the mechanical character of coir/epoxy composite. It was found that the hardness of the composite decreases by increasing length of fiber up to 20 mm and then after it increases. They also concluded that fiber length has a major influence on enhancing mechanical properties like tensile strength, flexural strength and impact strength. Romli et al. done a factorial study upon tensile strength of coir reinforced epoxy composite. In their study, the volume fraction, curing time and compression load during the solidification of composites were taken as parameters. From the results, they concluded that volume fraction influences the tensile strength of the composites. Authors also increased the percentage volume fraction of fiber and found that the tensile properties of composites increased to some extent. Curing time also showed some effects on the characteristics of composites meanwhile the influence of compression load on the properties of composites were not revealed properly.

Nam et al. studied the significance of alkali treatment on interfacial bonding and mechanical properties of coir fiber filled poly (butylene succinate) biodegradable composites. Composites with fiber concentration of 10-30% were prepared using 5% NaOH treated fibers. On comparing with untreated fiber composites authors found a remarkable improvement in the interfacial shear strength (IFSS) and mechanical properties of treated coir fiber/ polybutylene succinate (PBS) composites. The treated composites with 25% fiber content exhibits higher mechanical properties.

Comparison of Natural Coir Fibers with Synthetic Fibre Reinforced Composites

The development of natural fibre reinforced polymer composites has received widespread attention due to their environment friendly characteristics over the synthetic fibre based polymer composites. Although, different categories of natural fibre reinforced composites have been developed, their joining has not been explored extensively. In the present research initiative, natural fiber reinforced polypropylene composites have been prepared by microwave curing. The advantage of microwave processing is that it leads to significantly faster curing times compared to thermal processing. The fast heating rate encountered using microwave energy can thus lead to reduced processing time and consequent energy efficiency.

Applications of Coir/Coconut Fiber Reinforced Composites

Coir is abundant, versatile, renewable, inexpensive in addition to biodegradable lignocelluloses backbone used to construct a wide variety of products (Vijayan et al., 1982). Coconut backbone has also been checked as a filler or different re-inforcement composites (Vijayan et al., 1982) (Kumar et al. 2007) (Agnelli et al. 2006) (Geethamma, 1998). Furthermore, represents a non-food agribusiness as an additional feedstock (agribusiness and food industry waste) to be considered as raw materials for the formulation eco compatible complex. Coir is the almost everyone fascinating goods as it has the lowest thermal conductivity along with collection density. The coir adding together reduces the thermal conductivity of the composite fabric specimens and produced lightweight manufactured supplies. Expansion of combined materials for buildings with natural fibers such as coconut fiber through small thermal Conductivity is a fascinating alternative to explain environmental along by the concerns of energy (Czvikovszky et al., 1985) (Hirunlabh et al., 2002). Geethamma et al. (Geethamma, 1998) have studied the dynamics mechanical behavior of natural rubber and reinforced composites short coconut fibers. Coco backbone composites were tested as polyester hulls such as ceilings with mailboxes (Kalaprasad et al., 2005). These composites, loaded coir ranging from 9 to 15% by weight, have a flexural strength of about 38 MPa. Composites of coir-polyester treated and untreated fibers of coir fiber loading and 17 wt.% were experienced in tensile, flexural with notched Izod force (Pillai et al., 1986). The outcome obtained among the untreated fibers explain clear with signs of the presence of a long weak interface Output drawn resin without any fibers adhering to the fibers along with low mechanical Properties were found. While showing an improved mechanical performance, composites with treated fibers current, however, simply a moderate boost the values of the mechanical properties analyzed. Alkaline treatment is also reported for coconut fibers (Nayak et al., 2003).

Previous works on coconut fiber reinforced composites Slate (Slate, 1976) investigated compressive and flexural strength of coconut fiber reinforced mortar. Two cement-sand ratios by weight, 1:2.75 with water cement ratio of 0.54 and 1:4 with water cement ratio of 0.82 were considered. Fiber content was 0.08%, 0.16% and 0.32% by total weight of cement, sand and water. The mortars for both design mixes without any fibers were also tested as reference. Cylinders of 50 mm diameter and 100 mm height and beams of 50 mm width, 50 mm depth and 200 mm length were tested. The curing was done for 8 days only. It was found that, compared to that of plain mortar of both mix designs, all strengths were increased in the case of fiber reinforced mortar with all considered fiber contents. However, a decrease in strength of mortar with an increase of fiber content was also observed.

Cook et al. (Cook et al. 1978) reported the use of coconut fiber reinforced cement composites as low cost roofing materials. The parameters studied were fiber lengths (2.5, 3.75 and 6.35 cm), fiber volumes (2.5%, 5%, 7.5%, 10% and 15%) and casting pressure (from 1 to 2 MPa with an increment of 0.33 MPa). They concluded that the optimum composite consisted of fibers with a length of 3.75 cm, a fiber volume fraction of 7.5% and is casted under the pressure of 1.67 MPa.

A comparison revealed that this composite was much cheaper than locally available roofing materials. Aziz et al. (Aziz et al. 1981) cited the work of (Das Gupta et al., 1978) (Das Gupta et al., 1979) who studied the mechanical properties of cement paste composites for different lengths and volume fractions of coconut fibers. Aziz et al. concluded that the tensile strength and modulus of rupture of cement paste increased when fibers up to 38 mm fiber length and 4% volume fraction were used. A further increase in length or volume fraction could reduce the strength of composite. The tensile strength of cement paste composite was 1.9, 2.5, 2.8, 2.2 and 1.5 MPa when it was reinforced with 38 mm long coconut fiber and the volume fractions of 2%, 3%, 4%, 5% and 6%, respectively. The corresponding modulus of rupture was 3.6, 4.9, 5.45, 5.4 and 4.6 MPa, respectively. 4% volume fraction of coconut fibers gave the highest mechanical properties amongst all tested cases. With 4% volume fraction, they also studied the tensile strength of cement paste reinforced with different lengths of coconut fibers. With the fiber lengths of 25, 38 and 50 mm, the reported tensile strength was 2.3, 2.8 and 2.7 MPa, respectively. The results indicated that coconut fibers with a length of 38 mm and a volume fraction of 4% gave the maximum strength.

Paramasivam et al. (Paramasivam et al., 1984) conducted a feasibility study of coconut fiber reinforced corrugated slabs of 915 mm _ 460 mm _ 10 mm for low-cost housing. A cement–sand ratio of 1:0.5 and water–cement ratio of 0.35 were used. Test for flexural strength using third point loading was performed. For producing required slabs having a flexural strength of 22 MPa, a fiber length of 2.5 cm, a volume fraction of 3%, and a casting pressure of 0.15 MPa were recommended. The thermal conductivity and absorption coefficient for low frequency sound were comparable with those of asbestos boards.

Agopyan et al. (Agopyan et al., 2005) studied coir and sisal fibers as replacement of asbestos in roofing tiles. The dimensions of the tiles were 487 mm _ 263 mm _ 6 mm. Three-point bend test specimen with2% total fiber volume fraction, support span of 350 mm, deflection rate of 5 mm/min was employed for determination of the maximum load. After the ageing periods of 16 and 60 months, the corresponding maximum load taken by coir tile were 235 and 248 N, respectively while that by sisal tiles were 237 and 159 N, respectively. The major benefit of reinforced tiles was their at least 22% higher energy absorption than that of the unreinforced tiles which could help to avoid fragile rupture of tiles during transportation or installation.

John et al. (John et al., 2005) studied the coir fiber reinforced low alkaline cement mortar taken from the internal and external walls of a 12 year old house. The panel of the house was produced using 1:1.5:0.504 (cement: sand: water, by mass) mortar reinforced with 2% of coconut fibers by volume. Fibers removed from the old samples were reported to be undamaged. No significant difference was found in the lignin content of fibers removed from external and internal walls, confirming the durability of coconut fibers in cement composites.

Luisito et al. (Luisito et al., 2005) of PCA-Zamboanga Research Center in Philippines invented coconut fiber boards (CFB) for applications such as tiles, bricks, plywood and hollow blocks. It is used for internal and exterior walls, partitions and ceiling. CFB consisted of 70% cement and 30% fiber by weight. It has water absorption of 32%, water swelling of 4.2% and bending strength of 0.81 MPa, respectively.

Mohammad (Mohammad, 2005) tested wall panels made of gypsum and cement as binder and coconut fiber as reinforcement. Bending and compressive strength, moisture content, density and water absorption were investigated. As expected, coconut fibers did not contribute to bending strength of the tested wall panels. Compressive strength increased with the addition of coconut fibers. There was no considerable change of moisture content with coconut fibers. However, moisture content increased with time. Water absorption of panels was not significantly affected with an increase in fiber content.

Ramakrishna and Sundararajan (Ramakrishna and Sundararajan, 2005) carried out the experiments on impact resistance of slabs using a falling weight of 0.475 kg from a height of 200 mm. The slabs consisted of 1:3 cement–sand mortar with the dimension of 300 mm _ 300 mm _ 20 mm. They were reinforced with coconut, sisal, jute and hibiscus cannabinus fibers having four different fiber contents of 0.5%, 1.0%, 1.5% and 2.5% by weight of cement and three fiber lengths of 20, 30 and 40 mm. A fiber content of 2% and a fiber length of 40 mm of coconut fibers showed the best performance by absorbing 253.5 J impact energy. At ultimate failure all fibers, except coconut fibers, showed fiber fracture while coconut fiber showed fiber pull-out. The ultimate failure was determined based on the number of blows required to open a crack in the specimen sufficiently and for the propagation of the crack through the entire depth of the specimen.

Li et al. (Li et al., 2006) studied untreated and alkalized coconut fibers with the lengths of 20 mm and 40 mm as reinforcement in cementitious composites. Mortar was mixed in a laboratory mixer at a constant speed of 30 rpm, with cement: sand: water: super plasticizer ratio of 1:3:0.43:0.01 by weight, and fibers were slowly put into the running mixer. The resulting mortar had a better flexural strength (increased up to 12%), higher energy absorption ability (up to 1680%) and a higher ductility (up to 1740%), and is lighter than the conventional mortar.

Reis (Reis, 2006) performed third-point loading tests to investigate the flexural strength, fracture toughness and fracture energy of epoxy polymer concrete reinforced with coconut, sugarcane bagasse and banana fibers. The investigation revealed that fracture toughness and energy of coconut fiber reinforced polymer concrete were the highest, and an increase of flexural strength up to 25% was observed with coconut fibers.

Asasutjarit et al. (Asasutjarit et al., 2007) determined the physical (density, moisture content, water absorption and thickness swelling), mechanical (modulus of elasticity, modulus of rupture and internal bond) and thermal properties of coir-based light weight cement board after 28 days of hydration. The physical and mechanical properties were measured by Japanese Industrial Standard JIS A 5908-1994 and the thermal properties according to JIS R 2618. The parameters studied were fiber length, coir pre-treatment and mixture ratio. 6 cm long boiled and washed fibers with the optimum cement:fiber:water weight ratio of 2:1:2 gave the highest modulus of rupture and internal bond amongst the tested specimens. The board also had a thermal conductivity lower than other commercial flake board composite.

Baruah and Talukdar (Baruah and Talukdar, 2007) investigated the mechanical properties of plain concrete (PC) and fiber reinforced concrete (FRC) with different fiber volume fractions ranging from 0.5% to 2%. Steel, synthetic and jute and coconut fibers were used. Here, the discussion is limited to the coconut fibers reinforced concrete (CFRC) only. The cement: sand: aggregate ratio for plain concrete was 1:1.67:3.64, and the water cement ratio was 0.535. Coconut fibers having length of 4 cm and an average diameter of 0.4 mm with volume fraction of 0.5%, 1%, 1.5% and 2% were added to prepare CFRC. The sizes of specimens were (1) 150 mm diameter and 300 mm height for cylinders (2) 150 mm width, 150 mm depth and 700 mm length for beams, and (3) 150 mm cubes having a cut of 90 mm _ 60 mm in cross-section and 150 mm high for L-shaped shear test specimens. All specimens were cured for 28 days. The compressive strength r, splitting tensile strength (STS), modulus of rupture (MOR) using four point load test and shear strengths are shown in Table 1 for PC and CFRC. It can be seen that CFRC with 2% fibers showed the best overall performance amongst all volume fractions. The compressive strength, splitting tensile strength, modulus of rupture and shear strength of coir fiber reinforced concrete with 2% fibers by volume fraction were increased up to 13.7%, 22.9%, 28.0% and 32.7%, respectively as compared to those of plain concrete. Their research indicated that all these properties were improved as well for CFRC with other fiber volume fractions of 0.5%, 1% and 1.5%. Even for CFRC with small fiber volume fraction of 0.5% the corresponding properties were increased up to 1.3%, 4.9%, 4.0% and 4.7%, respectively.

Li et al. (Li et al., 2006) studied fiber volume fraction and fiber surface treatment with a wetting agent for coir mesh reinforced mortar using nonwoven coir mesh matting. They performed a four-point bending test and concluded that cementitious composites, reinforced by three layers of coir mesh with a low fiber content of 1.8%, resulted in a 40% improvement in the maximum flexural strength. The composites were 25 times stronger in flexural toughness and about 20 times higher in flexural ductility.

Nowadays, coir/coconut fiber-reinforced composites are mainly being used in the manufacture of building boards, roofing sheets, insulation boards, building panels, as a lightweight aggregate, cement board, geo-textile applications.

Yuhazri et al. (Yuhazri et al. 2007) utilized coconut fibers in the manufacturing of motor cycle helmet. They used epoxy resins from thermo set polymer as the matrix materials and coconut fibers as the reinforcement. After the development of helmet shells fabrication method, mechanical testing (dynamic penetration) was performed on this composite material to determine its performance. The result in the mechanical performance showed that coconut fibers performed well as a suitable reinforcement to the epoxy resin matrix.

A team of Baylor University researchers is trying to develop a technology to use coconut fiber as a replacement for synthetic polyester fibers in compression molded composites. Their aim is to use the coconut fibers to make trunk liners, floorboards and interior door covers on cars.

Apart from applications in engineering, coconut fibers are also used in yarn, ropes, mats, mattresses, brushes, sacking, caulking boats, rugs, geo-textiles, insulation panels and packaging.

SOLUTIONS AND RECOMMENDATIONS

Several investigations have been carried out to assess the potential of natural fibers as reinforcement in the polymers (De Rosa et al., 2010) (Mulinari et al., 2009) (Ochi, 2008) (Gu, 2009).

The results have shown that natural fibers present potential to be used as reinforcement for plastics, but generally they do not attain the full mechanical performance levels of glass fibers reinforced plastics. However, composite materials need a better understanding of their failure mechanisms. The stresses encountered during their service life are frequently cyclical in nature causing fatigue degradation and are responsible for the majority of failures of composite materials.

In many previous studies, it was observed some failure mechanism as fractured fibers and presence of pull out and poor bonding interfacial between coconut fibers and polymer matrix. This fact occurred most probably due to the treatment suggested on fibers surface. As a result, improvement of the treatment techniques on fibers surface is needed to obtain better mechanical performance of the coconut fiber reinforced polymer composites.

One of the main disadvantages of natural fibers in composites are the poor compatibility between fiber and matrix and the relative high moisture sorption. Therefore, chemical treatments should be considered in modifying the fiber surface properties.

FUTURE RESEARCH DIRECTIONS

There is a wide scope of future research to explore the area of coconut/coir fibers reinforced polymer composites and their applications. Further studies on coir/coconut fiber reinforced polymers should be extended to study other aspects of these composites like effect of fiber content, fiber orientation, loading pattern, fiber treatment on mechanical behavior of coconut coir based polymer composites.

Although there are several reports in the literature which discuss the mechanical behavior of coir fiber reinforced polymer composites, very limited work has been done on effect of fiber length on mechanical behavior of coir fiber reinforced polymer composites and studies on this effect should be considered in detail.

Development of composite materials for buildings using natural fiber as coconut coir with low thermal conductivity should be considered as an interesting alternative which would help to solve environment and energy concern (Khedari et al., 2002) (Asasutjarit et al., 2007).

CONCLUSION

This chapter has been written with an aim to explore the potential of the coconut/coir fiber reinforced polymer composites by considering previously done research, in terms of their mechanical performance, surface treatments/modifications and areas of application.

Fiber-reinforced polymer composites have played a significant role for a long time in a variety of applications, regarding their high specific strength and high modulus. Natural fibers in substitution of conventional fibers as reinforcement are major steps taken in promoting environmental sustainability. The use of natural fibers as reinforcement to polymer matrices is a way to recycle these fibers and produce new materials with high strength.

Reinforcement with natural fiber in composites has recently gained attention due to low cost, easy availability, low density, acceptable specific properties, ease of separation, enhanced energy recovery, C02 neutrality, biodegradability and recyclable in nature. There are many types of natural fibers and results from previously carried out indicate that coir can be used as a potential reinforcing material for many structural and non-structural applications of polymers.

Further research in the development of coir fiber-reinforced composites should be extended and increased in the future, regarding their enhanced performance, economical advantages and eco-friendly nature.

REFERENCES

Abdul Khalil, H. P. S., & Rozman, H. D. (2000). Acetylated Plant-Fiber Reinforced Polyester Composites. A Study of Mechanical, Hygrothermal and Aging Characteristics. *Polymer-Plastics Technology and Engineering*, *39*(4), 757–781. doi:10.1081/PPT-100100057

Agnelli, L. H., Corradini, M. F., Morais, E., Rosa, L. C., Mazzetto, S. E., & Mattoso, J. A. M. (2006). A preliminary study for the use of natural fibers as reinforcement in starch–gluten–glycerol matrix. In Macromolecular Symposia. Academic Press.

Agopyan, V., Savastano, H. Jr, John, V. M., & Cincotto, M. A. (2005). Developments on vegetable fiber-cement based materials in São Paulo, Brazil: An overview. *Cement and Concrete Composites*, *27*(5), 527–536. doi:10.1016/j.cemconcomp.2004.09.004

Arancon, R. N. (2007). *Natural Fiber Production and Food Securtiy: Coir in Asia and the Pacific*. Jakarta: Asian and Pacific Coconut Community.

Asasutjarit, C., Hirunlabh, J., Khedari, J., Charoenvai, S., Zeghmati, B., & Shin, U. C. (2007). Development of coconut coir-based lightweight cement board. *Construction & Building Materials*, *21*(2), 277–288. doi:10.1016/j.conbuildmat.2005.08.028

Asasutjarit, C., Hirunlabh, J., Khedari, J., Charoenvai, S., Zeghmati, S. B., & Shin, C. C. (2007). Development of coconut coir-based lightweight cement board. *Construction & Building Materials*, *21*(2), 277–288. doi:10.1016/j.conbuildmat.2005.08.028

Awboa, A. (2008). *Mechanical Property Evaluation of Coconut Fiber*. (Master thesis). Department of Mechanical Engineering, Blekinge Institute of Technology, Kariskrone, Sweden.

Ayrilmis, N., & Ashori, A. (2015). *Alternative Solutions for Reinforcement of Thermoplastic Composites*. Natural Fiber Composites. Taylor and Francis Group LLC. doi:10.1201/b19062-4

Aziz, M. A., Paramasivam, P., & Lee, S. L. (1981). Prospects for natural fiber reinforced concretes in construction. *Int J Cem Compos Lightweight Concr*, *3*(2), 123–132. doi:10.1016/0262-5075(81)90006-3

Baruah, P., & Talukdar, S. (2007). A comparative study of compressive, flexural, tensile and shear strength of concrete with fibers of different origins. *Indian Concrete Journal*, *81*(7), 17–24.

Biswas, S., Kindo, S., & Patnaik, A. (2011). Effect of fiber length on mechanical behavior of coir fiber reinforced epoxy composites. *Fibers and Polymers*, *12*(1), 73–78. doi:10.1007/s12221-011-0073-9

Bledzki, A. K., & Gassan, J. (1999). Composites reinforced with cellulose based fibers. *Progress in Polymer Science*, *24*(2), 221–274. doi:10.1016/S0079-6700(98)00018-5

Bledzki, A. K., & Gassan, J. (1999). Composites reinforced with cellulose based fibres. *Progress in Polymer Science*, *24*(2), 221–274. doi:10.1016/S0079-6700(98)00018-5

Brahmakumar, M., Pavithran, C., & Pillai, R. M. (2005). Coconut fibre reinforced polyethylene composites: Effect of natural waxy surface layer of the fibre on fibre/matrix interfacial bonding and strength of composites. *Composites Science and Technology*, *65*(3-4), 563–569. doi:10.1016/j.compscitech.2004.09.020

Calado, V., & Barreto, D. W. (2000). Article. *Journal of Materials Science Letters*, *23*(23), 2151–2153. doi:10.1023/A:1026743314291

Cook, D. J., Pama, R. P., & Weerasingle, H. L. S. D. (1978). Coir fiber reinforced cement as a low cost roofing material. *Building and Environment*, *13*(3), 193–198. doi:10.1016/0360-1323(78)90043-4

Czvikovszky, T., Owolabi, O., Kovacs, I., (1985). Coconut-fiber-reinforced thermosetting plastics. *Journal of Applied Polymer Science*, *30*, 1827–1836.

Das Gupta, N. C., Paramsivam, P., & Lee, S. L. (1978). Mechanical properties of coir reinforced cement pastes composites. *Housing Sci*, *2*(5), 391–406.

Das Gupta, N. C., Paramsivam, P., & Lee, S. L. (1979). Coir reinforced cement pastes composites. In *Conference proceedings of our world in concrete and structures.*

Dave, S., & Jadeja, M. (2015). A Survey on Mechanical Activities on Coir Fiber Resistant Polymer Matrix Composites. *International Journal of Advance Research and Innovative Ideas in Education*, *1*(1).

De Rosa, I. M., Santulli, C., & Sarasini, F. (2010). Mechanical and thermal characterization of epoxy composites reinforced with random and quasi-unidirectional untreated Phormium tenax leaf fibers. *Materials & Design*, *31*(5), 2397–2405. doi:10.1016/j.matdes.2009.11.059

Faruk, O., Bledzki, A. K., Fink, H.-P., & Sain, M. (2012). Biocomposites reinforced with natural fibers: 2000–2010. *Progress in Polymer Science*, *37*(11), 1552–1596. doi:10.1016/j.progpolymsci.2012.04.003

Geethamma, V.G., (1998). Composite of short coir fibers and natural rubber: effect of chemical modification, loading and orientation of fiber. *Polymer*, *39*(6–7), 1483–1497.

Geethamma, V. G., Thomas Mathew, K., Lakshminarayan, R., & Thomas, S. (1998). Composite of short coir fibers and natural rubber: Effect of chemical modification, loading and orientation of fiber. *Polymer*, *39*(6-7), 1483–1491. doi:10.1016/S0032-3861(97)00422-9

George, J., Sreekala, M. S., & Thomas, S. (2001). A review on interface modification and characterization of natural fiber reinforced plastic composites. *Polymer Engineering and Science*, *41*(9), 1471–1485. doi:10.1002/pen.10846

Gu, H. (2009). Tensile behaviours of the coir fiber and related composites after NaOH treatment. *Materials & Design*, *30*(9), 3931–3934. doi:10.1016/j.matdes.2009.01.035

Harisha, S., Peter Michaelb, D., Benselyb, A., Mohan Lalb, D., & Rajaduraic, A. (2009). Mechanical Property Evaluation of Natural Fiber Coir Composite. *Materials Characterization*, *60*(1), 44–49. doi:10.1016/j.matchar.2008.07.001

Hirunlabh, J., Pratintong, N., Khedari, J., & Suttisonk, B. (2002). *New lightweight composite construction materials with low thermal conductivity*. Cem Compos.

Idicula, M., Boudenne, A., Umadevi, L., Ibos, L., Candau, Y., & Thomas, S. (2006). Thermophysical properties of natural fibre reinforced polyester composites. *Composites Science and Technology*, *66*(15), 2719–2725. doi:10.1016/j.compscitech.2006.03.007

John, M. J., & Anandjiwala, R. D. (2008). Recent developments in chemical modification and characterization of natural fiber-reinforced composites. *Polymer Composites*, *29*(2), 187–207. doi:10.1002/pc.20461

John, V. M., Cincotto, M. A., Sjostrom, C., Agopyan, V., & Oliveira, C. T. A. (2005). Durability of slag mortar reinforced with coconut fiber. *Cement and Concrete Composites*, *27*(5), 565–574. doi:10.1016/j.cemconcomp.2004.09.007

Joseph, K., Mattoso, L. H. C., Toledo, R. D., Thomas, S., de Carvalho, L. H., Pothen, L., . . . James, B. (2000). Natural fiber reinforced thermoplastic composites. In Natural Polymers and Agrofibers Composites. Sãn Carlos, Brazil: Embrapa, USP-IQSC, UNESP.

Joseph, S., Sreekala, M. S., Ommen, Z., Kosy, P., & Thomas, S. (2002). A comparison of mechanical properties of phenol formaldehyde composites reinforced with banana fibers and glass fibers. *Composites Science and Technology*, *62*(14), 1857–1868. doi:10.1016/S0266-3538(02)00098-2

Kalaprasad G,,Geethamma V.G, Sabu T., Gabriel G. (2005). Dynamic mechanical behavior of short coir fiber reinforced natural rubber composites. *Composites, 36*, 1499–506.

Karnani, R., Krishnan, M., & Narayan, R. (1997). Biofiber-reinforced polypropylene composites. *Polymer Engineering and Science*, *37*(2), 476–483. doi:10.1002/pen.11691

Khedari, J., Suttisonk, B., Pratintong, N., & Hirunlabh, J. (2002). New lightweight composite construction materials with low thermal conductivity. *Cem Compos*, *23*(1), 65–70. doi:10.1016/S0958-9465(00)00072-X

Kumar, S., Adhikari, B., Choudhury, A., (2007). Recycled milk pouch and virgin low density. Polyethylene/linear low-density polyethylene based coir fiber composites. *Journal of Applied Polymer Science, 106*, 775–785.

Li, X. (2008). *Development of flax fiber-reinforced polyethylene biocomposites by injection molding*. (PhD dissertation). Department of agricultural and bioresources engineering, University of Saskatchewan.

Li, Z., Wang, L., & Wang, X. (2006). Flexural characteristics of coir fiber reinforced cementitious composites. *Fiber Polym*, *7*(3), 286–294. doi:10.1007/BF02875686

Lilholt, H., & Lawther, J. M. (2000). *Comprehensive Composite Materials*. Elsevier Ltd.

Luo, S., & Netravali, A. (1999). Mechanical and thermal properties of environment-friendly ''green'' composites made from pineapple leaf fibers and poly(hydroxybutyrate-co-valerate) resin. *Polymer Composites*, *20*(3), 367–378. doi:10.1002/pc.10363

Mohammad, H. B. M. H. (2005). *Coconut fiber reinforced wall panelling system*. (Master thesis). Faculty of Civil Engineering. Universiti Teknologi, Malaysia.

Mohanty, A. K., Misra, M., & Drzal, L. T. (2001). Surface modifications of natural fibers and performance of the resulting biocomposites: An overview. *Composite Interfaces*, *8*(5), 313–343. doi:10.1163/156855401753255422

Mohanty, A. K., Misra, M., & Drzal, L. T. (2005). Natural Fibers, Biopolymers and Biocomposites. Boca Raton, FL: Taylor & Francis.

Mohanty, A. K., Misra, M., & Hinrichsen, G. (2000). Biofibers biodegradable polymers and biocomposites: An overview. *Macromolecular Materials and Engineering*, *276-277*(1), 1–24. doi:10.1002/(SICI)1439-2054(20000301)276:1<1::AID-MAME1>3.0.CO;2-W

Mueller, D. H., & Krobjilowski, A. (2003). New Discovery in the Properties of Composites Reinforced with Natural Fibers. *Journal of Industrial Textiles*, *33*(2), 111–129. doi:10.1177/152808303039248

Mulinari, D. R., Voorwald, H. J. C., Cioffi, M. O. H., Da Silva, M. L. C. P., & Luz, S. M. (2009). Preparation and properties of HDPE/sugarcane bagasse cellulose composites obtained for thermokinetic mixer. *Carbohydrate Polymers*, *75*(2), 317–320. doi:10.1016/j.carbpol.2008.07.028

Munawar, S. S., Umemura, K., & Kawai, S. (2007). Characterization of the morphological, physical, and mechanical properties of seven non-wood plant fiber bundles. *Journal of Wood Science*, *53*(2), 108–113. doi:10.1007/s10086-006-0836-x

Murali, M. R. K., & Mohana, R. K. (2007). Extraction of the tensile properties of natural fibers, Vakka, Date and Bambo. *Journal of Composite Structures*, *77*(3), 288–295. doi:10.1016/j.compstruct.2005.07.023

Naidu, A.L, Sudarshan, B., Krishna, K.H. (2013). Study on Mechanical Behavior Of Groundnut Shell Fiber Reinforced Polymer Metal Matrix Composites. *International Journal of Engineering Research & Technology*.

Nam, T. H., Ogihara, S., Tung, N. H., & Kobayashi, S. (2011). Effect of alkali treatment on interfacial and mechanical properties of coir fiber reinforced poly (butylene succinate) biodegradable composites. *Composites. Part B, Engineering*, *42*(6), 1648–1656. doi:10.1016/j.compositesb.2011.04.001

Nayak, SK., Mohanty, A.K., Rout, J., Misra, M., Tripathy, SS. (2003). SEM observations of the fractured surfaces of coir composites. *J. Reinf. Plast. Compos.*

Nguong, C. W., Lee, S. N. B., & Sujan, D. (2013). *A Review on Natural Fibre Reinforced Polymer Composites*. World Academy of Science, Engineering and Technology.

Ochi, S. (2008). Mechanical properties of kenaf fibers and kenaf/PLA composites. *Mechanics of Materials*, *40*(4-5), 446–452. doi:10.1016/j.mechmat.2007.10.006

Owolabi, O., Czvikovszky, T., & Kovacs, I. (1985). Coconut fiber reinforced thermosetting plastics. *Journal of Applied Polymer Science*, *30*(5), 1827–1836. doi:10.1002/app.1985.070300504

Paramasivam, P., Nathan, G. K., & Das Gupta, N. C. (1984). Coconut fiber reinforced corrugated slabs. *Int J Cem Compos Lightweight Concr*, *6*(1), 19–27. doi:10.1016/0262-5075(84)90056-3

Paul, A., & Thomas, S. (1997). Electrical properties of natural-fiber-reinforced low density polyethylene composites: A comparison with carbon black and glass fiber-filled low density polyethylene composites. *Journal of Applied Polymer Science*, *63*(2), 247–266. doi:10.1002/(SICI)1097-4628(19970110)63:2<247::AID-APP12>3.0.CO;2-#

Pillai, S. G. K., Kulkarni, A. G., Sukumaran, K., Satyanarayana, P. K., & Rohatgi, K. G. (1986). Fabrication and properties of natural fiber-reinforced polyester composites. *Composites*, 17.

Prasad, S. V., Pavithran, C., & Rohatgi, P. K. (1983). Alkali treatment of coir fibers for coir-polyester composites. *Journal of Materials Science*, *18*(5), 1443–1454. doi:10.1007/BF01111964

Ràcz, I., & Hargitai, H. (2000). Influence of water on properties of cellulosic fiber reinforced polypropylene composites. *International Journal of Polymeric Materials*, *47*(4), 667–674. doi:10.1080/00914030008031321

Rahman, M. M., & Khan, M. A. (2007). Surface treatment of coir (Cocos nucifera) fibers and its influence on the fibers' physico–mechanical properties. *Composites Science and Technology*, *67*(11-12), 2369–2376. doi:10.1016/j.compscitech.2007.01.009

Ramakrishna, G., & Sundararajan, T. (2005). Studies on the durability of natural fibers and the effect of corroded fibers on the strength of mortar. *Cement and Concrete Composites*, *27*(5), 575–582. doi:10.1016/j.cemconcomp.2004.09.008

Ramakrishna, G., & Sundararajan, T. (2005, May). Impact strength of a few natural fiber reinforced cement mortar slabs: A comparative study. *Cement and Concrete Composites*, *27*(5), 547–553. doi:10.1016/j.cemconcomp.2004.09.006

Ramanaiah, K., Ratna Prasad, A. V., & Hema Chandra Reddy, K. (2012). Thermal and mechanical properties of waste grass broom fiber-reinforced polyester composites. *Materials & Design*, *40*, 103–108. doi:10.1016/j.matdes.2012.03.034

Ranjeth Kumar Reddy, T., Subba Rao, T., & Padma Suvarna, R. (2014). Studies on thermal characteristics of cow dung powder filled glass–polyester hybrid composites. *Composites. Part B, Engineering*, *56*, 670–672. doi:10.1016/j.compositesb.2013.08.059

Rao, K. M. M., & Rao, K. M. (2007). Extraction and Tensile Properties of Natural Fibers: Vakka, Date and Bamboo. *Composite Structures*, *77*(3), 288–295. doi:10.1016/j.compstruct.2005.07.023

Reis, J. M. L. (2006). Fracture and flexural characterization of natural fiber-reinforced polymer concrete. *Construction & Building Materials*, *20*(9), 673–678. doi:10.1016/j.conbuildmat.2005.02.008

Rijswijk, K. V., Brouwer, W. D., & Beukers, A. (2001). *Natural Fibre Composites*. Rome, Italy: FAO Economic and Social Development Department.

Romli, F. I., Alias, A. N., Rafie, A. S. M., & Majid, D. L. A. A. (2012). Factorial Study on the Tensile Strength of a Coir Fiber-Reinforced Epoxy Composite. *AASRI Procedia*, *3*, 242–247. doi:10.1016/j.aasri.2012.11.040

Rout, J., Misra, M., & Mohanty, A. K. (1999). Surface modification of coir fibers I: Studies on graft copolymerization of methyl methacrylate on to chemically modified coir fibers. *Polymers for Advanced Technologies*, *10*(6), 336–344. doi:10.1002/(SICI)1099-1581(199906)10:6<336::AID-PAT884>3.0.CO;2-O

Rout, J., Misra, M., Tripathy, S. S., Nayak, S. K., & Mohanty, A. K. (2001). The influence of fibre treatment on the performance of coir-polyester composites. *Composites Science and Technology*, *61*(9), 1303–1310. doi:10.1016/S0266-3538(01)00021-5

Rout, J., Tripathy, S. S., Nayak, S. K., Misra, M., & Mohanty, A. K. (2001). Scanning electron microscopy study of chemically modified coir fibers. *Journal of Applied Polymer Science*, *79*(7), 1169–1177. doi:10.1002/1097-4628(20010214)79:7<1169::AID-APP30>3.0.CO;2-Q

Samal, R. K., Panda, B. B., Rout, S. K., & Mohanty, M. (1995). Effect of chemical modification on FTIR spectra. I. Physical and chemical behavior of coir. *Journal of Applied Polymer Science*, *58*(4), 745–752. doi:10.1002/app.1995.070580407

Sanadi, A. R., Prasad, S. V., & Rohtgi, P. K. (1986). Sunhempfiber-reinforced polyester. *Journal of Materials Science*, *21*(12), 4299–4304. doi:10.1007/BF01106545

Satyanarayan, K. G., Kulkarni, A. G., & Rohatgi, P. K. (1981). Fabrication and properties of coir based composites. *Journal of Scientific and Industrial Research*, *40*, 222–229.

Satyanarayana, K. G., Pillai, C. K. S., Sukumaran, K., Pillai, S. G. K., Rohatgi, P. K., & Kalyani, V. (1982). Structure and properties of fibers from different parts of coconut tree. *Journal of Materials Science*, *17*, 2453–2472. doi:10.1007/BF00543759

Satyanarayana, K. G., Sukumaran, K., Mukherjee, P. S., Pavithran, C., & Pillai, S. G. K. (1990). Natural fiber–polymer composites. *Cement and Concrete Composites*, *12*(2), 117–136. doi:10.1016/0958-9465(90)90049-4

Silva, G. G., De Souza, D. A., Machado, J. C., & Hourston, D. J. (2000). Mechanical and thermal characterization of native brazilian coir fiber. *Journal of Applied Polymer Science*, *76*(7), 1197–1206. doi:10.1002/(SICI)1097-4628(20000516)76:7<1197::AID-APP23>3.0.CO;2-G

Slate, F. O. (1976). Coconut fibers in concrete. *Eng J Singapore*, *3*(1), 51–54.

Sreenivasan, S., Iyer, P. B., & Iyer, K. R. K. (1996). Influence of delignification and alkali treatment on the fine structure of coir fibres (Cocos Nucifera). *Journal of Materials Science*, *31*(3), 721–726. doi:10.1007/BF00367891

Varma, D. S., Varma, M., & Varma, I. K. (1985). Coir Fibers II: Evaluation as a Reinforcement in Unsaturated Polyester Resin Composites. *Journal of Reinforced Plastics and Composites*, *4*(4), 419–431. doi:10.1177/073168448500400406

Varma, D. S., Varma, M., & Varma, I. K. (1986). Coir fibers. 3. Effect of resin treatment on properties of fibers and composites. *Industrial & Engineering Chemistry Product Research and Development*, *25*(2), 282–289. doi:10.1021/i300022a027

Yuhazri, M.Y., & Dan, M.M.P. (2007) Helmet Shell Using Coconut Fiber (Deco-Helmet). *Journal of Advanced Manufacturing Technology, 1*(1), 23-30.

Compilation of References

Abdul Khalil, H. P. S., Bhat, A. H., & Ireana Yusra, A. F. (2012). Green composites from sustainable cellulose nanofibrils: A review. *Carbohydrate Polymers*, *87*(2), 963–979. doi:10.1016/j.carbpol.2011.08.078

Abdul Khalil, H. P. S., Kumar, R. N., Asri, S. M., Nik Fuaad, N. A., & Ahmad, M. N. (2007). Hybrid Thermoplastic Pre-preg Oil Palm Frond Fibers (OPF) Reinforced in Polyester Composites. *Polymer-Plastics Technology and Engineering*, *46*(1), 43–50. doi:10.1080/03602550600948749

Abdul Khalil, H. P. S., & Rozman, H. D. (2000). Acetylated Plant-Fiber Reinforced Polyester Composites. A Study of Mechanical, Hygrothermal and Aging Characteristics. *Polymer-Plastics Technology and Engineering*, *39*(4), 757–781. doi:10.1081/PPT-100100057

Abdul Khalil, H. P. S., Tehrani, M. A., Davoudpour, Y., Bhat, A. H., Jawaid, M., & Hassan, A. (2012). Natural fiber reinforced poly(vinyl chloride) composites: A review. *Journal of Reinforced Plastics and Composites*. doi:10.1177/0731684412458553

Agnelli, L. H., Corradini, M. F., Morais, E., Rosa, L. C., Mazzetto, S. E., & Mattoso, J. A. M. (2006). A preliminary study for the use of natural fibers as reinforcement in starch–gluten–glycerol matrix. In Macromolecular Symposia. Academic Press.

Agopyan, V., Savastano, H. Jr, John, V. M., & Cincotto, M. A. (2005). Developments on vegetable fiber-cement based materials in São Paulo, Brazil: An overview. *Cement and Concrete Composites*, *27*(5), 527–536. doi:10.1016/j.cemconcomp.2004.09.004

Ahmadi, M. A., Alidoust, O., Sadrinejad, I., & Nayeri, M. (2007). Development of Mechanical Properties of Self Compacting Concrete Contain Rice Husk Ash. *International Journal of Computer and Information Engineering*, *1*, 4.

Ahmed, K. S., Vijayarangan, S., & Rajput, C. (2006). Mechanical behavior of isothalic polyester-based untreated woven jute and glass fabric hybrid composites. *Journal of Reinforced Plastics and Composites*, *25*(15), 1549–1569. doi:10.1177/0731684406066747

Ahmed, N. M., & El-Sabbagh, S. H. (2014). The influence of kaolin and calcined kaolin on SBR composite properties. *Polymer Composites*, *35*(3), 570–580. doi:10.1002/pc.22697

Aji, I., Zainudin, E., Abdan, K., Sapuan, S., & Khairul, M. (2012). Mechanical properties and water absorption behavior of hybridized kenaf/pineapple leaf fiber-reinforced high-density polyethylene composite. *Journal of Composite Materials*, *47*(8), 979–990. doi:10.1177/0021998312444147

Akadiri, P. O., Olomolaiye, P. O., & Chinyio, E. (2013). Multi-criteria evaluation model for the selection of sustainable materials for building projects. *Automation in Construction*, *30*(0), 113–125. doi:10.1016/j.autcon.2012.10.004

Akil, H. M., Omar, M. F., Mazuki, A. A. M., Safiee, S., Ishak, Z. A. M., & Abu Bakar, A. (2011). Kenaf fiber reinforced composites: A review. *Materials & Design*, *32*(8–9), 4107–4121. doi:10.1016/j.matdes.2011.04.008

Akinci, A., Sen, S., & Sen, U. (2014). Friction and wear behavior of zirconium oxide reinforced PMMA composites. *Composites. Part B, Engineering*, *56*, 42–47. doi:10.1016/j.compositesb.2013.08.015

Ali, I., & Elleithy, R. (2011). Toughness of HDPE / CaCO. *Composite Structures*, *48*(1), 49–58.

Alves, C., Ferrão, P. M. C., Silva, A. J., Reis, L. G., Freitas, M., Rodrigues, L. B., & Alves, D. E. (2010). Ecodesign of automotive components making use of natural jute fiber composites. *Journal of Cleaner Production*, *18*(4), 313–327. doi:10.1016/j.jclepro.2009.10.022

Amico, S. C., Angrizani, C. C., & Drummond, M. L. (2008). Influence of the stacking sequence on the mechanical properties of glass/sisal hybrid composites. *Journal of Reinforced Plastics and Composites*.

Anupam, K., Lal, P. S., Bist, V., Sharma, A. K., & Swaroop, V. (2014). Raw material selection for pulping and papermaking using TOPSIS multiple criteria decision making design. *Environmental Progress and Sustainable Energy*, *33*(3), 1034–1041. doi:10.1002/ep.11851

Arai, T., Freddi, G., Innocenti, R., & Tsukada, M. (2003). Preparation of water-repellent silks by a reaction with octadecenylsuccinic anhydride. *Journal of Applied Polymer Science*, *89*(2), 324–332. doi:10.1002/app.12081

Arancon, R. N. (2007). *Natural Fiber Production and Food Securtiy: Coir in Asia and the Pacific*. Jakarta: Asian and Pacific Coconut Community.

Arato, C., Pye, E. K., & Gjennestad, G. (2005). The lignol approach to biorefining of woody biomass to produce ethanol and chemicals. *Applied Biochemistry and Biotechnology*, *123*(1-3), 871–882. doi:10.1385/ABAB:123:1-3:0871 PMID:15930566

Arib, R. M. N., Sapuan, S. M., Ahmad, M. M. H. M., Paridah, M. T., & Zaman, H. K. (2006). Mechanical properties of pineapple leaf fiber reinforced polypropylene composites. *Materials & Design*, *27*(5), 391–396. doi:10.1016/j.matdes.2004.11.009

Ariffin, A., Mansor, A. S., & Jikan, S. S. (2008). Mechanical, morphological, and thermal properties of polypropylene / kaolin composite. Part I. The effects of surface-treated kaolin and processing enhancement. *Journal of Applied Polymer Science*, *108*(6), 3901–3916. doi:10.1002/app.27950

Asasutjarit, C., Hirunlabh, J., Khedari, J., Charoenvai, S., Zeghmati, B., & Shin, U. C. (2007). Development of coconut coir-based lightweight cement board. *Construction & Building Materials, 21*(2), 277–288. doi:10.1016/j.conbuildmat.2005.08.028

Ashori, A. (2008). Wood–plastic composites as promising green-composites for automotive industries. *Bioresource Technology, 99*(11), 4661–4667. doi:10.1016/j.biortech.2007.09.043 PMID:18068352

Asim, M., Abdan, K., Jawaid, M., Nasir, M., Dashtizadeh, Z., Ishak, M. R., & Hoque, M. E. (2015). A Review on Pineapple Leaves Fibre and Its Composites. *International Journal of Polymer Science*, 2015.

Asyadi, F., Jawaid, M., Hassan, A., & Wahit, M. U. (2013). Mechanical Properties of Mica-Filled Polycarbonate/Poly (Acrylonitrile-Butadiene-Styrene) Composites. *Polymer-Plastics Technology and Engineering, 52*(7), 727–736. doi:10.1080/03602559.2012.762672

Athijayamani, A., Thiruchitrambalam, M., Manikandan, V., & Pazhanivel, B. (2010). Mechanical properties of natural fibers reinforced polyester hybrid composite. *International Journal of Plastics Technology, 14*(1), 104–116. doi:10.1007/s12588-009-0016-0

Attri, R., & Grover, S. (2013). Application of preference selection index method for decision making over the design stage of production system life cycle. *Journal of King Saud University -. Engineering and Science, 27*(0), 207–216.

Augier, L., Sperone, G., Vaca-Garcia, C., & Borredon, M. E. (2007). Influence of the wood fibre filler on the internal recycling of poly (vinyl chloride)-based composites. *Polymer Degradation & Stability, 92*(7), 1169–1176. doi:10.1016/j.polymdegradstab.2007.04.010

Aurrekoetxea, J., Sarrionandia, M., & Gomez, X. (2008). Effects of microstructure on wear behaviour of wood reinforced polypropylene composite. *Wear, 265*(5), 606–611. doi:10.1016/j.wear.2007.12.013

Awad, Z. K., Aravinthan, T., & Manalo, A. (2012). Geometry effect on the behavior of single and glue-laminated glass fibre reinforced polymer composite sandwich beams loaded in four-point bending. *Materials & Design, 39*, 93–103. doi:10.1016/j.matdes.2012.02.023

Awboa, A. (2008). *Mechanical Property Evaluation of Coconut Fiber*. (Master thesis). Department of Mechanical Engineering, Blekinge Institute of Technology, Kariskrone, Sweden.

Axen, N., Hutchings, I. M., & Jacobson, S. (1996). A model for the friction of multiphase materials in abrasion. *Tribological International, 29*(6), 467–475. doi:10.1016/0301-679X(95)00104-C

Ayrilmis, N., & Ashori, A. (2015). *Alternative Solutions for Reinforcement of Thermoplastic Composites*. Natural Fiber Composites. Taylor and Francis Group LLC. doi:10.1201/b19062-4

Aziz, M. A., Paramasivam, P., & Lee, S. L. (1981). Prospects for natural fiber reinforced concretes in construction. *Int J Cem Compos Lightweight Concr, 3*(2), 123–132. doi:10.1016/0262-5075(81)90006-3

Baillie, C. (2004). *Green Composites: Polymer Composites and the Environment*. New York: CRC Press.

Bakiyalakshmi, K., & Sudha, P. N. (2012). Adsorption of Copper (II) ion onto chitosan/sisal/banana fiber hybrid Composite. *International Journal Environmental Sciences*, 3.

Balakrishna, N. S., Ismail, H., & Othman, N. (2014). Polypropylene / Rattan Powder / Kaolin Hybrid Composites: Processing, Mechanical and Thermal Properties. *Polymer-Plastics Technology and Engineering*, *53*(5), 451–458. doi:10.1080/03602559.2013.844833

Banks, D. I., Griffin, N. J., Shackleton, C. M., Shackleton, S. E., & Mavrandonis, J. M. (1996). Wood supply and demand around two rural settlements in semi-arid Savanna, South Africa. *Biomass and Bioenergy*, *11*(4), 319–331. doi:10.1016/0961-9534(96)00031-1

Barkakaty, B. C. (1976). Some structural aspects of sisal fibers. *Journal of Applied Polymer Science*, *20*(11), 2921–2940. doi:10.1002/app.1976.070201101

Barone, J. R., Schmidt, W. F., & Liebner, C. F. E. (2005). Compounding and molding of polyethylene composites reinforced with keratin feather fiber. *Composites Science and Technology*, *65*(3–4), 683–692. doi:10.1016/j.compscitech.2004.09.030

Baruah, P., & Talukdar, S. (2007). A comparative study of compressive, flexural, tensile and shear strength of concrete with fibers of different origins. *Indian Concrete Journal*, *81*(7), 17–24.

Baxter, W. L., Aurisicchio, M., & Childs, P. R. N. (2015). (in press). Materials, use and contaminated interaction. *Materials & Design*, *0*(0). doi:10.1016/j.matdes.2015.04.019

Behzadian, M., Otaghsara, S. K., Yazdani, M., & Ignatius, J. (2012). A state-of the-art survey of TOPSIS applications. *Expert Systems with Applications*, *39*(17), 13051–13069. doi:10.1016/j.eswa.2012.05.056

Bello, S. R., & Adegbulugbe, T. A. (2010). Comparative Study on Utilization of Charcoal, Sawdust and Rice Husk in Biomass Furnace-Dryer. *Journal of Scientific Research and Development*, *12*, 1592.

Ben Difallah, B., Kharrat, M., Dammak, M., & Monteil, G. (2012). Mechanical and tribological response of ABS polymer matrix filled with graphite powder. *Materials & Design*, *34*, 782–787. doi:10.1016/j.matdes.2011.07.001

Benyus, J. M. (1997). *Biomimicry: Innovation Inspired by Nature*. New York: HarperCollins Publisher Inc.

Berndes, G., Hoogwijk, M., & Van den Broek, R. (2003). The contribution of biomass in the future global energy supply: A review of 17 studies. *Biomass and Bioenergy*, *25*(1), 1–28. doi:10.1016/S0961-9534(02)00185-X

Bhargavi, D. S., Suresh, Y., Naresh, C., & Upendra, N. (2014). An Investigation of Mechanical Properties of Hybrid Polymer Composites. *International Journal of Engine Research*, *3*, 116–123.

Bhowmick, M., Mukhopadhyay, S., & Alagirusamy, R. (2012). Mechanical properties of natural fibre reinforced composites. *Textile Progress*, *44*(2), 85–140. doi:10.1080/00405167.2012.676800

Binod, P., Sindhu, R., Singhania, R. R., Vikram, S., Devi, L., Nagalakshmi, S., & Pandey, A. et al. (2010). Bioethanol production from rice straw: An overview. *Bioresource Technology*, *101*(13), 4767–4774. doi:10.1016/j.biortech.2009.10.079 PMID:19944601

Bismarck, A., Baltazar-Y-Jimenez, A., & Sarlkakis, K. (2006). Green composites as Panacea? Socio-economic aspects of green materials. *Environment, Development and Sustainability*, *8*(3), 445–463. doi:10.1007/s10668-005-8506-5

Biswas, S., Ahsan, Q., Verpoest, I., & Hasan, M. (2011). Effect of Span Length on the Tensile Properties of Natural Fibers. *Advanced Materials Research*, *264-265*, 445–450. doi:10.4028/www.scientific.net/AMR.264-265.445

Biswas, S., Kindo, S., & Patnaik, A. (2011). Effect of fiber length on mechanical behavior of coir fiber reinforced epoxy composites. *Fibers and Polymers*, *12*(1), 73–78. doi:10.1007/s12221-011-0073-9

Bledzki, A. K. (2007). Abaca fibre reinforced PP composites and comparison with jute and flax fibre PP composites. *Express Polymer Letters*, *1*(11), 755–762. doi:10.3144/expresspolymlett.2007.104

Bledzki, A. K., & Gassan, J. (1999). Composites reinforced with cellulose based fibres. *Progress in Polymer Science*, *24*(2), 221–274. doi:10.1016/S0079-6700(98)00018-5

Bledzki, A. K., Reihmane, S., & Gassan, J. (1996). Properties and modification methods for vegetable fibers for natural fiber composites. *Journal of Applied Polymer Science*, *59*(8), 1329–1336. doi:10.1002/(SICI)1097-4628(19960222)59:8<1329::AID-APP17>3.0.CO;2-0

Bogoeva-Gaceva, G., Avella, M., Malinconico, M., Buzarovska, A., Grozdanov, A., Gentile, G., & Errico, M. E. (2007). Natural fiber eco-composites. *Polymer Composites*, *28*(1), 98–107. doi:10.1002/pc.20270

Brahmakumar, M., Pavithran, C., & Pillai, R. M. (2005). Coconut fibre reinforced polyethylene composites: Effect of natural waxy surface layer of the fibre on fibre/matrix interfacial bonding and strength of composites. *Composites Science and Technology*, *65*(3-4), 563–569. doi:10.1016/j.compscitech.2004.09.020

Brezet, H., & Van Hemel, C., & the United Nations Environment Programme. Industry and Environment (Paris), Rathenau Institute (TheHague), Delft University of Technology (Delft). (1997). Ecodesign: A promising approach to sustainable production and consumption. UNEP.

Brindha, D., Vinodhini, S., Alarmelumangai, K., & Malathy, N. S. (2012). Physico-Chemical Properties of Fibers rom Banana Varieties After Scouring. *Indian Journal of Fundamental and Applied Life Sciences*, *2*, 217–221.

Brooks, R. B. (2009). Soil stabilization with flyash and rice husk ash. *International Journal of Research and Review in Applied Science*, *1*, 210–217.

Burns, L. A., Mouritz, A. P., Pook, D., & Feih, S. (2012a). Bio-inspired design of aerospace composite joints for improved damage tolerance. *Composite Structures*, *94*(3), 995–1004. doi:10.1016/j.compstruct.2011.11.005

Burns, L. A., Mouritz, A. P., Pook, D., & Feih, S. (2012b). Strength improvement to composite T-joints under bending through bio-inspired design. *Composites. Part A, Applied Science and Manufacturing*, *43*(0), 1971–1980. doi:10.1016/j.compositesa.2012.06.017

Buzarovska, A. (2008). Potential use of rice straw as filler in eco-composite materials, *Australian Journal of Crop Science, 1*, 37-42.

Calado, V., & Barreto, D. W. (2000). Article. *Journal of Materials Science Letters*, *23*(23), 2151–2153. doi:10.1023/A:1026743314291

Çalişkan, H., Kurşuncu, B., Kurbanoĝlu, C., & Güven, Ş. Y. (2013). Material selection for the tool holder working under hard milling conditions using different multi criteria decision making methods. *Materials & Design*, *45*(0), 473–479. doi:10.1016/j.matdes.2012.09.042

Callister, W. D. Jr. (2006). *Materials science and engineering: an introduction* (7th ed.). New York: John Wiley & Sons.

Chand, N., Dwivedi, U. K., & Sharma, M. K. (2007). Development and tribological behaviour of UHMWPE filled epoxy gradient composites. *Wear*, *262*(1-2), 184–190. doi:10.1016/j.wear.2006.04.012

Chandrasegaran, S. K., Ramani, K., Sriram, R. D., Horváth, I., Bernard, A., Harik, R. F., & Gao, W. (2013). The evolution, challenges, and future of knowledge representation in product design systems. *Computer Aided Design*, *45*(2), 204–228. doi:10.1016/j.cad.2012.08.006

Chang, B. P., Akil, H. M., Nasir, R. B. M., Bandara, I. M. C. C. D., & Rajakpase, S. (2014). The effect of ZnO nanoparticles on the mechanical, tribological and antibacterial properties of ultra-high molecular weight polyethylene. *Journal of Reinforced Plastics and Composites*, 0731684413509426.

Chang, B. P., Akil, H. M., Nasir, R. M., & Nurdijati, S. (2011). Mechanical and antibacterial properties of treated and untreated zinc oxide filled UHMWPE composites. *Journal of Thermoplastic Composite Materials*, *24*(5), 653–667. doi:10.1177/0892705711399848

Chang, B. P., Md Akil, H., & Bt. Md Nasir, R. (2013). Comparative study of micro-and nano-ZnO reinforced UHMWPE composites under dry sliding wear. *Wear*, *297*(1), 1120–1127. doi:10.1016/j.wear.2012.11.083

Changduk, K., Park, H., & Lee, J. (2014). Study on structural design and analysis of flax natural fiber composite tank manufactured by vacuum assisted resin transfer molding. *Materials Letters*, *130*, 21–25. doi:10.1016/j.matlet.2014.05.042

Chan, J. W. K., & Tong, T. K. L. (2007). Multi-criteria material selections and end-of-life product strategy: Grey relational analysis approach. *Materials & Design*, *28*(5), 1539–1546. doi:10.1016/j.matdes.2006.02.016

Chattopadhyay, S. K., Singh, S., Pramanik, N., Niyogi, U., Khandal, R. K., Uppaluri, R., & Ghoshal, A. K. (2011). Biodegradability studies on natural fibers reinforced polypropylene composites. *Journal of Applied Polymer Science*, *121*(4), 2226–2232. doi:10.1002/app.33828

Chaudhary, V., & Gohil, P. P. (2015). Investigations on Drilling of Bidirectional Cotton Polyester Composite. *Materials and Manufacturing Processes*, 1–9. doi:10.1080/10426914.2015.1059444

Chen, H., Tu, J., & Guan, S. (2012). Applying the Theory of Problem-Solving and AHP to develop eco-innovative design. In M. Matsumoto et al. (Eds.), Design for Innovative Value Towards a Sustainable Society (pp. 489–494). Dordrecht: Springer Science+Business Media. doi: doi:10.1007/978-94-007-3010-6_93 doi:10.1007/978-94-007-3010-6_93

Chen, J., Feng, Y., Chen, N., Li, B., Chen, F., & Zhang, X. (2013). Fabrication and Properties of Silver-Doped Acrylonitrile Butadiene Styrene (ABS) Composites. *Polymer-Plastics Technology and Engineering*, *52*(11), 1081–1088. doi:10.1080/03602559.2013.769574

Chen, M., Zhao, J., & Xia, L. (2009). Comparison of four different chemical pretreatments of corn stover for enhancing enzymatic digestibility. *Biomass and Bioenergy*, *33*(10), 1381–1385. doi:10.1016/j.biombioe.2009.05.025

Cheung, H.-y., Ho, M.-, Lau, K.-, Cardona, F., & Hui, D. (2009). Natural fibre-reinforced composites for bioengineering and environmental engineering applications. *Composites. Part B, Engineering*, *40*(7), 655–663. doi:10.1016/j.compositesb.2009.04.014

Chiu, M. C., & Okudan, G. (2014). An investigation on the impact of product modularity level on supply chain performance metrics: An industrial case study. *Journal of Intelligent Manufacturing*, *25*(1), 129–145. doi:10.1007/s10845-012-0680-3

Chollakup, R., Tantatherdtam, R., Ujjin, S., & Sriroth, K. (2011). Pineapple leaf fiber reinforced thermoplastic composites: Effects of fiber length and fiber content on their characteristics. *Journal of Applied Polymer Science*, *119*(4), 1952–1960. doi:10.1002/app.32910

Ciccone-Nogueira, J. C., Borsatto, M. C., Souza-Zaroni, W. C., Ramos, R. P., & Palma-Dibb, R. G. (2007). Microhardness of composite resins at different depths varying the post-irradiation time. *Journal of Applied Oral Science*, *15*(4), 305–309. doi:10.1590/S1678-77572007000400012 PMID:19089149

Cook, D. J., Pama, R. P., & Weerasingle, H. L. S. D. (1978). Coir fiber reinforced cement as a low cost roofing material. *Building and Environment*, *13*(3), 193–198. doi:10.1016/0360-1323(78)90043-4

Cross, N. (1994). *Engineering Design Methods: Strategies for Product Design*. Chichester, UK: John Wiley & Sons.

Czvikovszky, T., Owolabi, O., Kovacs, I., (1985). Coconut-fiber-reinforced thermosetting plastics. *Journal of Applied Polymer Science*, *30*, 1827–1836.

Dabade, B. M., Reddy, G. R., Rajesham, S., & Kiran, C. U. (2006). Effect of fiber length and fiber weight ratio on tensile properties of sun hemp and palmyra fiber reinforced polyester composites. *Journal of Reinforced Plastics and Composites*, *25*(16), 1733–1738. doi:10.1177/0731684406068418

Danladi, A., & Shu'aib, J. (2014). Fabrication and Properties of Pineapple Fibre/High Density Polyethylene Composites. *American Journal of Materials Science*, *4*(3), 139–143.

Das Gupta, N. C., Paramsivam, P., & Lee, S. L. (1978). Mechanical properties of coir reinforced cement pastes composites. *Housing Sci*, *2*(5), 391–406.

Das Gupta, N. C., Paramsivam, P., & Lee, S. L. (1979). Coir reinforced cement pastes composites. In *Conference proceedings of our world in concrete and structures*.

Datschefski, E. (2001). *The Total Beauty of Sustainable Products*. New York: Rotovision.

Dave, S., & Jadeja, M. (2015). A Survey on Mechanical Activities on Coir Fiber Resistant Polymer Matrix Composites. *International Journal of Advance Research and Innovative Ideas in Education*, *1*(1).

Davoodi, M. M., Sapuan, S. M., Ahmad, D., Aidy, A., Khalina, A., & Jonoobi, M. (2011). Concept selection of car bumper beam with developed hybrid bio-composite material. *Materials & Design*, *32*(10), 4857–4865. doi:10.1016/j.matdes.2011.06.011

Davoodi, M. M., Sapuan, S. M., Ahmad, D., Ali, A., Khalina, A., & Jonoobi, M. (2010). Mechanical properties of hybrid kenaf/glass reinforced epoxy composite for passenger car bumper beam. *Materials & Design*, *31*(10), 4927–4932. doi:10.1016/j.matdes.2010.05.021

De Rosa, I. M., Santulli, C., & Sarasini, F. (2010). Mechanical and thermal characterization of epoxy composites reinforced with random and quasi-unidirectional untreated Phormium tenax leaf fibers. *Materials & Design*, *31*(5), 2397–2405. doi:10.1016/j.matdes.2009.11.059

Defoirdt, N., Biswas, S., De Vriese, L., Tran, L. Q. N., Van Acker, J., Ahsan, Q., & Verpoest, I. et al. (2010). Assessment of the tensile properties of coir, bamboo and jute fiber. *Composites. Part A, Applied Science and Manufacturing*, *41*(5), 588–595. doi:10.1016/j.compositesa.2010.01.005

Dickinson, W. C. (2000). *In integrative plant anatomy* (pp. 15–21). San Diego, CA: Harcourt Academic press.

Dieter, G. E. (2000). Engineering Design: A Materials and Processing Approach (3rd ed.). Boston: McGraw-Hill.

Dittenber, D. B., & GangaRao, H. V. S. (2012). Critical review of recent publications on use of natural composites in infrastructure. *Composites. Part A, Applied Science and Manufacturing*, *43*(8), 1419–1429. doi:10.1016/j.compositesa.2011.11.019

Dong, D. V., Huu, P. D., & Nguyen, N. L. (2008). Effect of rice husk ash on properties of high strength concrete. *The 3rd ACF International Conference- ACF/VCA 2008*.

Easterling, K. E. (1990). *Tomorrow's Materials*. London: The Institute of Materials.

Eichhorn, S. J., Dufresne, A., Aranguren, M., Marcovich, N. E., Capadona, J. R., Rowan, S. J., & Peijs, T. et al. (2010). Review: Current international research into cellulose nanofibres and nanocomposites. *Journal of Materials Science*, *45*(1), 1–33. doi:10.1007/s10853-009-3874-0

Eiss, N. S. Jr, & Czichos, H. (1986). Tribological studies on rubber-modified epoxies: Influence of material properties and operating conditions. *Wear*, *111*(4), 347–361. doi:10.1016/0043-1648(86)90132-8

El-Zeiny, R. M. A. (2012). Biomimicry as a problem solving methodology in interior architecture. *Procedia: Social and Behavioral Sciences*, *50*(0), 502–512. doi:10.1016/j.sbspro.2012.08.054

Engels, F. M., & Jungs, H. G. (1998). Alfalfa stem tissues: Cell-wall development and significa-tion. *Annals of Botany*, *82*(5), 561–568. doi:10.1006/anbo.1998.0705

Ephraim, M. E., Godwin, A. A., & Joseph, O. U. (2012). Compressive strength of concrete with rice husk ash aspartial replacement of ordinary Portland cement. *Scholarly Journal of Engineering Research*, *1*, 32–36.

Esthappan, S. K., Kuttappan, S. K., & Joseph, R. (2012). Thermal and mechanical properties of polypropylene/titanium dioxide nanocomposite fibers. *Materials & Design*, *37*, 537–542. doi:10.1016/j.matdes.2012.01.038

Faruk, O., Bledzki, A. K., Fink, H.-P., & Sain, M. (2012). Biocomposites reinforced with natural fibers: 2000–2010. *Progress in Polymer Science*, *37*(11), 1552–1596. doi:10.1016/j.progpolymsci.2012.04.003

Faruk, O., Bledzki, A. K., Fink, H.-P., & Sain, M. (2014). Progress Report on Natural Fiber Reinforced Composites. *Macromolecular Materials and Engineering*, *299*(1), 9–26. doi:10.1002/mame.201300008

Findik, F., & Turan, K. (2012). Materials selection for lighter wagon design with a weighted property index method. *Materials & Design*, *37*, 470–477. doi:10.1016/j.matdes.2012.01.016

Finkbeiner, M., & Hoffmann, R. (2006). Application of life cycle assessment for the environmental certificate of the Mercedes-Benz S-Class (7 pp). *The International Journal of Life Cycle Assessment*, *11*(4), 240–246. doi:10.1065/lca2006.05.248

Fiore, V., Botta, L., Scaffaro, R., Valenza, A., & Pirrotta, A. (2014). PLA based biocomposites reinforced with Arundo donax fillers. *Composites Science and Technology*, *105*(0), 110–117. doi:10.1016/j.compscitech.2014.10.005

Fisher, R. A. (1935). *The Design of Experiments* (1st ed.). Edinburgh, UK: Oliver and Boyd.

Flemming, M., Ziegmann, G., & Roth, S. (1995). Einfuhrung, Faserverbundbauweisen. In Fasern und Matrices. Springer-Verlag.

Gamelas, J. A. F., Ferraz, E., & Rocha, F. (2014). An insight into the surface properties of alcined kaolinitic clays: The grinding effect. Colloids and Surfaces A: *Physicochem. Eng. Aspects*, *455*, 49–57. doi:10.1016/j.colsurfa.2014.04.038

Gañán, P., Zuluaga, R., Cruz, J., Vélez, J., Retegí, A., & Mondragon, I. (2008). Elucidation of the fibrous structure of Musaceae maturate rachis. *Cellulose (London, England)*, *15*(1), 131–139. doi:10.1007/s10570-007-9150-z

Gassan, J., Chate, A., & Bledzki, A. (2001). Calculation of elastic properties of natural fibers. *Journal of Materials Science*, *36*(15), 3715–3720. doi:10.1023/A:1017969615925

Geethamma, V.G., (1998). Composite of short coir fibers and natural rubber: effect of chemical modification, loading and orientation of fiber. *Polymer*, *39*(6–7), 1483–1497.

Geethamma, V. G., Thomas Mathew, K., Lakshminarayan, R., & Thomas, S. (1998). Composite of short coir fibers and natural rubber: Effect of chemical modification, loading and orientation of fiber. *Polymer*, *39*(6-7), 1483–1491. doi:10.1016/S0032-3861(97)00422-9

George, J., Bhagawan, S. S., Prabhakaran, N., & Thomas, S. (1995). Short pineapple-leaf-fiber-reinforced low-density polyethylene composites. *Journal of Applied Polymer Science*, *57*(7), 843–854. doi:10.1002/app.1995.070570708

George, J., Bhagawan, S. S., & Thomas, S. (1998). Effects of environment on the properties of low-density polyethylene composites reinforced with pineapple-leaf fiber. *Composites Science and Technology*, *58*(9), 1471–1485. doi:10.1016/S0266-3538(97)00161-9

George, J., Sreekala, M. S., & Thomas, S. (2001). A Review on Interface Modification and Characterization. *Polymer Engineering and Science*, *41*(9), 1471–1485. doi:10.1002/pen.10846

Ghazilla, R. A. R., Taha, Z., Yusoff, S., Rashid, S. H. A., & Sakundarini, N. (2013). Development of decision support system for fastener selection in product recovery oriented design. *International Journal of Advanced Manufacturing Technology*, *70*(5-8), 1403–1413. doi:10.1007/s00170-013-5373-3

Ghofrani, M., Pishan, S., Mohammadi, M. R., & Omidi, H. (2012). Study on Rice-husk/Recycled High Density Polyethylene Composites – Their Physical and Mechanical properties. *Environmental Sciences*, *9*, 99–112.

Ghosh, R., Reena, G., Ramakrishna, A., Raju, B.L. (2011). Effect of fibre volume fraction on the tensile strength of Banana fibre reinforced vinyl ester resin composites. *International Journal of Advanced Engineering Sciences and Technologies, 4*, 89 – 91.

Gibbs, H. K., Brown, S., Niles, J. O., & Foley, J. A. (2007). Monitoring and estimating tropical forest carbon stocks: Making REDD a reality. *Environmental Research Letters*, *2*(4), 1–13. doi:10.1088/1748-9326/2/4/045023

Gidde, M. R., & Jivani, A. P. (2007). Waste to Wealth - Potential of Rice Husk in India: A Literature Review. *Proceedings of the International Conference on Cleaner Technologies and Environmental Management PEC.*

Girisha, C., Sanjeevamurthy, S., Gunti, R., & Manu, S. (2012). Mechanical performance of natural fiber reinforced epoxy hybrid composites. *International Journal of Engineering Research and Applications*, *2*(5), 615–619.

Girubha, R. J., & Vinodh, S. (2012). Application of fuzzy VIKOR and environmental impact analysis for material selection of an automotive component. *Materials & Design*, *37*(0), 478–486. doi:10.1016/j.matdes.2012.01.022

Goda, K., Sreekala, M. S., Gomes, A., Kaji, T., & Ohgi, J. (2006). Improvement of plant based natural fibers for toughening green composites- Effect of load application during mercerization of ramie fibers. *Composites. Part A, Applied Science and Manufacturing*, *37*(12), 2213–2220. doi:10.1016/j.compositesa.2005.12.014

Gomes, A., Matsuo, T., Goda, K., & Ohgi, J. (2007). Development and effect of alkali treatment on tensile properties of curaua fiber green composites. *Composites. Part A, Applied Science and Manufacturing*, *38*(8), 1811–1820. doi:10.1016/j.compositesa.2007.04.010

Grethlein, H. E. (1985). The effect of pore size distribution on the rate of enzymatic hydrolysis of cellulosic substrates. *Nature Biotechnology*, *3*(2), 155–160. doi:10.1038/nbt0285-155

Gritsch, C. S., Kleist, G., & Murphy, R. J. (2004, August19). Developmental changes in cell wall structure of phloem fibres of the bamboo Dendrocalamusasper. *Annals of Botany*, *94*(4), 497–505. doi:10.1093/aob/mch169 PMID:15319227

Gu, H. (2009). Tensile behaviours of the coir fiber and related composites after NaOH treatment. *Materials & Design*, *30*(9), 3931–3934. doi:10.1016/j.matdes.2009.01.035

Guimarães, J. L., Wypych, F., Saul, C. K., Ramos, L. P., & Satyanarayana, K. G. (2010). Studies of the processing and characterization of corn starch and its composites with banana and sugarcane fibers from Brazil. *Carbohydrate Polymers*, *80*(1), 130–138. doi:10.1016/j.carbpol.2009.11.002

Gupta, K. M. (2009). Experimental investigation of the Behavior of dual fiber hybrid composite under different stacking sequence. In *17th International Conference on Composite Materials (ICCM-17)*.

Habibi, Y., Zawawy, W. K., Ibrahim, M. M., & Dufresne, A. (2009). Processing andcharacterization of reinforced polyethylene composites made with lignocellulosic fibers from Egyptian agro-industrial residues. *Composites Science and Technology*, *68*(7-8), 1877–1885. doi:10.1016/j.compscitech.2008.01.008

Haefele, S. M., Konboon, Y., Wongboon, W., Amarante, S., Maarifat, A. A., Pfeiffer, E. M., & Knoblauch, C. (2011). Effects and fate of biochar from rice residues in rice-based systems. *Field Crops Research*, *121*(3), 430–440. doi:10.1016/j.fcr.2011.01.014

Hall, D. O. (1991) *Biomass for energy in the developing countries (current role, potential, problems and prospects)*. Pergamon Press. Retrieved from http://www.iea.org

Hambali, A., Sapuan, S. M., Ismail, N., & Nukman, Y. (2009a). Application of Analytical Hierarchy Process in the design concept selection of automotive composite bumper beam during the conceptual design stage. *Scientific Research and Essays*, *4*(4), 198–211.

Hambali, A., Sapuan, S. M., Ismail, N., & Nukman, Y. (2009b). Composite manufacturing process selection using Analytical Hierarchy Process. *International Journal of Mechanical and Materials Engineering*, *4*(1), 49–61.

Hamma, A., Kaci, M., Ishak, Z. A. M., & Pegoretti, A. (2014). Starch-grafted-polypropylene/kenaf fibres composites. Part 1: Mechanical performances and viscoelastic behaviour. *Composites. Part A, Applied Science and Manufacturing*, *56*(0), 328–335. doi:10.1016/j.compositesa.2012.11.010

Haneefa, A., Bindu, P., Arvind, I., & Thomas, S. (2008). Studies of Tensile and Flexural Properties of Short Banana/Glass Hybrid Fiber Reinforced Polystyrene Composites. *Journal of Composite Materials*, *42*(15), 1471–1489. doi:10.1177/0021998308092194

Harisha, S., Peter Michaelb, D., Benselyb, A., Mohan Lalb, D., & Rajaduraic, A. (2009). Mechanical Property Evaluation of Natural Fiber Coir Composite. *Materials Characterization*, *60*(1), 44–49. doi:10.1016/j.matchar.2008.07.001

Hassan, A., Akbari, A., Hing, N. K., & Ratnam, C. T. (2012). Mechanical and thermal properties of ABS / PVC composites: Effect of particles size and surface treatment of ground calcium carbonate. *Polymer-Plastics Technology and Engineering*, *51*(5), 473–479. doi:10.1080/03602559.2011.651242

He, & Wang. (2008). Study on Ramie Fiber Reinforced Polypropylene Composites (RF-PP) and its Mechanical Properties. *Advanced Materials Research, 41-42*.

Herrera-Estrada, L., Pillay, S. & Vaidya, U. (n.d.). *Banana Fibre Composites For Automotive and Transportation Applications*. Department of Material Science and Engineering, University of Alabama at Birmingham.

Herrera-Franco, P. J., & Valadez-Gonza'lez, A. (2004). Mechanical properties of continuous natural fibre-reinforced polymer composites. *Composites. Part A, Applied Science and Manufacturing*, *35*(3), 339–345. doi:10.1016/j.compositesa.2003.09.012

Hirunlabh, J., Pratintong, N., Khedari, J., & Suttisonk, B. (2002). *New lightweight composite construction materials with low thermal conductivity*. Cem Compos.

Hong, C. K., & Wool, R. P. (2005). Development of a bio-based composite material from soybean oil and keratin fibers. *Journal of Applied Polymer Science*, *95*(6), 1524–1538. doi:10.1002/app.21044

Hossain, M. K., Dewan, M. W., Hosur, M., & Jeelani, S. (2011). Mechanical performances of surface modified jute fiber reinforced biopol nanophased green composites. *Composites. Part B, Engineering*, *42*(6), 1701–1707. doi:10.1016/j.compositesb.2011.03.010

Huda, M. S., Drzal, L. T., Mohanty, A. K., & Misra, M. (2008). Effect of chemical modifications of the pineapple leaf fiber surfaces on the interfacial and mechanical properties of laminated biocomposites. *Composite Interfaces*, *15*(2-3), 169–191. doi:10.1163/156855408783810920

Huda, M. S., Mohanty, A. K., Drzal, L. T., Schut, E., & Mishra, M. (2005). Green composites from recycled cellulose and poly lactic acid: Physico-mechanical and Morphological Properties evaluation. *Journal of Materials Science*, *40*(16), 4221–4229. doi:10.1007/s10853-005-1998-4

Idicula, M., Boudenne, A., Umadevi, L., Ibos, L., Candau, Y., & Thomas, S. (2006). Thermophysical properties of natural fibre reinforced polyester composites. *Composites Science and Technology*, *66*(15), 2719–2725. doi:10.1016/j.compscitech.2006.03.007

Idicula, M., Neelakantan, N. R., Oommen, Z., Joseph, K., & Thomas, S. (2005). A study of the mechanical properties of randomly oriented short banana and sisal hybrid fiber reinforced polyester composites. *Journal of Applied Polymer Science*, *96*(5), 1699–1709. doi:10.1002/app.21636

Imihezri, S. S. S., Sapuan, S. M., Ahmad, M. M. H. M., & Sulaiman, S. (2005). A study of the comparison of `V' and `X' ribbing in a composite pedal using mold flow analysis software. *Materials & Design*, *26*(2), 157–166. doi:10.1016/j.matdes.2004.05.011

Imihezri, S. S. S., Sapuan, S. M., Sulaiman, S., Hamdan, M. M., Zainuddin, E. S., Osman, M. R., & Rahman, M. Z. A. (2006). Mould flow and component design analysis of polymeric based composite automotive clutch pedals. *Journal of Materials Processing Technology*, *171*(3), 358–365. doi:10.1016/j.jmatprotec.2005.06.077

Ishizaki, M. H., Visconte, L. L., Furtado, C. R., Oliveira, M. G., & Leblanc, J. L. (2008). Characterization of Green Coconut Fiber Composites. *International Journal of Polymer Analysis and Characterization*, *13*(5), 319–330. doi:10.1080/10236660802351409

Iyagba, , & Mangibo, , & Mohammad. (2009). The study of cow dung as co-substrate with rice husk in biogas production. *Scientific Research and Essays*, *9*, 861–866.

Jahan, A., Ismail, M. Y., Sapuan, S. M., & Mustapha, F. (2010). Material screening and choosing methods – A review. *Materials & Design*, *31*(2), 696–705. doi:10.1016/j.matdes.2009.08.013

Jahazi, M., & Hossein-Nejad, S. (2004). The development of an optimum manufacturing and material selection process for the fabrication of labyrinth seal strips. *Journal of Materials Processing Technology*, *152*(3), 272–275. doi:10.1016/j.jmatprotec.2003.11.035

Jain, S., Kumar, R., & Jindal, U. C. (1992). Mechanical behaviour of bamboo and bamboo composite. *Journal of Materials Science*, *27*(4), 598–604.

Janssen, J. J. (2000). *Designing and building with bamboo*. International Network for Bamboo and Rattan China.

Jarukumjorn, K., & Suppakarn, N. (2009). Effect of glass fiber hybridization on properties of sisal fiber–polypropylene composites. *Composites. Part B, Engineering*, *40*(7), 623–627. doi:10.1016/j.compositesb.2009.04.007

Jawaid, M., Khalil, H. A., & Bakar, A. A. (2011). Woven hybrid composites: Tensile and flexural properties of oil palm-woven jute fibres based epoxy composites. *Materials Science and Engineering A*, *528*(15), 5190–5195. doi:10.1016/j.msea.2011.03.047

Jayabal, S., Natarajan, U., & Sathiyamurthy, S. (2011). Effect of glass hybridization and staking sequence on mechanical behaviour of interply coir–glass hybrid laminate. *Bulletin of Materials Science*, *34*(2), 293–298. doi:10.1007/s12034-011-0081-9

Jindal, U. C. (1986). Development and testing of bamboo-fibers reinforced plastic composites. *Journal of Composite Materials*, *20*(1), 19–29. doi:10.1177/002199838602000102

Jinnah, M., Mariatti, M., & Abubacker, A. (2008). Effect of Chemical Surface Modification on the Properties of Woven Banana Reinforced Un-saturated Polyester Composites. *Journal of Reinforced Plastics and Composites*, *41*, 2371–2380.

John, K., & Naidu, S. V. (2004). Effect of fiber content and fiber treatment on flexural properties of sisal fiber/glass fiber hybrid composites. *Journal of Reinforced Plastics and Composites*, *23*(15), 1601–1605. doi:10.1177/0731684404039799

John, K., & Naidu, S. V. (2004). Sisal fiber/glass fiber hybrid composites: The impact and compressive properties. *Journal of Reinforced Plastics and Composites*, *23*(12), 1253–1258.

John, K., & Naidu, S. V. (2004). Tensile properties of unsaturated polyester-based sisal fiber–glass fiber hybrid composites. *Journal of Reinforced Plastics and Composites*, *23*(17), 1815–1819. doi:10.1177/0731684404041147

John, M. J., & Anandjiwala, R. D. (2008). Recent developments in chemical modification and characterization of natural fiber-reinforced composites. *Polymer Composites*, *29*(2), 187–207. doi:10.1002/pc.20461

John, M. J., Francis, B., Varghese, K. T., & Thomas, S. (2008). Effect of Chemical Modification on Properties of Hybrid Fiber Bio-composites. *Composites. Part A, Applied Science and Manufacturing*, *39*(2), 352–363. doi:10.1016/j.compositesa.2007.10.002

John, M. J., & Thomas, S. (2008). Biofibres and biocomposites. *Carbohydrate Polymers*, *71*(3), 343–364. doi:10.1016/j.carbpol.2007.05.040

Johnson, D. & Bhg, Y. Dato E. & Younus, N.B. (2009). Particle board from rice husk- A Brief Introduction to Renewable Materials of Construction. *JURUTERA*, 12-15.

John, V. M., Cincotto, M. A., Sjostrom, C., Agopyan, V., & Oliveira, C. T. A. (2005). Durability of slag mortar reinforced with coconut fiber. *Cement and Concrete Composites*, *27*(5), 565–574. doi:10.1016/j.cemconcomp.2004.09.007

Joseph, K., Mattoso, L. H. C., Toledo, R. D., Thomas, S., de Carvalho, L. H., Pothen, L., . . . James, B. (2000). Natural fiber reinforced thermoplastic composites. In Natural Polymers and Agrofibers Composites. Sãn Carlos, Brazil: Embrapa, USP-IQSC, UNESP.

Joseph, S., Sreekala, M. S., Oommen, Z., & Thomas, S. (2002). A Comparison of the Mechanical Properties of Phenol Formaldehyde Composites Reinforced with Banana Fibers and Glass Fibers. *Journal of Composite Science and Technology.*, *62*(14), 1857–1868. doi:10.1016/S0266-3538(02)00098-2

Joshi, S., Drzal, L., Mohanty, A., & Arora, S. (2004). Are natural fiber composites environmentally superior to glass fiber reinforced composites? *Composites. Part A, Applied Science and Manufacturing*, *35*(3), 371–376. doi:10.1016/j.compositesa.2003.09.016

Kalaprasad G,,Geethamma V.G, Sabu T., Gabriel G. (2005). Dynamic mechanical behavior of short coir fiber reinforced natural rubber composites. *Composites, 36*, 1499–506.

Kamireddy, S. R., Li, J., Abbina, S., Berti, M., Tucker, M., & Ji, Y. (2013). Converting forage sorghum and sunn hemp into biofuels through dilute acid pretreatment. *Industrial Crops and Products*, *49*, 598–609.

Karana, E., Hekkert, P., & Kandachar, P. (2008). Material considerations in product design: A survey on crucial material aspects used by product designers. *Materials & Design*, *29*(6), 1081–1089. doi:10.1016/j.matdes.2007.06.002

Karana, E., Hekkert, P., & Kandachar, P. (2010). A tool for meaning driven materials selection. *Materials & Design*, *31*(6), 2932–2941. doi:10.1016/j.matdes.2009.12.021

Karnani, R., Krishnan, M., & Narayan, R. (1997). Biofiber-reinforced polypropylene composites. *Polymer Engineering and Science*, *37*(2), 476–483. doi:10.1002/pen.11691

Kartini, K., Mahmud, H. B., & Hamidah, M. S. (2010). Absorption and permeability performance of selangor rice husk ash blended grade 30 concrete. *Journal of Engineering Science and Technology*, *5*, 1–16.

Kaseem, M., Hamad, K., & Deri, F. (2013). On-line rheological measurements and mechanical properties of acrylonitrile-butadiene-styrene / corn starch composite. *International Journal of Polymeric Materials and Polymeric Biomaterials*, *62*(5), 260–264. doi:10.1080/00914037.2011.641698

Kelly, A. (1967). The Nature of Composite Materials. *Scientific American Magazine*, *217*(B), 161.

Kengkhetkit, N., & Amornsakchai, T. (2014). A new approach to "greening" plastic composites using pineapple leaf waste for performance and cost effectiveness. *Materials & Design*, *55*, 292–299. doi:10.1016/j.matdes.2013.10.005

Ketty, B. (2007). Study of banana and coconut fibers: Botanical composition, thermal degradation and textural observations. *Bioresource Technology*, *98*(1), 58–68. doi:10.1016/j.biortech.2005.11.030 PMID:16442281

Khalil, H. A., Kang, C. W., & Adawi, T. O. (2008). The effect of different laminations on mechanical and physical properties of hybrid composites. *Journal of Reinforced Plastics and Composites*.

Khalil, H. P. S. A., Hanida, S., Kang, C. W., & Fuaad, N. A. N. (2007). Agro-hybrid Composite: The Effects on Mechanical and Physical Properties of Oil Palm Fiber (EFB)/Glass Hybrid Reinforced Polyester Composites. *Journal of Reinforced Plastics and Composites*, *26*(2), 203–218. doi:10.1177/0731684407070027

Khedari, J., Suttisonk, B., Pratintong, N., & Hirunlabh, J. (2002). New lightweight composite construction materials with low thermal conductivity. *Cem Compos*, *23*(1), 65–70. doi:10.1016/S0958-9465(00)00072-X

Kikuchi, T., Tani, Y., Takai, Y., Goto, A., & Hamada, H. (2014). Mechanical Properties of Jute Composite by Spray up Fabrication Method. *Energy Procedia*, *56*, 289–297. doi:10.1016/j.egypro.2014.07.160

Kim, S. (2009). Incombustibility, physico-mechanical properties and TVOC emission behavior of the gypsum–rice husk boards for wall and ceiling materials for construction. *Industrial Crops and Products, 29*, 381–387.

Kim, H. J., & Gueneom, Y. (2001). Thermogravimetric Analysis of Rice Husk Flour for a New Raw Material of Lignocellulosic Fiber Thermoplastic Polymer composite. *Mokchae Konghak*, *29*, 59–67.

Kim, H. S., Kim, S., Kim, H. J., & Yang, H. S. (2006). Thermal properties of bio-flour-fliied polyolefin composite with different compatibilizing agent type and content. *Thermochimica Acta*, *451*(1-2), 181–188. doi:10.1016/j.tca.2006.09.013

Kim, H. S., Kim, S., & Kim, H.-J. (2006). Enhanced Interfacial Adhesion of Bioflour-Filled Poly(propylene) Biocomposites by Electron-Beam Irradiation. *Macromolecular Materials and Engineering*, *291*(7), 762–772. doi:10.1002/mame.200600121

Kim, H. S., Yang, H.-S., Kim, H.-J., & Park, H.-J. (2004). Thermogravimetric analysis of rice husk flour filled thermoplastic polymer composites. *Journal of Thermal Analysis and Calorimetry*, *76*(2), 395–404. doi:10.1023/B:JTAN.0000028020.02657.9b

Kim, H.-J., Son, J., Park, H.-J., Lee, B.-J., & Hwang, T.-S. (2004). Rice-husk flour filled polypropylene composites; mechanical and morphological study. *Composite Structures*, *63*(3-4), 305–312. doi:10.1016/S0263-8223(03)00179-X

Kim, H.-S., Kim, H.-J., Lee, J.-W., & Choi, I.-G. (2006). Biodegradability of bio-flour filled bio degradable poly(butylene succinate) bio-composites in natural and compost soil. *Polymer Degradation & Stability*, *91*(5), 1117–1127. doi:10.1016/j.polymdegradstab.2005.07.002

Kim, J. S., Eom, Y. G., Kim, S., & Kim, H.-J. (2007). Effect of natural-resource-based scavengers on the adhesion properties and formaldehyde emission of engineered flooring. *Journal of Adhesion Science and Technology*, *21*(3-4), 211–225. doi:10.1163/156856107780684602

Kim, S., Kim, H. J., Kim, H. S., & Lee, H. Y. (2006). Effect of Bio-Scavengers on the Curing Behavior and Bonding Properties of Melamine-Formaldehyde Resins. *Macromolecular Materials and Engineering*, *291*(9), 1027–1034. doi:10.1002/mame.200600213

Klemm, D., Heublein, B., Fink, H.-P., & Bohn, A. (2005). Cellulose: Fascinating Biopolymer and Sustainable Raw Material. *Angewandte Chemie International Edition*, *44*(22), 3358–3393. doi:10.1002/anie.200460587 PMID:15861454

Kocserha, I., & Gömze, L. A. (2010). Friction properties of clay compounds. *Applied Clay Science*, *48*(3), 425–430. doi:10.1016/j.clay.2010.01.017

Koren, H., & Bisesi, M. S. (2003). *Handbook of environmental health: pollutant interactions in air, water, and soil.* Boca Raton, FL: Lewis Publishers.

Koronis, G., Silva, A., & Fontul, M. (2013). Green composites: A review of adequate materials for automotive applications. *Composites. Part B, Engineering*, *44*(1), 120–127. doi:10.1016/j.compositesb.2012.07.004

Kranthi, G., & Satapathy, A. (2010). Evaluation and prediction of wear response of pine wood dust filled epoxy composites using neural computation. *Computational Materials Science*, *49*(3), 609–614. doi:10.1016/j.commatsci.2010.06.001

Krishnan, A. K., George, T. S., Anjana, R., & George, K. E. (2013). Effect of modified kaolin clays on the mechanical properties of polypropylene / polystyrene blends. *Journal of Applied Polymer Science*, *127*(2), 1409–1415. doi:10.1002/app.38043

Ku, H., Wang, H., Pattarachaiyakoop, N., & Trada, M. (2011). A review on the tensile properties of natural fiber reinforced polymer composites. *Composites. Part B, Engineering*, *42*(4), 856–873. doi:10.1016/j.compositesb.2011.01.010

Kulkarni, A. G., Satyanarayana, K. G., Rohtagi, P. K., & Vijayam, K. (1983). Mechanical Properties of Banana Fibers. *Journal of Materials Science*, *18*(8), 2290–2296. doi:10.1007/BF00541832

Kumar, S., Adhikari, B., Choudhury, A., (2007). Recycled milk pouch and virgin low density. Polyethylene/linear low-density polyethylene based coir fiber composites. *Journal of Applied Polymer Science, 106*, 775–785.

Kumar, A., & Mohanta, K. (2012). Properties and Industrial Applications of Rice husk: A review. *International Journal of Emerging Technology and Advanced Engineering*, *2*, 86–90.

Kumar, R., Mishra, S., & Varma, I. K. (2008). Banana fiber-reinforced biodegradable soy protein composites. *Frontiers of Chemistry in China*, (September), 243–250.

Laly, P. A., Thomas, S., & George, J. (2012).Tensile and impact properties of bananafibre/glass fibre hybrid polyester composites. ICCM 2012-Eupore.

Leão, A., Cherian, B., Narine, S., Souza, S., Sain, M., & Thomas, S. (2015). The use of pineapple leaf fibers (PALFs) as reinforcements in composites. Biofiber Reinforcements. *Composite Materials*, 211-235.

Lee, S. M., Cho, D., Park, W. H., Lee, S. G., Han, S. O., & Drzal, L. T. (2005). Novel silk/poly(butylene succinate) biocomposites: The effect of short fibre content on their mechanical and thermal properties. *Composites Science and Technology*, *65*(3–4), 647–657. doi:10.1016/j.compscitech.2004.09.023

Lee, Y-K., Kim, S., Yang, H.S., & Kim, H.J. (2003). Mechanical Properties of Rice Husk Flour-Wood Particleboard by Urea-Formaldehyde Resins. *MokchaeKonghak*, *31*, 42–49.

Lewin, L. M. E. M. P. (1985). Handbook of Fiber Science and Technology: Fiber Chemistry (Vol. 4). New York: CRC Press.

Li, X. (2008). *Development of flax fiber-reinforced polyethylene biocomposites by injection molding*. (PhD dissertation). Department of agricultural and bioresources engineering, University of Saskatchewan.

Liese, W. (2010). *The anatomy of bamboo culms*. INBAR Technical Report No. 18. International Network for Bamboo and Rattan.

Liese, D. G. W. (1971). On the Anatomy of Asian Bamboos, with Special Reference to their Vascular Bundles. *Wood Science and Technology*, *5*(4), 290–312. doi:10.1007/BF00365061

Lilholt, H., & Lawther, J. M. (2000). *Comprehensive Composite Materials*. Elsevier Ltd.

Lin, C.-S., Chen, L.-S., & Hsu, C.-C. (2011). An innovative approach for RFID product functions development. *Expert Systems with Applications*, *38*(12), 15523–15533. doi:10.1016/j.eswa.2011.06.007

Li, S. H., Zeng, Q. Y., Xiao, Y. L., Fu, S. Y., & Zhou, B. L. (1995). Biomimicry of bamboo bast fiber with engineering composite materials. *Materials Science and Engineering C*, *3*(0), 125–130. doi:10.1016/0928-4931(95)00115-8

Liu, H., Dong, L., Xie, H., & Xiong, C. (2012). Novel-modified kaolin for enhancing the mechanical and thermal properties of poly (vinyl chloride). *Polymer Engineering and Science*, *52*(10), 2071–2077. doi:10.1002/pen.23161

Liu, W., Misra, M., Askeland, P., Drzal, L. T., & Mohanty, A. K. (2005). 'Green' composites from soy based plastic and pineapple leaf fiber: Fabrication and properties evaluation. *Polymer*, *46*(8), 2710–2721. doi:10.1016/j.polymer.2005.01.027

Liu, W., Mohanty, A. K., Drzal, L. T., & Misra, M. (2005). Novel Biocomposites from Native Grass and Soy Based Bioplastic: Processing and Properties Evaluation. *Industrial & Engineering Chemistry Research*, *44*(18), 7105–7112. doi:10.1021/ie050257b

Li, X., Cai, Z., Winandy, J. E., & Basta, A. H. (2010). Selected properties of particleboard panels manufactured from rice straws of different geometries. *Bioresource Technology*, *101*(12), 4662–4666. doi:10.1016/j.biortech.2010.01.053 PMID:20153181

Li, Z. H., & Chen, T. (2012). The mechanical and tribological properties of MAH-grafted HDPE-filled clay / PEI composites. *Polymer-Plastics Technology and Engineering*, *51*(8), 776–779. doi:10.1080/03602559.2012.663848

Li, Z., Wang, L., & Wang, X. (2006). Flexural characteristics of coir fiber reinforced cementitious composites. *Fiber Polym*, *7*(3), 286–294. doi:10.1007/BF02875686

Lodha, P., & Netravali, N. (2002). Characterization of interfacial and mechanical properties of green composite with soy protein isolate and ramie fiber. *Journal of Materials Science*, *37*(17), 3657–3665. doi:10.1023/A:1016557124372

London, X., Camayo, G., Rian, N., & Lopez, Y. (2002). Characterization of the anatomy of Guadua angustifolia (Poaceae: Bambusoideae) culms. *Bamboo Sci Cult: J Am Bamboo Soc*, *16*, 18–31.

Lopattananon, N., Panawarangkul, K., Sahakaro, K., & Ellis, B. (2006). Performance of pineapple leaf fiber–natural rubber composites: The effect of fiber surface treatments. *Journal of Applied Polymer Science*, *102*(2), 1974–1984. doi:10.1002/app.24584

Luduena, L., Fasce, D., Alvarez, V. A., & Stefani, P. M. (2011). Nanocellulose from rice husk. *BioResources*, *6*, 1440–1453.

Luo, S., & Netravali, A. N. (1999). Mechanical and thermal properties of environmental-friendly" green" composites made from pineapple leaf fibers and poly (hydroxybutyrate-co-valerate) resin. *Polymer Composites*, *20*(3), 367–378. doi:10.1002/pc.10363

Luttropp, C., & Lagerstedt, J. (2006). EcoDesign and The Ten Golden Rules: Generic advice for merging environmental aspects into product development. *Journal of Cleaner Production*, *14*(15-16), 1396–1408. doi:10.1016/j.jclepro.2005.11.022

Luz, S. M., Caldeira-pires, A., & Ferrão, P. M. C. (2010). Environmental benefits of substituting talc by sugarcane bagasse fibers as reinforcement in polypropylene composites: Ecodesign and LCA as strategy for automotive components. *Resources, Conservation and Recycling*, *54*(12), 1135–1144. doi:10.1016/j.resconrec.2010.03.009

Madhusudan, S., Rao, D. S. S., & Madhukiran, J. (2013). Fabrication and testing of natural fiber reinforced hybrid composites banana/pineapple. *International Journal of Modern Engineering and Research*, *3*, 2239–2243.

Majhi, S. K., Nayak, S. K., Mohanty, S., & Unnikrishnan, L. (2010). Mechanical and Fracture Behavior of Banana Fiber Reinforced Polylactic Acid Biocomposites. *International Journal of Plastics Technology*, *14*(S1), 57–75. doi:10.1007/s12588-010-0010-6

Ma, L., He, H., Jiang, C., Zhou, L., Luo, Y., & Jia, D. (2012). Effect of alkali treatment on structure and mechanical properties of acrylonitrile-butadiene-styrene / bamboo fiber composites. *Journal of Macromolecular Science, Part B: Physics*, *51*(11), 2232–2244. doi:10.1080/00222348.2012.669688

Maleque, M. A., Belal, E. Y., & Sapuan, S. A. (2006). Mechanical Properties of Study of Pseudo-stem Banana Fiber Reinforced Epoxy Composite. *The Arabian Journal for Science and Engineering*, *32*, 359–363.

Malkapuram, R., Kumar, V., & Negi, Y. S. (2008). Recent development in natural fiber reinforced polypropylene composites. *Journal of Reinforced Plastics and Composites*.

Maniya, K., & Bhatt, M. G. (2010). A selection of material using a novel type decision-making method: Preference selection index method. *Materials & Design*, *31*(4), 1785–1789. doi:10.1016/j.matdes.2009.11.020

Mansor, M. R., Sapuan, S. M., Hambali, A., Zainudin, E. S., & Nuraini, A. A. (2015). Conceptual design of kenaf polymer composites automotive spoiler using TRIZ and Morphology Chart methods. *Applied Mechanics and Materials*, *761*(0), 63–67. doi:10.4028/www.scientific.net/AMM.761.63

Mansor, M. R., Sapuan, S. M., Zainudin, E. S., Nuraini, A. A., & Hambali, A. (2013). Hybrid natural and glass fibers reinforced polymer composites material selection using Analytical Hierarchy Process for automotive brake lever design. *Materials & Design*, *51*(0), 484–492. doi:10.1016/j.matdes.2013.04.072

Mansor, M. R., Sapuan, S. M., Zainudin, E. S., Nuraini, A. A., & Hambali, A. (2014a). Application of integrated AHP-TOPSIS Method in hybrid natural fiber composites materials selection for automotive parking brake lever component. *Australian Journal of Basic and Applied Sciences*, *8*(5), 431–439.

Mansor, M. R., Sapuan, S. M., Zainudin, E. S., Nuraini, A. A., & Hambali, A. (2014b). Conceptual design of kenaf fiber polymer composite automotive parking brake lever using integrated TRIZ – Morphological Chart – Analytic Hierarchy Process method. *Materials & Design*, *54*(0), 473–482. doi:10.1016/j.matdes.2013.08.064

Mathews, F. L., & Rawlings, R. D. (1994). Engineering Composites. Chapman and Hall.

Memon, S. A., Shaikh, M. A., & Akbar, H. (2008). Production of Low Cost Self Compacting Concrete Using Rice Husk Ash. *First International Conference on Construction In Developing Countries*.

Min, S. H., Kim, H. Y., Kwon, Y. J., & Sohn, S. Y. (2011). Conjoint analysis for improving the e-book reader in the Korean market. *Expert Systems with Applications*, *38*(10), 12923–12929. doi:10.1016/j.eswa.2011.04.087

Mirmehdi, S. M., Zeinaly, F., & Dabbagh, F. (2014). Date palm wood flour as filler of linear low-density polyethylene. *Composites. Part B, Engineering*, *56*, 137–141. doi:10.1016/j.compositesb.2013.08.008

Mishra, S., Mohanty, A. K., Drzal, L. T., Misra, M., & Hinrichsen, G. (2004). A review on pineapple leaf fibers, sisal fibers and their biocomposites. *Macromolecular Materials and Engineering*, *289*(11), 955–974. doi:10.1002/mame.200400132

Mishra, S., Mohanty, A. K., Drzal, L. T., Misra, M., Parija, S., Nayak, S. K., & Tripathy, S. S. (2003). Studies on mechanical performance of biofibre/glass reinforced polyester hybrid composites. *Composites Science and Technology*, *63*(10), 1377–1385. doi:10.1016/S0266-3538(03)00084-8

Mishra, S., Naik, J. B., & Patil, Y. P. (2000). The Compatibilising Effect of Maleic Anhydride on Swelling and Mechanical Properties of Plant Fiber Reinforced Novolac Composites. *Journal of Composite Science & Technology*, *60*(9), 1729–1735. doi:10.1016/S0266-3538(00)00056-7

Mitra, B. C. (2014). Environment Friendly composite materials: Biocomposites and Green composites. *Defence Science Journal*, *64*(3), 244–261. doi:10.14429/dsj.64.7323

Mohammad, H. B. M. H. (2005). *Coconut fiber reinforced wall panelling system.* (Master thesis). Faculty of Civil Engineering. Universiti Teknologi, Malaysia.

Mohanta, N., & Acharya, S. K. (2014). Investigation of mechanical properties of luffa cylindrica fibre reinforced epoxy hybrid composite. *International Journal of Engineering Science and Technology*, *7*(1), 1–10. doi:10.4314/ijest.v7i1.1

Mohanty, A. K., Misra, M., & Drzal, L. T. (2005). Natural Fibers, Biopolymers and Biocomposites. Boca Raton, FL: Taylor & Francis.

Mohanty, A. K., Mishra, M., & Drzal, L. T. (2001). Surface modifications of natural fibers and performance of the resulting biocomposites: An overview. *Composite Interfaces*, *8*(5), 313–343. doi:10.1163/156855401753255422

Mohanty, A. K., Misra, M., & Drzal, L. T. (Eds.). (2005). *Natural fibers, biopolymers, and biocomposites*. CRC Press. doi:10.1201/9780203508206

Mohanty, A. K., Misra, M., & Hinrichsen, G. (2000). Biofibres, biodegradable polymers and biocomposites: An overview. *Macromolecular Materials and Engineering*, *276-277*(1), 1–24. doi:10.1002/(SICI)1439-2054(20000301)276:1<1::AID-MAME1>3.0.CO;2-W

Molding, C. (n.d.). Retrieved from: http://www.coremt.com/processes/compression-molding/

Montgomery, D. C. (1997). *Design and analysis of experiments* (Vol. 7). New York: Wiley.

Mubashirunnisa, A., Vijayalakshmi, K., Gomathi, T., & Sudha, P. N. (2012). Development of banana/glass short hybrid fiber reinforced nano-chitosan polymer composities. *Scholars Research Library Der Pharmacia Lettre*, *4*, 1162–1168.

Mueller, D. H., & Krobjilowski, A. (2003). New Discovery in the Properties of Composites Reinforced with Natural Fibers. *Journal of Industrial Textiles*, *33*(2), 111–129. doi:10.1177/152808303039248

Mukhopadhyay, S., Fanguerio, R., Arpac, Y., & Sentrik, U. (2008). Banana Fibers _ Variability and Fracture Behavior. *Journal of Engineered Fibers and Fabrics*, *3*, 39–45.

Mulinari, D. R., Voorwald, H. J. C., Cioffi, M. O. H., Da Silva, M. L. C. P., & Luz, S. M. (2009). Preparation and properties of HDPE/sugarcane bagasse cellulose composites obtained for thermokinetic mixer. *Carbohydrate Polymers*, *75*(2), 317–320. doi:10.1016/j.carbpol.2008.07.028

Munawar, S. S., Umemura, K., & Kawai, S. (2007). Characterization of the morphological, physical, and mechanical properties of seven non-wood plant fiber bundles. *Journal of Wood Science*, *53*(2), 108–113. doi:10.1007/s10086-006-0836-x

Munirah, M., Rahmat, A. R., & Hassan, A. (2007). *Characterization and Treatments of Pineapple Leaf Fiber Thermoplastic Composite for Construction Application*. Universiti Teknologi Malaysia.

Murali, M. R. K., & Mohana, R. K. (2007). Extraction of the tensile properties of natural fibers, Vakka, Date and Bambo. *Journal of Composite Structures*, *77*(3), 288–295. doi:10.1016/j.compstruct.2005.07.023

Mwaikambo, L. Y., & Ansell, M. P. (2003). Hemp fiber reinforced cashew nut shell liquid composites. *Composites Science and Technology*, *63*(9), 1297–1305. doi:10.1016/S0266-3538(03)00101-5

Mylsamy, K., & Rajendran, I. (2011). The mechanical properties, deformation and thermo mechanical properties of alkali treated and untreated Agave continuous fibre reinforced epoxy composites. *Materials & Design*, *32*(5), 3076–3084. doi:10.1016/j.matdes.2010.12.051

Naidu, A.L, Sudarshan, B., Krishna, K.H. (2013). Study on Mechanical Behavior Of Groundnut Shell Fiber Reinforced Polymer Metal Matrix Composites. *International Journal of Engineering Research & Technology*.

Nam, S., & Netravali, A. N. (2002). Green Composites ii. Environment-friendly, Biodegradable composites using Ramie Fibers and Soy Protein Concentrate (SPC) Resin. *Fibers and Polymers*, *7*(4), 380–388. doi:10.1007/BF02875770

Nam, T. H., Ogihara, S., Tung, N. H., & Kobayashi, S. (2011). Effect of alkali treatment on interfacial and mechanical properties of coir fiber reinforced poly (butylene succinate) biodegradable composites. *Composites. Part B, Engineering*, *42*(6), 1648–1656. doi:10.1016/j.compositesb.2011.04.001

Nascimento, D. C., Ferreira, A. S., Monteiro, S. N., Aquino, R. C., & Kestur, S. G. (2012). Studies on the characterization of piassava fibers and their epoxy composites. *Composites. Part A, Applied Science and Manufacturing*, *43*(3), 353–362. doi:10.1016/j.compositesa.2011.12.004

Nassif. (2010). Effect of Chemical Treatment on The Some Electrical And Thermal Properties For Unsaturated Polyester Composites Using Banana Fibers. *Eng. & Tech. Journal*, *28*, 10.

Nayak, SK., Mohanty, A.K., Rout, J., Misra, M., Tripathy, SS. (2003). SEM observations of the fractured surfaces of coir composites. *J. Reinf. Plast. Compos.*

Netravali, A. N., & Chabba, S. (2003). Composites get greener. *Materials Today*, *6*(4), 22–29. doi:10.1016/S1369-7021(03)00427-9

Netravali, A. N., Huang, X., & Mizuta, K. (2007). Advanced 'green' composites. *Advanced Composite Materials*, *16*(4), 269–282. doi:10.1163/156855107782325230

Nguong, C. W., Lee, S. N. B., & Sujan, D. (2013). *A Review on Natural Fibre Reinforced Polymer Composites*. World Academy of Science, Engineering and Technology.

Nogata, F., & Takahashi, H. (1995). Intelligent functionally graded material: Bamboo. *Compos Eng*, *5*(7), 43–51. doi:10.1016/0961-9526(95)00037-N

Nunna, S., Chandra, P. R., Shrivastava, S., & Jalan, A. K. (2012). A review on mechanical behavior of natural fiber based hybrid composites. *Journal of Reinforced Plastics and Composites*, *31*(11), 759–769. doi:10.1177/0731684412444325

Nyambo, C., Mohanty, A. K., & Misra, M. (2010). Polylactide-Based Renewable Green Composites from Agricultural Residues and Their Hybrids. *Biomacromolecules*, *11*(6), 1654–166. doi:10.1021/bm1003114 PMID:20499931

O'Sullivan, A. (1997). Cellulose: The structure slowly unravels. *Cellulose (London, England)*, *4*(3), 173–207. doi:10.1023/A:1018431705579

Ochi, S. (2008). Mechanical properties of kenaf fibers and kenaf/PLA composites. *Mechanics of Materials*, *40*(4-5), 446–452. doi:10.1016/j.mechmat.2007.10.006

Okafor, F. O., & Okonkwo, U. N. (2009). Effects of Rice Husk Ash on Some Geotechnical Properties of Lateritic Soil. *Leonardo Electronic Journal of Practices and Technologies*, *15*, 67–74.

Okubo, K., Fujii, T., & Yamamoto, Y. (2004). Development of bamboo-based polymer composites and their mechanical properties. *Composites. Part A, Applied Science and Manufacturing*, *35*(3), 377–383. doi:10.1016/j.compositesa.2003.09.017

Omofuma, F. E., Adeniye, S. A., & Adeleke, A. E. (2011). A Case Study of Rice Husk and Wood Flour. *World Applied Sciences Journal*, *14*, 1347–1352.

Openshaw, K. (2011). Supply of woody biomass, especially in the tropics: Is demand outstripping sustainable supply? *International Review (Steubenville, Ohio)*, *13*, 487–499.

Osorio, E. T. L., Van Vuure, A. W., & Verpoest, I. (2010). Morphological aspect and mechanical properties of single bamboo fibers and flextural characterization of bamboo/epoxy composites. *Journal of Reinforced Plastics and Composites*, *30*(5), 396–408. doi:10.1177/0731684410397683

Owolabi, O., Czvikovszky, T., & Kovacs, I. (1985). Coconut fiber reinforced thermosetting plastics. *Journal of Applied Polymer Science*, *30*(5), 1827–1836. doi:10.1002/app.1985.070300504

Pahl, G., Beitz, W., Feldhusen, J., & Grote, K. H. (2007). Engineering Design: A Systematic Approach (3rd ed.). London: Springer-Verlag London Limited.

Pandey, J. K., Ahn, S. H., Lee, C. S., Mohanty, A. K., & Misra, M. (2010). Recent Advances in the Application of Natural Fiber Based Composites. *Macromolecular Materials and Engineering*, *295*(11), 975–989. doi:10.1002/mame.201000095

Pan, P., Zhu, B., Kai, W., Serizawa, S., & Ji, M. (2007). Crystallization behavior and mechanical properties of bio-based green composites based on poly(L-lactide) and kenaf fiber. *Journal of Applied Polymer Science*, *105*(3), 1511–1520. doi:10.1002/app.26407

Paramasivam, P., Nathan, G. K., & Das Gupta, N. C. (1984). Coconut fiber reinforced corrugated slabs. *Int J Cem Compos Lightweight Concr*, *6*(1), 19–27. doi:10.1016/0262-5075(84)90056-3

Parasmewaran, N., & Liese, W. (1976). On the fine structure of bamboo fibers. *Wood Science and Technology*, *10*, 231–246.

Parasmewaran, N., & Liese, W. (1980). Ultrastructural aspects of bamboo cells. *Cellul Chem Technol*, *14*, 587–609.

Park, , Han, Kim, Yang, & Kim. (2005). Evaluation of the impact on manufacturing temperature and time in the production process of bio-composites. *MokchaeKonghak*, *33*, 29–37.

Paul, A., & Thomas, S. (1997). Electrical properties of natural-fiber-reinforced low density polyethylene composites: A comparison with carbon black and glass fiber-filled low density polyethylene composites. *Journal of Applied Polymer Science*, *63*(2), 247–266. doi:10.1002/(SICI)1097-4628(19970110)63:2<247::AID-APP12>3.0.CO;2-#

Paul, S. A., Boudene, A., Joseph, K., & Thomas, S. (2008). Effect of Fiber Loading and Chemical Treatments on Thermo Physical Properties of Banana Fiber/Polypropylene Commingled Composite Materials. *Composites. Part A, Applied Science and Manufacturing*, *39*(9), 1582–1588. doi:10.1016/j.compositesa.2008.06.004

Peng, A. H., & Xiao, X. M. (2013). Material selection using PROMETHEE combined with analytic network process under hybrid environment. *Materials & Design*, *47*(0), 643–652. doi:10.1016/j.matdes.2012.12.058

Pereira, M. A. R., & Beraldo, A. L. (2007). Bambu de Corpo e Alma. Bauru: Canal.

Pervaiz, M., & Sain, M. M. (2003). Carbon storage potential in natural fiber composites. *Resources, Conservation and Recycling*, *39*(4), 325–340. doi:10.1016/S0921-3449(02)00173-8

Pickering, K. (Ed.). (2008). *Properties and performance of natural-fibre composites*. Elsevier.

Pillai, S. G. K., Kulkarni, A. G., Sukumaran, K., Satyanarayana, P. K., & Rohatgi, K. G. (1986). Fabrication and properties of natural fiber-reinforced polyester composites. *Composites*, 17.

Poggio, C., Lombardini, M., Gaviati, S., & Chiesa, M. (2012). Microhardness of composite resins at different depths varying the post-irradiation time. *Journal of Conservative Dentistry*, *15*(3), 237–241. doi:10.4103/0972-0707.97946 PMID:22876009

Pothan, L. A., George, C. N., Jacob, M., & Thomas, S. (2007). Effect of Chemical Modification on the Mechanical and Electrical Properties of Banana Fiber Polyester Composites. *Composite Material*, *41*(19), 2371–2386. doi:10.1177/0021998307075456

Pothan, L. A., Oommen, Z., & Thomas, S. (2003). Dynamic Mechanical Analysis of Banana Fiber Reinforced with Polyester Composites. *Journal of Composite Science and Technology*, *63*(2), 283–293. doi:10.1016/S0266-3538(02)00254-3

Pothan, L. A., Thomas, S., & George, J. (2012). Tensile and Impact Properties of Banana Fibre/ Glass Fibre Hybrid Polyester Composites.*ICCM Conference*.

Pothan, L. A., Thomas, S., & Groeninckx, G. (2006). The Role of Fibre/Matrix Interactions on the Dynamic Mechanical Properties of Chemically Modified Banana Fiber/Polyester Composites. *Composite Part A, 36*(9), 1260–1269. doi:10.1016/j.compositesa.2005.09.001

Pothan, L. A., Thomas, S., & Habler, R. (2005). The Static and Dynamic Mechanical Properties of Banana and Glass Fiber Woven Fabric Reinforced Polyester Composites. *Journal of Composite Materials*, *39*(11), 1007–1025. doi:10.1177/0021998305048737

Pothan, L. A., Thomas, S., & Neelakantan, N. R. (1997). Short Banana Fiber Reinforced Polyester Composites: Mechanical Properties, Failure and Aging Characteristics. *Journal of Reinforced Plastics and Composites*, *16*, 1012–1025.

Prachayawarakorn, J., & Yaembunying, N. (2005). Effect of recycling on properties of rice husk-filled-polypropylene*; songklanakarinJ. Sci.Technol*, *27*, 343–352.

Prasad, S. V., Pavithran, C., & Rohatgi, P. K. (1983). Alkali treatment of coir fibers for coir-polyester composites. *Journal of Materials Science*, *18*(5), 1443–1454. doi:10.1007/BF01111964

Prasetyoko, D., Ramli, Z., Endud, S., Hamdan, H., & Sulikowski, B. (2006). Conversion of rice husk ash to zeolite beta. *Waste Management (New York, N.Y.)*, *26*(10), 1173–1179. doi:10.1016/j.wasman.2005.09.009 PMID:16274981

Praveen, R. N., Muruli, D., & Nayak, S. S. (2013). Sliding wear behavior of Acrylonitrile Butadiene Styrene (ABS) thermoplastic composites. *International Journal of Mechanical Engineering and Robotics Research*, *2*(4), 361–372.

Prukkaewkanjana, K., Thanawan, S., & Amornsakchai, T. (2015). High Performance Hybrid Reinforcement of Nitrile Rubber using Short Pineapple Leaf Fiber and Carbon Black. *Polymer Testing*, *45*, 76–82. doi:10.1016/j.polymertesting.2015.05.004

Pugh, S. (1991). *Total Design: Integrated Methods for Successful Product Engineering*. Wokingham, UK: Addison-Wesley Publishing.

Ràcz, I., & Hargitai, H. (2000). Influence of water on properties of cellulosic fiber reinforced polypropylene composites. *International Journal of Polymeric Materials*, *47*(4), 667–674. doi:10.1080/00914030008031321

Radetzki, M. (1997). The economics of biomass in industrialized countries: An overview. *Energy Policy*, *25*(6), 545–554. doi:10.1016/S0301-4215(97)00043-8

Rahman, M. M., & Khan, M. A. (2007). Surface treatment of coir (Cocos nucifera) fibers and its influence on the fibers' physico–mechanical properties. *Composites Science and Technology*, *67*(11-12), 2369–2376. doi:10.1016/j.compscitech.2007.01.009

Rai, A., & Jha, C. N. (n.d.). Natural fibre composite and its potential as building material. *BM-TPC Seminar.*

Ramakrishna, G., & Sundararajan, T. (2005). Studies on the durability of natural fibers and the effect of corroded fibers on the strength of mortar. *Cement and Concrete Composites*, *27*(5), 575–582. doi:10.1016/j.cemconcomp.2004.09.008

Ramakrishna, G., & Sundararajan, T. (2005, May). Impact strength of a few natural fiber reinforced cement mortar slabs: A comparative study. *Cement and Concrete Composites*, *27*(5), 547–553. doi:10.1016/j.cemconcomp.2004.09.006

Ramanaiah, K., Ratna Prasad, A. V., & Hema Chandra Reddy, K. (2012). Thermal and mechanical properties of waste grass broom fiber-reinforced polyester composites. *Materials & Design*, *40*, 103–108. doi:10.1016/j.matdes.2012.03.034

Ramezanianpour, A., Mahdi Khani, M., & Ahmadibeni, G. (2009). The Effect of Rice Husk Ash on Mechanical Properties and Durability of Sustainable Concretes. *International Journal of Civil Engineering*, *7*, 83–91.

Randjbaran, E., Zahari, R., Majid, D. L., Jalil, N. A. A., Vaghei, R., & Ahmadi, R. (2014). Experimental Study of the Influence of Stacking Order of the Fibrous Layers on Laminated Hybrid Composite Plates Subjected to Compression Loading. *Journal of Science and Engineering*, *4*(1), 1-8.

Ranjeth Kumar Reddy, T., Subba Rao, T., & Padma Suvarna, R. (2014). Studies on thermal characteristics of cow dung powder filled glass–polyester hybrid composites. *Composites. Part B, Engineering*, *56*, 670–672. doi:10.1016/j.compositesb.2013.08.059

Rao, K., Pranav, D., & Anusha, P. R. T. (2011). Stabilization of expansive soil with rice husk ash, lime and gypsum –an experimental study *International Journal of Engineering Science and Technology*, *3*, 8076–8085.

Rasheva, Z., Zhang, G., & Burkhart, T. (2010). A correlation between the tribological and mechanical properties of short carbon fibers reinforced PEEK materials with different fiber orientations. *Tribology International*, *43*(8), 1430–1437. doi:10.1016/j.triboint.2010.01.020

Rashid, M.H., & Molla, M.K., & Ahmed, T. (2010). Mortar Incorporating Rice Husk Ash: Strength and Porosity. *European Journal of Scientific Research*, *40*, 471–477.

Rawat, R., Kumbhar, B., & Tewari, L. (2013). Optimization of alkali pretreatment for bioconversion of poplar (Populus deltoides) biomass into fermentable sugars using response surface methodology. *Industrial Crops and Products*, *44*, 220–226. doi:10.1016/j.indcrop.2012.10.029

Reddy, D. V. (2006) Durability Characteristics of Rice Husk Ash-Modified Reinforced Concrete. *Fourth LACCEI International Latin American and Caribbean Conference for Engineering and Technology.*

Reddy, N., & Yang, Y. (2005a). Properties and potential applications of natural cellulose fibers from cornhusks. *Green Chemistry*, *7*(4), 190–195. doi:10.1039/b415102j

Reddy, N., & Yang, Y. (2005b). Structure and properties of high quality natural cellulose fibers from cornstalks. *Polymer*, *46*(15), 5494–5500. doi:10.1016/j.polymer.2005.04.073

Reis, J. M. L. (2006). Fracture and flexural characterization of natural fiber-reinforced polymer concrete. *Construction & Building Materials*, *20*(9), 673–678. doi:10.1016/j.conbuildmat.2005.02.008

Richard, P. W., Shrikant, N. K., John, J. L., Shana, P. B., Jue, L., Wim, T., & George, I. W. et al. (2002). Affordable Composites and Plastics from Renewable Resources: Part II: Manufacture of Composites *Advancing Sustainability through Green Chemical Engineering (Albany, N.Y.)*, *823*, 205–224.

Rijswijk, K. V., Brouwer, W. D., & Beukers, A. (2001). *Natural Fibre Composites*. Rome, Italy: FAO Economic and Social Development Department.

Rodrigo, O. T., Alberto, J., Talens, P., & Chiralt, A. (2014). Properties of starch–hydroxyl propyl methylcellulose based films obtained by compression molding. *Carbohydrate Polymers*, *109*, 155–165. doi:10.1016/j.carbpol.2014.03.059 PMID:24815412

Romli, F. I., Alias, A. N., Rafie, A. S. M., & Majid, D. L. A. A. (2012). Factorial Study on the Tensile Strength of a Coir Fiber-Reinforced Epoxy Composite. *AASRI Procedia*, *3*, 242–247. doi:10.1016/j.aasri.2012.11.040

Rosa, S.M.L., Santos, E.F., Ferreira, C.A., Marli, S. & Nachtigall, B. (2009). Thermal and Dynamic mechanical characterization of rice husk filled Polypropylene Composites. *Macromolecular Reaearch, 17*, 8-13.

Rosa, S. M. L., Santos, E. F., Ferreira, C. A., Marli, S., & Nachtigall, B. (2009). Studies on properties of rice husk-filled-PP composite. *Materials Research*, *12*(3), 333–338. doi:10.1590/S1516-14392009000300014

Roselman, I. C., & Tabor, D. (1976). The friction of carbon fibres. *Journal of Physics. D, Applied Physics*, *9*(17), 2517–2532. doi:10.1088/0022-3727/9/17/012

Rose, S. K., Ahammad, H., Eickhout, B., Fisher, B., Kurosawas, A., Rao, S., & van Vuuren, D. P. et al. (2012). Land-based mitigation in climate stabilization. *Energy Econ*, *34*(1), 365–380. doi:10.1016/j.eneco.2011.06.004

Rout, J., Misra, M., & Mohanty, A. K. (1999). Surface modification of coir fibers I: Studies on graft copolymerization of methyl methacrylate on to chemically modified coir fibers. *Polymers for Advanced Technologies*, *10*(6), 336–344. doi:10.1002/(SICI)1099-1581(199906)10:6<336::AID-PAT884>3.0.CO;2-O

Rout, J., Misra, M., Tripathy, S. S., Nayak, S. K., & Mohanty, A. K. (2001). The influence of fibre treatment on the performance of coir-polyester composites. *Composites Science and Technology*, *61*(9), 1303–1310. doi:10.1016/S0266-3538(01)00021-5

Rout, J., Tripathy, S. S., Nayak, S. K., Misra, M., & Mohanty, A. K. (2001). Scanning electron microscopy study of chemically modified coir fibers. *Journal of Applied Polymer Science*, *79*(7), 1169–1177. doi:10.1002/1097-4628(20010214)79:7<1169::AID-APP30>3.0.CO;2-Q

Rozainee, M., Ngo, S.P., & Arshad, A., Salema, & Tan, K.G. (2010). Effect of feeding methods on the rice husk ashquality in a fluidised bed combustor. *Emirates Journal for Engineering Research*, *15*, 1–12.

Sabat, A.K. & Nanda, R.P. (2011). Effect of marble dust on strength and durability of Rice husk ashstabilised expansive soil. *International Journal of Civil and Structural Engineering*, *1*, 939-948.

Safa, Y., & Bhatti, H. N. (2011). Adsorptive removal of direct dyes by low cost rice husk: Effect of treatments and modifications. *African Journal of Biotechnology*, *10*, 3128–3142.

Saheb, D. N., & Jog, J. P. (1999). Natural fiber polymer composites: A review. *Advances in Polymer Technology*, *18*(4), 351–363. doi:10.1002/(SICI)1098-2329(199924)18:4<351::AID-ADV6>3.0.CO;2-X

Sahoo, S., Misra, M., & Mohanty, A. K. (2011). Enhanced properties of lignin-based biodegradable polymer composites using injection moulding process. *Composites. Part A, Applied Science and Manufacturing*, *42*(11), 1710–1718. doi:10.1016/j.compositesa.2011.07.025

Samal, R. K., Panda, B. B., Rout, S. K., & Mohanty, M. (1995). Effect of chemical modification on FTIR spectra. I. Physical and chemical behavior of coir. *Journal of Applied Polymer Science*, *58*(4), 745–752. doi:10.1002/app.1995.070580407

Sanadi, A. R., Prasad, S. V., & Rohtgi, P. K. (1986). Sunhempfiber-reinforced polyester. *Journal of Materials Science*, *21*(12), 4299–4304. doi:10.1007/BF01106545

Sandberg, D., Haller, P., & Navi, P. (2013). Thermo-hydro and thermo-hydro-mechanical wood processing: An opportunity for future environmentally friendly wood products. *Wood Material Science and Engineering*, *8*(1), 64–88. doi:10.1080/17480272.2012.751935

Sangthong, S., Pongprayoon, T., & Yanumet, N. (2009). Mechanical property improvement of unsaturated polyester composite reinforced with admicellar-treated sisal fibers. *Composites. Part A, Applied Science and Manufacturing*, *40*(6–7), 687–694. doi:10.1016/j.compositesa.2008.12.004

San, Y. T., Jin, Y. T., & Li, S. C. (2009). *TRIZ: Systematic Innovation in Manufacturing*. Selangor: Firstfruits Sdn Bhd.

Sapuan, S. M., Mohamed, A. R., Siregar, J. P., & Ishak, M. R. (2011). Pineapple Leaf Fibers and PALF-Reinforced Polymer Composites. In Cellulose Fibers: Bio-and Nano-Polymer Composites (pp. 325-343). Springer Berlin Heidelberg.

Sapuan, S. M. (2005). Concurrent design and manufacturing process of automotive composite components. *Assembly Automation*, *25*(2), 146–152. doi:10.1108/01445150510590514

Sapuan, S. M. (2006). Using the morphological chart technique for the design of polymeric-based composite automotive pedals. *Discovery and Innovation*, *18*, 311–317.

Sapuan, S. M. (2010). *Concurrent Engineering for Composites*. Serdang: UPM Press.

Sapuan, S. M., & Abdalla, H. S. (1998). A prototype knowledge-based system for the material selection of polymeric-based composites for automotive components. *Composites. Part A, Applied Science and Manufacturing*, *29*(7), 731–742. doi:10.1016/S1359-835X(98)00049-9

Sapuan, S. M., Leenie, A., Harimi, M., & Beng, Y. K. (2006). Mechanical properties of woven banana fibre reinforced epoxy composites. *Materials & Design*, *27*(8), 689–693. doi:10.1016/j.matdes.2004.12.016

Sapuan, S. M., & Mansor, M. R. (2014). Concurrent engineering approach in the development of composite products: A review. *Materials & Design*, *58*, 161–167. doi:10.1016/j.matdes.2014.01.059

Sassner, P., Galbe, M., & Zacchi, G. (2008). Techno-economic evaluation of bioethanol production from three different lignocellulosic materials. *Biomass and Bioenergy*, *32*(5), 422–430. doi:10.1016/j.biombioe.2007.10.014

Satyanarayana, K. G., Pillai, C. K. S., Sukumaran, K., Rohatgi, P. K., & Vijayan, K. (1982). Structure property studies of fibres from various parts of the coconut tree. *Journal of Materials Science*, *17*(8), 2453–2462.

Satyanarayana, K. G., Arizaga, G. G. C., & Wypych, F. (2009). Biodegradable composites based on lignocellulosic fibers—An overview. *Progress in Polymer Science*, *34*(9), 982–1021. doi:10.1016/j.progpolymsci.2008.12.002

Satyanarayana, K. G., Pillai, C. K. S., Sukumaran, K., Pillai, S. G. K., Rohatgi, P. K., & Kalyani, V. (1982). Structure and properties of fibers from different parts of coconut tree. *Journal of Materials Science*, *17*, 2453–2472. doi:10.1007/BF00543759

Satyanarayana, K. G., Sukumaran, K., Mukherjee, P. S., Pavithran, C., & Pillai, S. G. K. (1990). Natural fiber–polymer composites. *Cement and Concrete Composites*, *12*(2), 117–136. doi:10.1016/0958-9465(90)90049-4

Satyanarayan, K. G., Kulkarni, A. G., & Rohatgi, P. K. (1981). Fabrication and properties of coir based composites. *Journal of Scientific and Industrial Research*, *40*, 222–229.

Savastano, H. Jr, Warden, P. G., & Coutts, R. S. P. (2000). Brazilian waste fibres as reinforcement for cement-based composites. *Cement and Concrete Composites*, *22*(5), 379–384. doi:10.1016/S0958-9465(00)00034-2

Savastano, H. Jr, Warden, P. G., & Coutts, R. S. P. (2003). Potential of alternative fibre cements as building materials for developing areas. *Cement and Concrete Composites*, *25*(6), 585–592. doi:10.1016/S0958-9465(02)00071-9

Savastano, H. Jr, Warden, P. G., & Coutts, R. S. P. (2005). Microstructure and mechanical properties of waste fibre–cement composites. *Cement and Concrete Composites*, *27*(5), 583–592. doi:10.1016/j.cemconcomp.2004.09.009

Schmehl, M., Mussig, J., Schonfeld, U., & Von Buttlar, H.-B. (2008). Life cycle assessment on a bus body component based on hemp and PTP. *Journal of Polymers and the Environment*, *16*(1), 51–60. doi:10.1007/s10924-008-0103-x

Schön, J. (1996). Coefficient of friction of composite delamination surfaces. *Wear*, *237*(1), 77–89. doi:10.1016/S0043-1648(99)00315-4

Sedjo, R. (1997). The economics of forest-based biomass supply. *Energy Policy*, *25*(6), 559–566. doi:10.1016/S0301-4215(97)00045-1

Selvam, B., Marimuthu, P., Narayanasamy, R., Anandakrishnan, V., Tun, K. S., Gupta, M., & Kamaraj, M. (2014). Dry sliding wear behaviour of zinc oxide reinforced magnesium matrix nano-composites. *Materials & Design*, *58*, 475–481.

Sen, T., Reddy, H.N.J., (2011). Application of Sisal, Bamboo, Coir and Jute Natural Composites in Structural Upgradation. *International Journal of Innovation, Management and Technology*, *2*(3).

Shah, D. U. (2013). Developing plant fibre composites for structural applications by optimising composite parameters: A critical review. *Journal of Materials Science*, *48*(18), 6083–6107. doi:10.1007/s10853-013-7458-7

Shah, D. U. (2014). Natural fibre composites: Comprehensive Ashby-type materials selection charts. *Materials & Design*, *62*, 21–31. doi:10.1016/j.matdes.2014.05.002

Shanian, A., & Savadogo, O. (2006). A material selection model based on the concept of multiple attribute decision making. *Materials & Design*, *27*(4), 329–337. doi:10.1016/j.matdes.2004.10.027

Shao, Z., & Vollrath, F. (2002). Materials: Surprising strength of silkworm silk. *Nature*, *418*(6899), 741–741. doi:10.1038/418741a PMID:12181556

Shibata, M., Oyamada, S., Kobayashi, S., & Yaginuma, D. (2004). Mechanical properties and biodegradability of green composites based on biodegradable polyesters and lyocell fabric. *Journal of Applied Polymer Science*, *92*(6), 3857–3863. doi:10.1002/app.20405

Shih, Y.-F. (2007). A study of the fiber obtained from the water bamboo husks. *Bioresource Technology*, *98*(4), 819–828. doi:10.1016/j.biortech.2006.03.025 PMID:16759852

Shi, J., Sharma-Shivappa, R. R., Chinn, M., & Howell, N. (2009). Effect of microbial pretreatment on enzymatic hydrolysis and fermentation of cotton stalks for ethanol production. *Biomass and Bioenergy*, *33*(1), 88–96. doi:10.1016/j.biombioe.2008.04.016

Silva, G. G., De Souza, D. A., Machado, J. C., & Hourston, D. J. (2000). Mechanical and thermal characterization of native brazilian coir fiber. *Journal of Applied Polymer Science*, *76*(7), 1197–1206. doi:10.1002/(SICI)1097-4628(20000516)76:7<1197::AID-APP23>3.0.CO;2-G

Silvester, S., Beella, S. K., Van Timmeren, A., Bauer, P., Quist, J., & Van Dijk, S. (2013). Exploring design scenarios for large-scale implementation of electric vehicles: The Amsterdam Airport Schiphol case. *Journal of Cleaner Production*, *48*(0), 211–219. doi:10.1016/j.jclepro.2012.07.053

Singh, M., Khan, I. A., & Grover, S. (2011). Selection of manufacturing process using graph theoretic approach. *International Journal of System Assurance Engineering and Management*, *2*(4), 301–311. doi:10.1007/s13198-012-0083-z

Singh, V. K, & Gope, P. C., Sakshi, C., & Bisht, D. S. (2012). Mechanical Behaviour of Banana Fiber Based Hybrid Bio Composites. *J. Mater. Environ. Sci.*, *3*, 185–194.

Siregar, J. P., Sapuan, S. M., Rahman, M. Z. A., & Zaman, H. M. D. K. (2012). Effects of Alkali Treatments on the Tensile Properties of Pineapple Leaf fibre Reinforced High Impact Polystyrene Composites. Editorial Board, 409.

Sisman, C. B., & Gezer, E. (2011). Effects of rice husk ash on characteristics of the briquette produced for masonry units. *Scientific Research and Essays*, *6*, 984–992.

Slate, F. O. (1976). Coconut fibers in concrete. *Eng J Singapore*, *3*(1), 51–54.

Smitthipong, W., Tantatherdtam, R., & Chollakup, R. (2015). Effect of pineapple leaf fiber-reinforced thermoplastic starch/poly (lactic acid) green composite Mechanical, viscosity, and water resistance properties. *Journal of Thermoplastic Composite Materials*, *28*(5), 717–729. doi:10.1177/0892705713489701

Sohlenius, G. (1992). Concurrent engineering. *CIRP Annals-Manufacturing Technology*, *41*(2), 645–655. doi:10.1016/S0007-8506(07)63251-X

Song, H. J., Zhang, Z. Z., Men, X. H., & Luo, Z. Z. (2010). A study of the tribological behavior of nano-ZnO-filled polyurethane composite coatings. *Wear*, *269*(1), 79–85. doi:10.1016/j.wear.2010.03.011

Song, X., Zheng, S., Huang, J., Zhu, P., & Guo, Q. (2000). Miscibility, morphology and fracture toughness of tetrafunctional epoxy resin/poly (styrene-co-acrylonitrile) blends. *Journal of Materials Science*, *35*(22), 5613–5619. doi:10.1023/A:1004824628535

Sreekala, M. S., George, J., Kumaran, M. G., & Thomas, S. (2002). The mechanical performance of hybrid phenol-formaldehyde-based composites reinforced with glass and oil palm fibres. *Composites Science and Technology*, *62*(3), 339–353. doi:10.1016/S0266-3538(01)00219-6

Sreenivasan, S., Iyer, P. B., & Iyer, K. R. K. (1996). Influence of delignification and alkali treatment on the fine structure of coir fibres (Cocos Nucifera). *Journal of Materials Science*, *31*(3), 721–726. doi:10.1007/BF00367891

Sukmawan, R., Takagi, H., & Nakagaito, A. N. (2015). Strength evaluation of cross-ply green composite laminates reinforced by bamboo fiber. *Composites. Part B, Engineering*, *84*, 9–16. doi:10.1016/j.compositesb.2015.08.072

Summerscales, J., Dissanayake, N. P. J., Virk, A. S., & Hall, W. (2010). A review of bast fibres and their composites. Part 1 – Fibres as reinforcements. *Composites. Part A, Applied Science and Manufacturing*, *41*(10), 1329–1335. doi:10.1016/j.compositesa.2010.06.001

Summerscales, J., & Grove, S. (2014). Manufacturing Methods for Natural Fiber Composites. *Natural Fiber Composites: Materials. Processes and Properties*, *7*, 176–215.

Takagi, H., & Asano, A. (2008). Effects of processing conditions on flexural properties of cellulose Nano fiber reinforced "green" composites. *Composites. Part A, Applied Science and Manufacturing*, *39*(4), 685–689. doi:10.1016/j.compositesa.2007.08.019

Thakur, V. K. (2013). *Green composites from natural resources*. CRC Press. doi:10.1201/b16076

Thakur, V. K., & Singha, A. S. (2013). *Biomass-based biocomposites*. Smithers Rapra Technology.

Thakur, V. K., Singha, A. S., & Thakur, M. K. (2012). Green Composites from Natural Fibers: Mechanical and chemical aging properties. *International Journal of Polymer Analysis and. Characterization*, *17*, 401–407.

Thakur, V. K., Singha, A. S., & Thakur, M. K. (2012a). Green Composites from Natural Fibers: Mechanical and Chemical Aging Properties. *International Journal of Polymer Analysis and Characterization*, *17*(6), 401–407. doi:10.1080/1023666X.2012.668665

Thakur, V. K., Singha, A. S., & Thakur, M. K. (2012b). In-Air Graft Copolymerization of Ethyl Acrylate onto Natural Cellulosic Polymers. *International Journal of Polymer Analysis and Characterization*, *17*(1), 48–60. doi:10.1080/1023666X.2012.638470

Thakur, V. K., & Thakur, M. K. (2014). Processing and characterization of natural cellulose fibers/thermoset polymer composites. *Carbohydrate Polymers*, *109*, 102–117. doi:10.1016/j.carbpol.2014.03.039 PMID:24815407

Thiruchitrambalam, M., Alavudeen, A., Athijayamani, A., Venkateshwaran, N., & Elaya Perumal, A. (2009). Improving Mechanical Properties of Banana/Kenaf Polyester Hybrid Composites using Sodium Laulryl Sulfate Treatment. *Materials Physics and Mechanics*, *8*, 165–173.

Thomas, S., Idicula, M., & Joseph, K. (2009). Mechanical performance of short banana/sisal hybrid fibre reinforced polyester composites. *Journal of Reinforced Plastics and Composites*.

Threepopnatkul, P. K., & Athipongarporn, N. (2008). Polycarbonate with Piņeapple Leaf Fiber to Produce Functional Composites. *Advanced Materials Research*, *47*(50).

Threepopnatkul, P., Kaerkitcha, N., & Athipongarporn, N. (2009). Effect of surface treatment on performance of pineapple leaf fiber–polycarbonate composites. *Composites. Part B, Engineering*, *40*(7), 628–632. doi:10.1016/j.compositesb.2009.04.008

Trujillo, E. E., Osorio, L., Fuentes, C., Van Vuure, A., & Verpoest, I. (2010) Bamboo fiber thermoplastic for transport applications. In *Proceedings of SAMPE Europe 31 International Technical Conference and Forum*.

Turmanova, S. Ch., & Genieva, S. D. (2008). Non isothermal degradation kinetics of filled with rise husk ash polypropene composites. *Express Polymer Letters*, *2*(2), 133–146. doi:10.3144/expresspolymlett.2008.18

Twidell, J. (1998). Biomass energy. *Renew Energy World*, *1*, 38–39.

Uday Kiran, C., Ramachandra Reddy, G., Dabade, B. M., & Rajesham, S. (2007). Tensile Properties of Sun Hemp, Banana and Sisal Fiber Reinforced with Polyester Composites. *Journal of Reinforced Plastics and Composites*, *26*(10), 1043–1050. doi:10.1177/0731684407079423

Ulrich, K. T., & Eppinger, S. D. (2004). Product Design and Development (3rd ed.). Boston: McGraw-Hill Irwin.

van Vuure, A. W., Osorio, L., Trujillo, E., Fuentes, C., & Verpoest, I. (2011). Long bamboo fibre composites. In *Proceedings of the ICCM 18*.

Vara Prasad, V. (2011). Tensile and flexural properties of Glass fiber-reinforced polyester hybrid composites. *International Journal of Materials and Biomaterials Applications*, *1*(1), 14–16.

Varma, D. S., Varma, M., & Varma, I. K. (1985). Coir Fibers II: Evaluation as a Reinforcement in Unsaturated Polyester Resin Composites. *Journal of Reinforced Plastics and Composites*, *4*(4), 419–431. doi:10.1177/073168448500400406

Varma, D. S., Varma, M., & Varma, I. K. (1986). Coir fibers. 3. Effect of resin treatment on properties of fibers and composites. *Industrial & Engineering Chemistry Product Research and Development*, *25*(2), 282–289. doi:10.1021/i300022a027

Velmurugan, R., & Manikandan, V. (2007). Mechanical properties of palmyra/glass fiber hybrid composites. *Composites. Part A, Applied Science and Manufacturing*, *38*(10), 2216–2226. doi:10.1016/j.compositesa.2007.06.006

Venkateshwaran, N., Elayaperumal, A., & Sathiya, G. K. (2012). Prediction of tensile properties of hybrid-natural fiber composites. *Composites. Part B, Engineering*, *43*(2), 793–796. doi:10.1016/j.compositesb.2011.08.023

Verma, D., Gope, P. C., Singh, I., & Jain, S. (2015). *Composites from Bagasse Fibers, Its Characterization and Applications. In Agricultural Biomass Based Potential Materials* (pp. 91–119). Springer International Publishing.

Versos, C. A. M., & Coelho, D. A. (2011). Biologically inspired design: Methods and validation. In D. A. Coelho (Ed.), *Industrial Design – New Frontiers* (pp. 101–120). Rijeka, Croatia: InTech. doi:10.5772/20326

Vivekanandan, S., & Kamaraj, G. (2011). Effect of co-digestion of cow dung with rice chaff, rice straw and rice husk in biogas production using anaerobic digestion. *International Research Journal of Biotechnology*, *2*, 114–118.

Wambua, P., Ivens, J., & Verpoest, I. (2003). Natural fibers: Can they replace glass in fiber reinforced plastics? *Composites Science and Technology*, *63*(9), 259–1264. doi:10.1016/S0266-3538(03)00096-4

Wang, F.-K., & Chen, K.-S. (2010). Applying Lean Six Sigma and TRIZ methodology in banking services. *Total Quality Management & Business Excellence*, *21*(3), 301–315. doi:10.1080/14783360903553248

Wang, L. L., Zhang, L. Q., & Tian, M. (2012). Mechanical and tribological properties of acrylonitrile–butadiene rubber filled with graphite and carbon black. *Materials & Design*, *39*, 450–457. doi:10.1016/j.matdes.2012.02.051

Wan, Y. Z., Chen, G. C., Raman, S., Xin, J. Y., Li, Q. Y., Huang, H., & Luo, H. L. et al. (2006). Friction and wear behavior of three-dimensional braided carbon fiber/epoxy composites under dry sliding conditions. *Wear*, *260*(9-10), 933–941. doi:10.1016/j.wear.2005.06.007

Waswa-Sabuni, B., Syagga, P.M., Dulo, S.O., & Kamau, G.N. (2002). Rice Husk Ash Cement – An alternative pozzolanacement for Kenyan building Industry. *Journal of Civil Engineering*, *8*, 13 – 26.

Woodhouse, I. H. (2006a). *Introduction to microwave remote sensing*. New York: CRC Press.

Woodhouse, I. H. (2006b). Predicting backscatter-biomass and height-biomass trends using a macroecology model. *IEEE Transactions on Geoscience and Remote Sensing*, *44*(4), 871–877. doi:10.1109/TGRS.2006.872356

Wötzel, K., Wirth, R., & Flake, M. (1999). Life cycle studies on hemp fibre reinforced components and ABS for automotive parts. *Die Angewandte Makromolekulare Chemie*, *272*(1), 121–127. doi:10.1002/(SICI)1522-9505(19991201)272:1<121::AID-APMC121>3.0.CO;2-T

Wu, W., Chiang, C., & Lin, C. (1998). Comparing the aggregation methods in the analytic hierarchy process when uniform distribution. *WSEAS Transactions on Business and Economics*, *5*(3), 82–87.

Xie, G. Y., Zhuang, G. S., Sui, G. X., & Yang, R. (2010). Tribological behavior of PEEK / PTFE composites reinforced with potassium titanate whiskers. *Wear*, *268*(3), 424–430. doi:10.1016/j.wear.2009.08.032

Yaacob, Q. (2015). *The effect of dispersion on the mechanical properties of PLA-based polymer composites*. (Unpublished undergraduate dissertation). Universiti Teknikal Malaysia Melaka.

Yahaya, R., Sapuan, S. M., Jawaid, M., Leman, Z., & Zainudin, E. S. (2014). Mechanical performance of woven kenaf-Kevlar hybrid composites. *Journal of Reinforced Plastics and Composites*, *33*(24), 2242–2254. doi:10.1177/0731684414559864

Yahaya, R., Sapuan, S. M., Jawaid, M., Leman, Z., & Zainudin, E. S. (2015). *Effect of fibre orientations on the mechanical properties of kenaf–aramid hybrid composites for spall-liner application*. Defence Technology.

Yamashina, H., Ito, T., & Kawada, H. (2002). Innovative product development process by integrating QFD and TRIZ. *International Journal of Production Research*, *40*(5), 1031–1050. doi:10.1080/00207540110098490

Yang, H-S. (2004). *Role of rice husk flour as reinforcing filler for thermoplastic polymer bio composite*. Doctoral Thesis.

Yang, H. S., Kim, H. J., Park, H. J., Lee, B. J., & Hwang, T. S. (2007). Effect of compatibilizing agents on rice-husk flour reinforced polypropylene composite. *Composite Structures*, *77*(1), 45–55. doi:10.1016/j.compstruct.2005.06.005

Yang, H. S., Wolcon, M. P., Kim, H.-S., & Kim, H.-J. (2005). Thermal properties of lignocellulosic filler-thermoplastic polymer bio-composites. *Journal of Thermal Analysis and Calorimetry*, *82*(1), 157–160. doi:10.1007/s10973-005-0857-5

Yang, H. S., Wolcott, M. P., Kim, H. S., Kim, S., & Kim, H. J. (2006). Properties of lignocellulosic material filled polypropylene bio-composite made with different manufacturing process. *Polymer Testing*, *25*(5), 668–676. doi:10.1016/j.polymertesting.2006.03.013

Yao, F., Wu, Q., Lei, Y., & Xu, Y. (2008). Rice straw fiber-reinforced high-density polyethylene composite: Effect of fiber type and loading. *Industrial Crops and Products*, *28*(1), 63–42. doi:10.1016/j.indcrop.2008.01.007

Yeh, C. H., Huang, J. C. Y., & Yu, C. K. (2011). Integration of four-phase QFD and TRIZ in product R&D: A notebook case study. *Research in Engineering Design*, *22*(3), 125–141. doi:10.1007/s00163-010-0099-9

Yen, S. B., & Chen, J. L. (2005). An eco-innovative tool by integrating FMEA and TRIZ methods. In *Proceeding of the 4th International Symposium on Environmentally Conscious Design and Inverse Manufacturing* (pp. 678–683). Tokyo, Japan: IEEE.

You, Y. L., Li, D. X., Si, G. J., Lv, R. Y., Deng, X. (2014). Improvement in the tribological properties of polyamide 6: Talc, glass fiber, graphite, and ultrahigh-molecular-weight polyethylene. *Journal of Thermoplastic Composite Materials.*

Yu, C. (2015). *Natural Textile Fibers: Vegetable Fibers. In Textiles and Fashion* (pp. 29–56). Woodhead Publishing.

Yuhazri, M.Y., & Dan, M.M.P. (2007) Helmet Shell Using Coconut Fiber (Deco-Helmet). *Journal of Advanced Manufacturing Technology, 1*(1), 23-30.

Yulianto, F. E., & Mochtar, N. E. (2010). Mixing of Rice Husk Ash (RHA) and lime for peat soil stabilization. *Proceedings of the First Makassar International Conference on Civil Engineering* (MICCE2010).

Yurtkuran, S., Kırlı, G., & Taneli, Y. (2013). Learning from nature: Biomimetic design in architectural education. *Procedia: Social and Behavioral Sciences*, *89*(0), 633–639. doi:10.1016/j.sbspro.2013.08.907

Yusof, Y., Yahya, S. A., & Adam, A. (2015). Novel Technology for Sustainable Pineapple Leaf Fibers Productions. *Procedia CIRP*, *26*, 756–760. doi:10.1016/j.procir.2014.07.160

Zainuddin, E. S., Sapuan, S. M., Abdan, K., & Mohamad, M. T. M. (2008). Thermal Degradation of Banana Pseudo-stem Filled Unplasticized Polyvinyl Chloride Composites. *Journal of Materials and Design*, *30*(3), 557–562. doi:10.1016/j.matdes.2008.05.060

Zakikhani, P., Zahari, R., Sultan, M.T.H, & Majid, D.L. (2014). Bamboo Fiber Extraction and Its Reinforced Polymer Composite Material. *International Journal of Chemical, Molecular, Nuclear, Materials and Metallurgical Engineering*, *8*(4).

Zeng, A., Zheng, Y., Guo, Y., Qiu, S., & Cheng, L. (2012). Effect of tetra-needle-shaped zinc oxide whisker (T-ZnOw) on mechanical properties and crystallization behavior of isotactic polypropylene. *Materials & Design*, *34*, 691–698. doi:10.1016/j.matdes.2011.05.028

Zhang, X., Yang, W., & Blasiak, W. (2011). Modeling Study of Woody Biomass: Interactions of Cellulose, Hemicellulose, and Lignin. *Energy & Fuels*, *25*(10), 4786–4795. doi:10.1021/ef201097d

Zhong, R., Burk, D. H., & Ye, Z. (2001). Fibers. A model for studying cell differentiation, cell elongation, and cell wall biosynthesis. *Plant Physiology*, *126*(2), 477–479. doi:10.1104/pp.126.2.477 PMID:11402177

Zhou, S. F., Reekie, L., Chan, H. P., Chow, Y. T., Chung, P. S., & Man Luk, K. (2012). Characterization and modelling of Bragg gratings written in polymer fiber for use as filters in the THz region. *Optics Express*, *20*(9), 9564–9571. doi:10.1364/OE.20.009564 PMID:22535048

Zhu, W., Zhu, J., Gleisner, R., & Pan, X. (2010). On energy consumption for size-reduction and yields from subsequent enzymatic saccharification of pretreated lodgepole pine. *Bioresource Technology*, *101*(8), 2782–2792. doi:10.1016/j.biortech.2009.10.076 PMID:20006490

Zini, E., & Scandola, M. (2011). Green composites: An overview. *Polymer Composites*, *32*(12), 1905–1915. doi:10.1002/pc.21224

Zou, Y., Xu, H., & Yang, Y. (2010). Lightweight Polypropylene Composites Reinforced by Long Switchgrass Stems. *Journal of Polymers and the Environment*, *18*(4), 464–473. doi:10.1007/s10924-010-0165-4

Zuluaga, R., Putaux, J. L., Cruz, J., Vélez, J., Mondragon, I., & Gañán, P. (2009). Cellulose micro fibrils from banana rachis: Effect of alkaline treatments onstructural and morphological features. J. *Carbohydrate Polymers*, *76*(1), 51–59. doi:10.1016/j.carbpol.2008.09.024

About the Contributors

Deepak Verma is an Assistant Professor in the Department of Mechanical Engineering, Graphic Era Hill University Dehradun, India. He received his M.Tech in Mechanical Engineering from College of Technology, G.B., Pant University of Agriculture and Technology Pantnagar India. He is having around 8 years of Teaching Experience and taught different courses at UG and PG level. He has a good background in Novel Processing of Green Composite Materials, their fabrication, Characterization and Mechanical Properties Evaluation.

Siddharth Jain received the M.Tech from IIT Roorkee and PhD from IIT Roorkee. He did his Post-Doctoral Fellowship at the National University of Singapore. He is a Scientist and Visiting Professor at the University of Alberta and an Associate Professor at the College of Engineering Roorkee.

Xiaolei Zhang is currently a Lecturer at Queen's University Belfast, she gained her PhD degree from Royal Institute of Technology in Sweden and had 1.5 years' of Post-Doctoral experience at University of Alberta in Canada. She has particular expertise in renewable energy utilization and power generation.

* * *

Mohd Zaid Bin Akop is currently a senior lecturer and researcher at the Faculty of Mechanical Engineering, Universiti Teknikal Malaysia Melaka, Malaysia. He obtained his PhD in Mechanical Engineering from Gunma University, Japan in 2014. His research interest includes diesel spray, solar energy (renewable energy), HVAC and Thermal Stress which led to publications of more than 30 peer-reviewed journals and conference proceedings. He also successfully contribute in several research projects from the Malaysian and Japan government acting both principle researcher and co-researcher.

Sumit Bhowmik received his B. E. in Mechanical Engineering from Tripura Engg. College (Tripura University), India, Master in Mechanical Engineering and PhD from Jadavpur University, Kolkata. He has over 15 years of teaching and research experience. Presently he is an Assistant Professor at the Department of Mechanical Engineering, National Institute of Technology, Silchar, Assam, India. His areas of interests are Fracture Mechanics, Materials properties, Composites, and Optimization etc. He has 9 International Journal and 6 International Conference publications to his credit. He is on the editorial board and review panel of 2 International Journals of repute.

Vijaykumar Chaudhary obtained his B.E. (Mechanical) degree in August- 1999 from BVM Engineering College, Vallabh Vidyanagar – Gujarat. He received his M.E. Mechanical (Machine Design) degree in October- 2005 from BVM Engineering College, Sardar Patel University. He is pursuing his Ph.D. from Charotar University of Science & Technology, Changa. He has written more than 28 research papers. His areas of interest include machining of composite materials. He has completed three MODROB projects of AICTE-New Delhi INDIA. Prof. Chaudhary is a life member of ISTE & AMM.

J. Paulo Davim received his PhD degree in Mechanical Engineering from the University of Porto in 1997 and the Aggregate title from the University of Coimbra in 2005. Currently, he is Professor at the Department of Mechanical Engineering of the University of Aveiro. He has more 29 years of teaching and research experience in manufacturing, materials and mechanical engineering with special emphasis in machining and tribology. Recently, he has also interest in management/industrial engineering and higher education for sustainability. He is the Editor-in-Chief of six international journals, guest editor of journals, books editor, book series editor and scientific advisory for many international journals and conferences. Presently, he is an editorial board member of 30 international journals and acts as reviewer for than 80 prestigious ISI Web of Science journals. In addition, he has also published in his field of research as author and co-author more than 50 book chapters and 350 articles in journals and conferences (more 200 articles in ISI Web of Science, h-index 34+).

Piyush P. Gohil did his Ph.D. in August- 2010 from SVNIT-Surat (Gujarat) INDIA. His area of research includes Design & Manufacturing of Composites. He is a life member of ISTE, TSI, ISAMPE and annual member of SAE. He has more than 60 research papers in referred journals & national & international conferences. He has delivered more than 25 expert lectures. He is acting as reviewer in many national and international journals. He has completed two MODROB projects of

AICTE-New Delhi INDIA. He was awarded travel grants as a Young Scientist from DST INDIA in march 2011 for Malaysia & under UGC Merged scheme 2014-2015 in March 2015 for London.

Kaushik Kumar received his BTech in Mechanical Engineering from REC (Now NIT), Warrangal, India, MBA from IGNOU and PhD from Jadavpur University, Kolkata. He has over 25 years of teaching, research and industrial experience. Presently he is an Associate Professor at the Department of Mechanical Engineering, Birla Institute of Technology, Mesra, Ranchi, India. His areas of interests are Optimization, Non-conventional machining, CAD/CAM, rapid prototyping and composites. He has 3 Patents, 4 Book Chapters, 83 international Journal, 18 International and 8 National Conference publications to his credit. He is on the editorial board and review panel of 7 International and 1 National Journals of repute. He is Mentor of Organising body of Institute for Research and Development India (IRD India) and has been felicitated with many awards and honours.

Muhd Ridzuan Mansor is currently a lecturer and researcher at the Faculty of Mechanical Engineering, Universiti Teknikal Malaysia Melaka, Malaysia. He obtained his PhD in Mechanical Engineering from Universiti Putra Malaysia, Malaysia in 2015. His research interest includes concurrent design and product development, biocomposites and renewable energy which led to publications of 3 book chapters, 21 peer-reviewed journals and 39 conference proceedings. He also successfully contributed in several research projects from the Malaysian government acting both principle researcher and co-researcher.

M. T. Musthafah currently is a Deputy Director, Center for Sustainability & Environment, Universiti Teknikal Malaysia Melaka. Completed PhD at Nagaoka University of Technology, Japan in 2010 with thesis title of "Thermodynamical analysis of expander for low-temperature Rankine cycle". Became a Lecturer since 2005 and is a researcher in Green Technology Vehicle Research Group, CARe, UTeM. Current research interest are about CNG engine analysis, renewable energy and heat recovery energy.

Kundan Patel obtained his B.E. (Mechanical) degree in July - 2009 from A. D. Patel Institute of Technology, Vallabh Vidyanagar – Gujarat. He received his M.Tech Mechanical (CAD/CAM) degree in July- 2012 from Charotar University of Science & Technology, Changa. He is pursuing his Ph.D. from Charotar University of Science & Technology, Changa. His areas of interest include machining of polymer matrix composite materials. Prof. Patel is a life member of ISTE.

Ronak Ramjiyan has done B.Tech from S S College of Engineering, Udaipur in July 2012. He has completed his M.Tech in Advanced Manufacturing Technology from Charotar University of Science & Technology in June 2015.

Supriyo Roy, M. Sc (Statistics), M. Tech (OR&IBM), Ph. D (Business Model Optimization) has been working as Associate Professor, BIT-Mesra, Ranchi since 1st July, 2012. A Post Doctoral Fellowship from EADS-SMI Center for Supply Chain, IIM - Bangalore, Dr. Roy has an experience of 4 years in Manufacturing Industry and 15 years of post graduate teaching. His teaching interest comprises Interdisciplinary areas of Management Science, Production and Operations Management, Decision Science, Soft Computing for manufacturing, etc. His publications have appeared in leading journals in INFOR-Canada, AMSE-France, BJO&PM - Brazil, IGI-Global, MDI journals, etc. He is impaneled reviewer of IJPR (UK), ITOR (Australia), BJO&PM (Brazil) and many National Journals. His current research interest covers development of Optimization models relating to Management Science problems via Structural Equation Modeling.

Mohd Azli Salim is currently senior lecturer and researcher at the Centre for Advanced Research on Energy, Faculty of Mechanical Engineering, Universiti Teknikal Malaysia Melaka. He is currently a PhD student (waiting for VIVA) from Universiti Teknikal Malaysia Melaka and PhD Visiting Student from Institute of Sound and Vibration Research (ISVR), University of Southampton, United Kingdom. His research interest includes vibration and control, noise and acoustics, structural dynamics, finite element analysis, control system and mathematical modelling, where he published over 50 international journals and 50 international conference. Additionally, he also contributed a lot of research projects from Ministry of Higher Education, Malaysia.

Irem Sanal is an assistant professor in Civil Engineering Department, Atilim University, Turkey. Her research interests are mainly composite materials, their behaviors and applications.

S. M. Sapuan was born on 25 September 1965, in Teluk Intan, Perak, Malaysia and earned his B.Eng degree in Mechanical Engineering from University of Newcastle, Australia in 1990. Afterward he continued his study and obtained his MSc from Loughborough University, UK in 1994 and finally Ph.D from De Montfort University, UK in 1998. Later on, he has managed to pursue other qualifications including Life Fellow, International Biographical Association (LFIBA), Fellow, Society of Automotive Engineers Inc. (SAE), Fellow, Institute of Materials Malaysia (FIMM), Fellow, Malaysian Scientific Association (FMSA), Life Member, Institute of Energy, Malaysia (LMInTeM), Member, International Association of Engineers, Fellow, Plastics, Rubber Institute Malaysia (FPRIM) and Honorary Life Member, Asian Polymer Association (HLMAPA), Member, International Network on Engineering Education and Research (iNEER), Member, National Professor Council, and Professional Engineer, Board of Engineers, Malaysia (PEng). Professor Sapuan is a Head, Laboratory of Biocomposite Technology, Institute of Tropical Forestry and Forest Products, Universiti Putra Malaysia (from August 2014) and Professor of Composite Materials in Department of Mechanical and Manufacturing Engineering, Universiti Putra Malaysia (UPM); the position he took up since May 2007. Prof Sapuan's contribution to his research field is evidenced by his publications. To date he has authored or co-authored more than 1000 publications on mechanical engineering, specializing mainly in composite materials, natural fibre composites, materials selection, concurrent engineering, total design and design methods including in national and international journals (450 papers published/accepted), books (11), edited books (5) and conference proceedings/seminars (385). He was invited to present keynote lectures at 11 conferences organized by various organizations such as UPM, Universiti Kuala Lumpur, Universitas Malahayati, Indonesia, and STRIDE, Ministry of Defense, Malaysia. In addition he presented 55 invited lectures in conferences and seminars. The books that he authored titled Engineering Design, Polymer-Based Composites, Product Design and Concurrent Engineering, Industrial Management, Glossary of Composite Materials, Concurrent Engineering for Composites, Materials Selection and Design, and Tropical Natural Fibre Composites. He also edited five books titled Research in Natural Fibre Reinforced Polymer Composites, Composite Materials Technology: Neural Network Applications, published by CRC Press, USA, Simulation for Undergraduates, Engineering Composites: Properties and Applications and Manufacturing of Natural Fibre Composites by Springer. His current h-index is 28 with 2917 citations.Professor Sapuan has successfully supervised 21 PhD as main supervisor and 20 PhD students as co-supervisor and is now supervising 13 PhD students. He has also successfully supervised 16 MS as main supervisors and 29 MS students as co-supervisor and is now supervising 6 MS students. In addition, he has supervised six postdoctoral researchers. Over the years, Prof. Sapuan has received numerous awards and honours, among others, 2001-2006 and 2008, 2009, 2010 and 2011 (ten times), Excellence Service Awards, UPM; 2002, AnugerahKaryawan Putra Cemerlang, UPM ((Excellent Putra Publication Award); 2002, Who's Who Award of achievement; 2005, Excellence Researcher Award (the highest number of papers published citation indexed journals); 2005, Excellence Award, science Publications, New York; 2006, Silver Medal, Malaysia Technology Expo; 2007 and 2008, Excellence Researcher Award, UPM, (publication incentive award; the highest in the category of Professor); 2007, Certificate of Excellence Award in Teaching 2007, Faculty of Engineering, UPM; 2008, Honorary Recent Advances in Environment, Ecosystems and Development ISBN: 978-1-61804-301-6 11, Life Member and Vice President, Asian Polymer Association, 2008, Vice Chancellor Fellowship Prize, UPM and 2008 ISESCO Science Prize in Technology, 2008 where he received a gold medal, a certificate and cash prize of USD 5,000. For 2009, Professor Sapuan is awarded with UPM Excellence Researcher Award 2008 (Special nternational Award and Publication incentive award; the highest in the category of professor). In 2010, Prof Sapuan is awarded with Plastic and Rubber Institute, Malaysia (PRIM) Fellowship Award, First Prize, Forest Research Institute of Malaysia (FRIM) Publication Award (Category: Semi/Non Technical Publication), 2010 and Excellent Researcher Award, Faculty of Engineering, UPM, 2008, 2010 and 2011. In 2008, his paper won the best oral presentation award in the 8th National Symposium on Polymeric Materials held in Penang, and in March 2011, his paper won the Best Poster Award in the Eight International Conference on Composite Science and Technology held in Kuala Lumpur. Also, in December 2011, his paper won the Best Poster Award in the International Conference on Innovation in Polymer Science and Technology 2011 held in Bali Indonesia. In February 2012, his paper won the Best Scientific Paper and Oral Presenter Award in UPM-UniKL Symposium on Polymeric Materials 2012 held in Melaka and the Second Place in Poster Presentation Award in SAMPE 2012 Asia Conference held in Kuala Lumpur. Professor Sapuan is listed among the ISESCO Science Laureattes and is ranked the 34th in UPM's Top 100 researchers in 2011. Professor Sapuan is the recipient of Rotary Research Gold Medal Award 2012. Professor Sapuan was awarded The Alumni Medal for Professional Excellence Finalist, 2012 Alumni Awards, University of Newcastle, NSW, Australia. In 2013, professor Sapuan was awarded Khwarizmi International Award (KIA). In 2013 he was awarded with 5 Star Role Model Supervisor award by UPM. He has been awarded "Outstanding Reviewer" by Elsevier for his contribution in reviewing journal papers.

M. A. Shaharuzaman received his B.Eng. (Hons.) and M.Eng. in Mechanical Engineering from Universiti Teknologi Malaysia (UTM) in 2007 and 2009, respectively. His professional life at the Universiti Teknikal Malaysia Melaka (UTeM) began in 2007 as a lecturer in the Department of Automotive, Faculty of Mechanical Engineering. He is the member of Green Technology Vehicle (GTeV) group in UTeM.

Sanjay Sharma is an Associate Professor and Head of the Department of Mechanical Engineering, Graphic Era Hill University, Dehradun, India. He is having around 21 years of teaching experience in the technical education and also around 5 years of Industrial Experience. His research interest are mainly composite materials and their processing by Friction Stir Methods.

Index

A

Agricultural wastes 6, 24, 166-167, 193
applied load 99, 111
ash (RHA). 169, 184

B

Bamboo Fiber 39, 47, 80, 83, 122, 228-229, 233, 237-238, 241-244, 246
bamboo fibers 17, 38-39, 192, 229, 232-244, 246
Banana fiber 201, 212, 216, 222, 224-226
Bast Fiber 4-5, 7, 23, 57, 71
bicycle frame 50, 62-66
bio-based polymer matrix 149
Biocomposite 13, 69, 147, 171-174, 191, 228, 247
Biodegradable composites 17, 22, 164, 221, 261, 272
biomass 3, 23, 47, 168, 183, 185, 187, 193-200
bumper beam 49, 58, 69-70

C

candidate materials 51, 59-60, 62, 126
CE approach 49-52, 62, 66-67
Cellulose 1, 4, 7-11, 13, 16, 18-23, 25, 43, 98, 100, 118, 125, 127-129, 133-134, 143, 146-147, 150-152, 158, 164, 168, 170, 172-174, 185, 205-211, 213, 222, 227, 234-235, 238, 247, 250, 252, 255, 258, 269, 272
Cement composites 16, 201, 216, 226, 263-264
Coconut Fiber 98, 121, 192, 247, 249, 252, 254-256, 258, 260, 262-268, 270-271, 273, 275
Coir Fiber 247, 252, 257-259, 261-262, 264-265, 267, 269-272, 275
Composite material 20, 58, 66, 77, 99, 106, 109-110, 147, 192, 202, 218, 225, 229, 234, 244, 246, 249-250, 266
Conceptual Design 48-55, 60-67, 70-72, 74
Concurrent conceptual design 52-54, 60-62, 66
Concurrent Design 48-50, 54-59, 72, 74
cycle assessment 2, 48, 50, 52, 61, 69, 73-74, 144, 192-193

D

design stage 48, 50-52, 54, 57, 60-63, 67, 70
dimensional stability 4, 53, 98, 107, 186, 241-242, 260
DMT 148

E

Economic Analysis 198
Economic Factors 193
energy production 194-196
Epoxy 16, 61, 69, 80-82, 92-94, 96-100, 114-115, 117, 120-123, 137, 216, 218-219, 224, 226, 240-241, 243, 246, 261, 265-266, 269-270, 274
Equilibrium Moisture Content 14, 23
Eucalyptus grandis 216-218
extraction process 39, 131-133, 204, 240

F

Fabrication Methods 24, 138
fiber matrix 127
Fiber Reinforced Polymers 228, 231, 247, 267
Fiber Type 1, 4, 191
friction coefficient 96, 99, 114-115, 118
Fruit Fiber 23

G

glass 2-3, 14-15, 17, 20, 25, 28, 34, 46, 53, 69, 71-72, 76-79, 81-83, 87-88, 91-95, 97, 99, 101, 123, 132, 140-141, 149, 154, 158, 199, 204, 219-220, 223-225, 230-231, 233, 235, 240-242, 246, 249, 251, 258, 260, 266, 271, 273-274
Green Composite 1, 23-24, 38, 47-50, 52-54, 57-61, 66-67, 96, 146, 148, 154-155, 159, 163-164, 193
green technology 2

H

Hardness 96, 107-109, 114, 118, 185-186, 218, 220, 234, 258, 260-261
Hemicellulose 1, 7-11, 23, 25, 43, 127, 129, 152, 169-170, 172-174, 234
high moisture 4, 14, 18, 210, 232, 259, 267
husk ash 166, 169, 174, 179, 184-191
Hybridization 76, 80, 91-93, 140, 144, 220
Hydrophilic 3, 10, 14, 23, 42, 125, 132, 147, 174, 249, 260
Hydrophobic 14, 42, 78, 125, 132-133, 147, 174, 239, 249, 251

I

indoor air 194-195
industrial applications 39, 189, 204, 230, 240
interfacial adhesion 16, 128, 133, 163, 173-174, 188, 251
interfacial bonding 42, 125, 133, 143-144, 151, 158, 160, 171-172, 175-176, 178, 239, 255-256, 261, 269
izod impact 160, 171-173, 175, 177-178

L

Leaf Fiber 16-17, 22-23, 125-126, 129-130, 132-134, 137, 144-147, 180
leaf fibers 6, 16, 58, 125, 129, 131-132, 141, 143, 145-147, 251, 270-271
life cycle 2, 48-50, 52, 61, 67, 69, 73-74, 97, 144, 192-193, 231
Life Cycle Assessment 2, 48, 50, 52, 61, 69, 73-74, 144, 192-193
life forms 229
Lignin 1, 7-9, 11, 23, 25, 38, 42-43, 100, 118, 127, 129, 152, 169-170, 172-173, 185, 205-206, 211, 219, 234-235, 240-241, 247, 252, 255-256, 259, 264
lignocellulosic 1-3, 8-9, 15, 22-23, 163, 169-170, 173, 175-178, 186, 188, 191, 199, 211, 223
load bearing 53, 62, 251
Long Fiber 125, 140-141
low density 15, 26, 128, 149, 165, 178, 201, 204, 211, 229, 233, 235, 244, 249, 251, 259, 267, 271, 273

M

material science 2, 122, 223
materials selection 48, 50, 52, 58-60, 69, 71-73
Mechanical Behavior 16, 76, 81, 92, 94, 262, 267, 269, 271-272
Mechanical Property 22, 148, 268, 270
Mercerization 133, 147, 152, 163, 255, 257
micro fibrils 205, 207, 209-211, 227
moisture absorption 4, 10, 15, 18, 79, 144, 182, 232, 235, 239-240, 248, 255
Morphological Chart 55-56, 64-65, 72, 74
Morphological Study 24, 46, 109, 111, 116, 137, 142, 188

N

NaOH solution 206, 210-211, 219

Natural Fiber 2-3, 5-6, 9-11, 14-15, 18-21, 23, 25, 40, 46, 48, 50, 55-56, 62, 71-72, 75, 77-80, 82, 92-94, 125-127, 132, 143-144, 146, 199, 229, 231, 245, 248-252, 259-261, 267-271, 273, 275
natural resources 2, 22, 58, 62, 75, 97, 126, 231, 243
normal load 99-100, 106, 109, 114-115, 118

O

Optical Property 148

P

particle size 100, 169, 182, 185-186
PDS requirements 51-52
performance engineering 2
Physical Properties 15-16, 20, 35, 93, 97, 110, 125, 132, 219, 225, 243
pineapple leaf 6, 16-17, 22, 58, 62, 125-126, 128-135, 137-138, 141, 143-147, 221, 271
pineapple leaf fibers 6, 16, 58, 125, 131-132, 141, 143, 145-147, 271
Polymer matrix 42, 79, 98, 115, 119, 134, 149, 174, 177, 183, 193, 201, 229, 232, 248-249, 255, 260, 266, 269
polymer network 77
polyvinyl chloride 14-15, 216, 226
poor interfacial bonding 171-172, 178, 239, 255-256
potential candidate materials 51, 59, 62
Product development 48-53, 55, 58-59, 61, 67, 71, 74, 105
pseudo stem 204, 216

R

raw material 5, 20, 35, 61, 67, 167, 169, 188, 241, 248
Recyclability 1, 3, 16, 18, 23, 96-97, 201, 229, 249
reinforced polyester 16, 20, 92-94, 145, 219-220, 224-225, 261, 268, 270
reinforced polymer 39-40, 46, 72, 80, 93, 166, 229, 231, 241, 246-247, 249, 260-261, 265, 267, 272
Reinforcement 2, 4, 6, 13-17, 25, 27, 29, 39, 55, 60, 71, 77-78, 80, 83, 96-98, 127-128, 135-137, 139, 141, 146, 148-150, 165, 169, 180, 201-204, 218, 226, 230-231, 240, 242-243, 247-249, 252, 262, 264, 266-268, 275
reinforcing filler 170-171, 173-175, 178, 186, 191
renewable assets 77
Rice Husk 6, 165-175, 178-184, 186-193
rice-husk flour 170, 173-175, 178, 186, 188, 191

S

Scouring 133, 147, 210-211, 222
SEM 44-45, 109-111, 115-118, 138, 141-142, 148, 160-163, 176, 182, 207-208, 214-215, 220-221, 257-258, 272
Short Fiber 125, 135-136, 138, 141, 247
slurry vacuum 217-218
Sodium hydroxide 133-134, 141, 218-219, 236, 255
specific gravity 78, 170, 184, 186
specific modulus 15-16, 77
specific wear 99, 109, 114-115, 118
stacking sequence 76, 79, 81, 83-85, 89-92
Staking Sequence 76, 93
starch (TPS) 128, 140
Stiffness 15, 18, 53, 59, 62, 135, 150-151, 165, 183, 202, 229-230, 233, 237, 244, 247-248, 256
Strength 7, 9-10, 15-18, 22, 24-25, 28, 37, 39-40, 42-43, 45, 47, 53, 57, 59, 62, 66, 68, 77, 79-82, 86-91, 96-97, 99, 107, 118, 131-134, 138, 140-141, 150-160, 165, 170-178, 180-182, 184-188, 190, 193, 202, 204, 206-207, 209, 211, 216, 218-222, 229-240, 242-244, 247-248, 251, 254-256, 260-267, 269, 273-274
Sustainability 1, 3, 18, 21, 23, 25, 48, 50, 52, 61-63, 68, 139, 174, 229, 235, 247, 267

Synthetic fiber 11, 62, 77, 79, 193, 241, 249

T

technical performance 51, 59
thermal stability 13, 16-17, 134, 170, 173, 175, 178, 180, 212, 216, 230, 258
Tribological Properties 96, 114, 121, 123
TRIZ 54-56, 63-64, 66, 71-72, 74-75
tropical fruit 126, 128

V

Void Content 32, 96, 106, 110-111, 117
volume fraction 29, 40-41, 60, 79-81, 99, 141, 206, 216, 219-222, 238, 255, 260-261, 263, 265-266

W

water absorption 13, 16, 144, 172, 184, 186, 207, 220, 248, 255, 260-261, 264-265
weight fraction 76, 79-81, 83-88, 91, 141
Wood 3-4, 8, 13-14, 25-26, 28, 37, 67, 96-100, 114-115, 117, 119, 121-122, 126, 148-149, 165, 169-170, 172, 175, 179-180, 185-186, 189, 193, 195-196, 198, 209, 234, 245-246, 251, 259, 272

X

XRD 148, 212

Y

Young's modulus 39-40, 44, 81, 150, 154-157, 209, 216, 220, 254

www.ingramcontent.com/pod-product-compliance
Lightning Source LLC
Chambersburg PA
CBHW081213180926
51613CB00003B/227
9781522504245